*On the Organic Law of Change*

# On the Organic Law of Change

*A Facsimile Edition and Annotated Transcription of*
*Alfred Russel Wallace's Species Notebook of 1855–1859*

Annotated by

James T. Costa

Harvard
University Press
Cambridge, Massachusetts
London, England
2013

*Library of Congress Cataloging-in-Publication Data*

Wallace, Alfred Russel, 1823–1913.
    On the organic law of change : a facsimile edition and annotated transcription of Alfred Russel Wallace's Species notebook of 1855–1859 / Alfred Russel Wallace ; annotated by James T. Costa.
        pages cm
    Includes bibliographical references and index.
    Summary: Notes from Wallace's Malay expedition.
    ISBN 978-0-674-72488-4 (alk. paper)
    1. Wallace, Alfred Russel, 1823–1913—Travel—Malay Archipelago.
2. Natural history—Malay Archipelago 3. Natural selection. 4. Evolution
(Biology) I. Costa, James T., 1963– II. Title. III. Title: Species notebook
of 1855–1859.
    QH375.W35 2013
    508.598—dc23         2013010172

*For Leslie,*
*tiger swallowtail mom,*
*with love*

# Contents

# Preface

Alfred Russel Wallace (1823–1913) would have appreciated the improbable and happenstance manner in which this project got its start: aboard the Oxford tube en route to London one characteristically wet English morning, I was reading a paper by the late historian of science H. Lewis McKinney. Wallace might, in fact, have considered it providential that I happened to be reading McKinney's account of Wallace's field notebook dubbed the "Species Notebook" not merely en route to London, but in fact on my way to the very locale where, according to the paper, this wondrous notebook resided: the library and archives of the venerable Linnean Society of London. Andrew Berry and I make an annual pilgrimage to the Linnean with our students in Harvard's Oxford-based summer program on Darwin and evolutionary biology, and we had never realized that most of Wallace's notebooks and journals from his travels in Southeast Asia sit demurely on a paneled bookshelf scant meters from the library where we have the privilege of viewing precious works on natural history. The sedate and stately Linnean is a far cry from the thatched huts and praus of the Malay Archipelago where they were written some 150 years ago, but it is a most fitting home considering that Wallace's notebooks and journals are now among the crown jewels of biology.

It was, therefore, with much anticipation that we approached Lynda Brooks, head librarian of the Linnean, about seeing the Species Notebook at the end of our program that day. Lynda kindly proffered it, and as we pored over its pages—some packed with Wallace's neat script, others with dense tables recording his daily catch and collecting triumphs and failures, and still others with lovely drawings and diagrams—I suddenly wondered, why has this notebook not been published? Biologists, students of evolutionary biology and its history, evolution mavens, Wallace and Darwin enthusiasts—all, I knew, would be keenly interested in its contents.

Concurring with McKinney, I saw that Wallace's Species Notebook was well named: it is a window into Wallace's early evolutionary thinking, encompassing the period in which his fertile mind was churning out paper after insightful paper as he searched for clues to the origin of species. The period of the notebook is, notably, punctuated by the publication of two watershed papers in the history of the field—Wallace's Sarawak Law paper (1855) and Ternate essay (1858)—among others. In its notes, sketches, and narratives we see Wallace the philosopher and Wallace the collector: the arguments for his planned book on transmutation are found side by side with specimen label designs; his discussions of the nature of species and varieties, the meaning of fossils, struggle in nature, or the branching history of life all share space with notes on the cost of rice, lists of books to read, a proposed remedy for the proliferation of taxonomic synonyms, and a practical scheme for a library of natural history. Then there are the myriad collection notes, from the hunting of orangutans and birds of paradise to his catches of curious beetles and beautiful butterflies.

Although historians have long been aware of the Species Notebook, and some, like McKinney, have studied it to good effect, beyond a small circle of Wallace scholars it has been a well-kept secret. Andrew and I wondered: could we make it more widely available? And perhaps not merely as a transcription but as an *annotated* transcription, to help readers more fully understand and appreciate the remarkable depth and breadth of Wallace's thinking? That last point is important. Wallace is decidedly underappreciated relative to his illustrious colleague Charles Darwin, all but unknown outside of scholarly circles and even little known within, and this despite his being not only the co-discoverer of one of humanity's greatest insights into the workings of nature—evolution by natural selection—but also the founder of the modern discipline of zoogeography. He made fundamental contributions in evolutionary biology, biogeography, ornithology, and entomology, and weighed in on a host of social issues of his day in a continuous stream of scientific papers, essays, interviews, and books. One does not have to be a conspiracy theorist to appreciate that history, in some respects, has not been just to Wallace.

That his gifts and contributions are not more widely appreciated today is our loss, for Wallace's story is nothing short of epic, his achievements all the more poignant for their realization against all odds. Wallace's unique blend of pluck, perseverance, and creativity were combined with the temperament of a philosopher, the sense of wonder of a child, and more than a dash of genius. This centennial year marking the naturalist's death in 1913 offers an opportunity for both celebration and reflection on the man and his accomplishments, and there is so much to celebrate and reflect upon. In this regard Wallace's Species Notebook is an unparalleled lens through which the scope of his early evolutionary thought is seen to advantage, in the context of life in the field where the ideas were conceived. It is in the spirit of enabling the reader to accompany Wallace on his journeys both physical and intellectual that my annotations are offered.

In the annotations, I also highlight the many points of parallel thinking, intersection, and occasional divergence between Wallace and Darwin during the notebook period and beyond. These should be read as an effort not to measure Wallace against Darwin but to show how very resonant was the thinking of these giants, a point that underscores the importance, even justice, of restoring Wallace's unwarrantably dimmed star to its former brilliance. The Species Notebook is an excellent starting point for appreciating just how profound Wallace's insights were in those years, pre–*Origin of Species,* when he labored in solitude.

We are fortunate indeed that his Species Notebook did not meet the fate of those from his South American journey, lost in the Atlantic as his homebound ship went first up in flames and then down into the depths. Considering the rigors of Wallace's peripatetic life in Southeast Asia, covering, by his reckoning, some 14,000 miles in eight years of travels, the survival of the notebook is itself cause for celebration. Through it we hear Wallace's voice loud and clear from the steamers and praus, huts and houses, jungle trails and trading villages across that infinite archipelago.

Berlin
FEBRUARY 10, 2013

# Note on the Text

## *Species Notebook Transcription*

The A. R. Wallace Species Notebook, Linnean Society of London manuscript number 180, was used "tête-bêche" by Wallace, with text running in opposite directions inward from each cover, flipped with respect to each other. These pages are here designated recto (going forward from the front of the book to the back) and verso (the reverse, running from back to front). This notebook transcription is based on (but not identical to) the Manuscript Transcription Protocol of the A. R. Wallace Correspondence Project at the Natural History Museum, London (www.nhm.ac.uk/wallacelettersonline).

The Species Notebook is paginated following Wallace's page numbers. Prefatory pages and endleaves with entries on them are designated [a], [a-1], [b], [b-1], and so on. The following notations are used in the transcription:

Every effort was made to reflect the layout of Wallace's original text in the transcription. Transcribed lines that for reasons of space wrap to the next line are indented on the wrapped line.

| | |
|---|---|
| Regular square brackets [ ] | Enclose editorial insertions, as when completing abbreviated terms: e.g., "coll." is completed as "coll.[ected]"; "coll.$^{d}$" as "coll.[ecte]$^{d}$" (exception: in lists and tables on the verso pages, words are left uncompleted as a space-saving measure) |
| Italic square brackets *[ ]* | Enclose editorial comments, which are themselves italicized |
| *[?]* | Indicates uncertain interpretation of text |
| *[illeg.]* | Indicates text is missing or illegible |
| Struck ~~text~~ | Text crossed out in original |
| Super- and subscript insertions in original | Shown in superscript or subscript |
| A full stop that appears as a dash in original | Transcribed as a period (.) |
| Wallace's spelling variants (e.g., Aru, Arru) | Not corrected |

## Annotations

### Place Names

The names of many of the locales (towns, islands) referenced by Wallace have changed over time. The place names Wallace used in the Species Notebook and in publications quoted by him are given; modern versions of these names are added parenthetically, except in quoted passages.

### Editions of Key Books

In his Southeast Asian travels Wallace used the fourth (1835) edition of Charles Lyell's *Principles of Geology* (four volumes); all quotations from Lyell's *Principles* come from this edition unless otherwise noted.

Much material from the Species Notebook was later incorporated into Wallace's acclaimed travel memoir *The Malay Archipelago,* first published in 1869 in two volumes by Macmillan and Company (London) and in the same year in a single volume by Harper & Brothers (New York). The single-volume American edition (see Wallace 1869a in the References) is used for all quotations and page numbers reported in the Species Notebook annotations; it is cited simply as *Malay Archipelago.*

Wallace referenced the second (1845) edition of Darwin's *Journal of Researches* in the Species Notebook. All references to Darwin's *On the Origin of Species* in the annotations pertain to the first (1859) edition unless otherwise noted.

*On the Organic Law of Change*

Alfred Russel Wallace (age thirty-nine), Singapore, February 1862. This photograph, the only one known of Wallace from the Malay Archipelago, was taken just before his departure for Britain. Copyright A. R. Wallace Memorial Fund and G. W. Beccaloni. Photograph generously provided by G. W. Beccaloni.

# Introduction

The year 1844 was momentous in the history of evolutionary biology. That fateful year saw the publication of *Vestiges of the Natural History of Creation,* the sensational manifesto on transmutation—what we now term biological evolution—published anonymously by the Scottish author and publisher Robert Chambers, as well as the drafting of a lengthy private essay on transmutation by the thirty-five-year-old rising naturalist Charles Darwin. That same year, twenty-one-year-old Alfred Russel Wallace, an unemployed sometime apprentice builder and surveyor, moved to the town of Leicester, England, to take a teaching post in a local school. Wallace spent little more than a year in Leicester, but in his own assessment his short stay there had a "determining influence" on his future. He concluded his discussion of Leicester in his autobiography, *My Life,* by declaring that his year there "must . . . be considered as perhaps the most important in my early life" (Wallace 1905, 1: 240).

. . .

The eighth of nine children, Wallace was born in 1823 into a genteel but impoverished family in Usk, Wales. The family moved often, as the finances of Wallace's father, lawyer and erstwhile publisher Thomas Vere Wallace, waned even as the size of his family waxed. When the young Wallace was five the family moved to Hertford, the hometown of his mother, Mary Ann Greenell, where he received his only formal education at the local grammar school. Continued deterioration of the family's finances compelled him to leave school at fourteen, and he then moved to London to live with his brother John for a time before becoming apprenticed to his eldest brother, William, a land surveyor (and thereafter, briefly, to a clockmaker). He returned to surveying with William in 1839, as Great Britain's burgeoning railway network and the controversial new enclosure laws, which required new surveys for the enclosure of common lands, had created a surveying boom. Living in Neath, on Swansea Bay, Wales, the brothers also became involved in a host of architectural and civil-engineering projects (Wallace would eventually design a new building for the Neath Mechanics' Institute, later the Free Library, erected in 1847 and still extant). Busts tend to follow on the heels of booms, and so it was with land surveying in Wales. A downturn forced Wallace to leave his brother's employment in early 1844, but he put his six and a half years of practical surveying experience to good use in securing a job as a schoolteacher in Leicester, hired by the local clergyman to teach surveying, mapping, and drawing.

Keen of mind and intensely curious, Wallace frequented the town and workingmen's libraries and mechanic's institutes wherever he lived, reading voraciously and attending lectures. He was a regular in the Leicester library, and the subjects he chose to explore reflected his somewhat iconoclastic personality: reading the great revolutionary Thomas Paine in Leicester and listening to utopian socialist Robert Owen in London were two formative experiences that tidily represent Wallace's openness to heterodox ideas (see Claeys 2008). It was in Leicester that Wallace first encountered mesmerism and phrenology, and read works scientific, philosophical, and social that were to have a lasting influence on him: Malthus's *Essay on Population* (1798), Combe's *Constitution of Man* (1828), Humboldt's *Personal Narrative* (1818), Darwin's *Journal of Researches* (1845), and Lyell's *Principles of Geology* (1830–1833), to name some of the more important authors. In Leicester he also met a kindred spirit in Henry Walter Bates, two years his junior, an apprentice hosier and ardent amateur entomologist who expanded Wallace's interest in nature, in particular inspiring him to become a keen beetle collector. A new world, that of natural history, fired his imagination (see Berry 2008).

Wallace had to leave Leicester in early 1845 when his brother William unexpectedly succumbed to pneumonia. After moving back to Neath to help sort out his brother's affairs, Wallace kept in touch with Bates, and they exchanged ideas and news of their latest insect catches. It was in Neath that Wallace first read the anonymously published *Vestiges of the Natural History of Creation* (1844). Victorian England was abuzz with the scandalous *Vestiges,* a book Wallace would eagerly devour, immediately convinced that its author was on to something. In a November 1845 letter to Bates Wallace wrote, "Have you read 'Vestiges of the Natural History of Creation,' or is it out of your line?" (Wallace Correspondence Project, WCP345). Bates evidently was not as taken with *Vestiges* as Wallace was, provoking Wallace to write a month later:

> I have rather a more favorable opinion of the "Vestiges" than you appear to have—I do not consider it as a hasty generalisation, but rather as an ingenious hypothesis strongly supported by some striking facts and analogies but which remains to be proved by more facts & the additional light which future researches may throw upon the subject.—it at all events furnishes a subject

for every observer of nature to turn his attention to; every fact he observes must make either for or against it, and it thus furnishes both an incitement to the collection of facts & an object to which to apply them when collected—I would observe that many eminent writers gave great support to the theory of the progressive development of species in Animals & plants. (Wallace Correspondence Project, WCP346)

The two kept up their correspondence, much of it spent reveling in their favorite topic: collecting insects, especially beetles. But clearly their interest in the diversity of insects extended to a broader philosophical interest in the nature of species and variation: "I begin to feel rather dissatisfied with a mere local collection; little is to be learnt by it. I should like to take some one family to study thoroughly, principally with a view to the theory of the origin of species. By that means I am strongly of opinion that some definite results might be arrived at," Wallace wrote in an 1847 letter to Bates. The letter concluded: "There is a work published by the Ray Society I should much like to see, Okens Elements of Physiophilosophy. There is a review of it in the Athenaeum. It contains some remarkable views on my favourite subject—the variations, arrangements, distribution, etc., of species" (Wallace Correspondence Project, WCP348). Wallace later quoted this letter in his autobiography to make the point that the question of species origins was very much on his and Bates' minds in those early days. "I was not satisfied with the more or less vague solutions [concerning species origins] at that time offered," he wrote. "I believed the conception of evolution through natural law so clearly formulated in the 'Vestiges' to be, so far as it went, a true one; and . . . I firmly believed that a full and careful study of the facts of nature would ultimately lead to a solution of the mystery" (Wallace 1905, 1: 257).

Bates visited Wallace in Neath in the autumn of 1847, and it was then that an audacious plan the two had been vaguely devising began to actually take shape: a journey to the Amazon, a locale inspired by W. H. Edwards's romantic book *A Voyage up the River Amazon* (1847). They arranged to meet in London to visit the entomological collections of the British Museum and the Royal Botanic Garden at Kew. Fortuitously, they met Edwards himself, who proved to be helpful and encouraging, and received advice and assistance from a host of others as well. Crucially, the pair got a crash course in the processing, preservation, and shipping of all manner of specimens (including insects, birds, mammals, reptiles, and fish), and were referred to Samuel Stevens, the highly respected natural-history agent of Bloomsbury Street, London. Himself a knowledgeable naturalist and the brother of a natural-history specimen auctioneer, Stevens took Wallace and Bates on as clients and under his wing, and he proved to be an able and enthusiastic promoter as well as supporter of their endeavors.

Seemingly against all odds, these earnest young men managed to pull off their great adventure, finding themselves collectors deep in Amazonia before the year was out. Wallace would spend the next four years there, Bates eleven. "The more I see of the country, *the more I want to,* and I can see *no end* of, the species of butterflies when the whole country is well explored," Wallace wrote to Stevens in 1849 (Wallace Correspondence Project, WCP 4267; original emphasis). The great motivation for Wallace's travels was the species question, and collecting was both the means to this end—by paying his way—and had the added bonus of providing ample raw material for the minute study of species diversity and variation. A steady stream of letter extracts and short articles by Wallace were published back in London, demonstrations of his industry and powers of observation; papers with titles like "On the Umbrella Bird *(Cephalopterus ornatus)*" (1850), "On the Monkeys of the Amazon" (1852), "On the Habits of the Butterflies of the Amazon Valley" (1854), and more.

The triumphs and tragedies of that first great expedition of Wallace's life have been compellingly recounted in several works published over the past decade or so (see, e.g., Berry 2002, 9–18; Fichman 2004, 24–29; Raby 2001, 34–82; Shermer 2002, 60–76; Slotten 2004, 47–83). Suffice it to say here that the tragedies—the death of his younger brother, Herbert, who had come from England to help him but succumbed to disease in Amazonia, followed by the loss of most of his private collection and all the specimens he had collected during the last two (and most interesting) years of his trip, when the ship he was sailing home on burned and sank in the Atlantic—were somehow balanced by the triumphs: the grandeur, sublimity, and excitement of the tropics, where, Wallace felt, one must be closer to the great engine of species creation, whatever that might be. He had already, too, built a reputation as a diligent and able collector and observer. Wallace was not back in England long before he began to feel the pull of the tropics, the collecting life, and of course the species question; there were observations to be made and data to be collected (and not incidentally, a living to be made as well). The solution to the mystery of how species change was waiting out there, he believed, a mystery that, as he had written nine years previously, "furnishes both an incitement to the collection of facts & an object to which to apply them when collected" (Wallace Correspondence Project, WCP346). Impressed by his achievements, the Royal Geographical Society agreed to facilitate his travels to another far-flung locale. He decided to head to the eastern tropics this time, to the vast, 2,500-mile-wide Malay Archipelago.

## The Species Notebook

Wallace, age thirty-one, arrived in Singapore on April 20, 1854, accompanied by a sixteen-year-old assistant, Charles Allen. The two spent their first few months exploring the immediate environment and preparing for extended journeys to collect specimens in the more remote and little-traveled regions of the archipelago. During his subsequent eight years of travels Wallace kept a series of journals and notebooks (described in the References section of this volume). Two of the notebooks (both registers of specimen collections: a bird and insect register and an insect, bird, and mammal register) are now in the collection of the Natural History Museum, London. Another, earlier notebook covering his initial collections in Singapore and peninsular Malaysia appears to have been lost sometime after his return to England. The remaining four notebooks and four travel journals were presented in 1936 to the Linnean Society by Wallace's son William.

The journals mainly consist of narrative accounts of Wallace's travels, and many of his descriptions of people, places, journeys, and other experiences served as primary material for *The Malay Archipelago,* his tour-de-force memoir published in 1869, seven years after his return. The notebooks, in contrast, are mainly devoted to records and descriptions of species and specimens, brief memoranda,

and notes taken on passages from a diversity of books, journals, and magazines. One notebook stands out as containing significant narrative on a range of subjects, from hunting orangutans in Borneo to extended critiques of naturalists on topics relevant to the species question (notably the eminent geologist Charles Lyell, and the noted comparative anatomists and paleontologists Richard Owen and Louis Agassiz). This notebook is thus of special interest, opening a window on Wallace's early thinking on transmutation. Dubbed the "Species Notebook" by the late American historian of science H. Lewis McKinney (1966, 1972), who first recognized its significance for understanding the development of Wallace's thinking, the period covered in this notebook saw the publication of Wallace's most revolutionary and insightful early papers written while he was in the field. The first of these was "On the Law Which Has Regulated the Introduction of New Species" (known as the "Sarawak Law" paper), which came out in 1855, followed by "Attempts at a Natural Arrangement of Birds" (notable for Wallace's evolutionary-tree device) in 1856. The brief "Note on the Theory of Permanent and Geographical Varieties" was published in 1858, the same year that the landmark "On the Tendency of Varieties to Depart Indefinitely from the Original Type"—the "Ternate essay"—was published, in which Wallace triumphantly announced his discovery of the mechanism of species change. One of his earliest studies of biogeography, "On the Zoological Geography of the Malay Archipelago," was published in the following year, 1859.

In mid-October 1854 Wallace departed Singapore for Sarawak, in northern Borneo, at the invitation of Sir James Brooke, the "White Rajah" who ruled Sarawak. It was there, albeit some months after his arrival, that the Species Notebook was begun. Wallace arrived in Borneo on November 1, 1854. In the first couple of months he collected at the mouths of the Sarawak and Santubong Rivers and upriver as far as the village of Bau, then spent Christmas as Brooke's guest. He was free to make use of Brooke's extensive library (which included *Vestiges,* among many other works of interest to Wallace), and he enjoyed nightly after-dinner philosophical discussions in the company of the Rajah and his circle. He eventually moved into a small bungalow provided by Brooke that sat at the mouth of the Sarawak River, at the foot of the striking Santubong Mountain, where he waited out the last of the rainy season. During that wait Wallace wrote his Sarawak Law paper, which he mailed off in February 1855; it appeared in the *Annals and Magazine of Natural History* in September of that year. The earliest date mentioned in the Species Notebook is March 12, 1855, when Wallace commenced a set of entomological notes after arriving at a new collecting locality on the Simunjon River in Sarawak. There, capitalizing on the fact that swarms of beetles were attracted to the downed timber cleared for the coal mining operations under development nearby, he had a small dwelling built with two rooms and a veranda. This would be his base of operations for the next nine months.

· · ·

The Species Notebook (manuscript 180 at the Linnean Society of London), is 11 by 17.8 centimeters in size and, like others among Wallace's notebooks, was used tête-bêche, with entries running in both directions from the ends of the book, inverted with respect to each other. The forward or recto direction consists of 181 pages including endleaves, while the opposite (verso) direction consists of about 80 pages that have entries. There is a rough chronology to the entries, but it is evident from the different inks used and varying penciled notes that Wallace edited and added information at various times, and it is likely that many changes were made later as he mined the notebook for material to incorporate into his voluminous writings. In fact, although by convention this notebook is labeled as spanning the years 1855 to 1859, there are several entries, both recto and verso, that date to as late as 1862, some perhaps later.

The subjects covered in the Species Notebook are varied, as would be expected in a working field notebook. The entries largely consist of extracts, notes, and memoranda on a diversity of books and articles; musings on practical matters like arranging taxonomic catalogs and formatting specimen labels; species lists; ethnological observations; accounts of orangutans; and observations of various birds and insects. In some ways this notebook is more like a commonplace book; it stands out from the other notebooks in containing significant narrative on a range of subjects. These more lengthy entries decrease in frequency after mid-1856. In June of 1856, on the island of Bali, Wallace decided to commence a detailed travel narrative that might serve as the basis for an eventual memoir. Thus was opened the first of what grew into four "Malay Journals" (Linnean Society manuscripts 178a–d). From the time of his Bali visit onward, these journals replaced the notebook as the primary running record of his experiences, and they became the basis for *The Malay Archipelago* (1869). Many notes and observations found in the Species Notebook, however, were also incorporated into *Malay Archipelago* and other writings.

The journals and *Malay Archipelago* lack Wallace's musings on the nature and transmutation of species. Entries on these subjects, which were never far from Wallace's thoughts, are scattered liberally throughout the Species Notebook and reflect the mode of Wallace's pro-transmutation reasoning and his broad interests as he amassed evidence for species change. His astonishingly far-ranging notes, observations, and arguments bear on many facets of the species problem: critiques of arguments for "proofs of design"; discussion of transitions between species; the geological record; morphology and affinity of species; relationships of island species to continental species; the nature of instinct; domestication; speculations on the nature of knowledge, authority, what constitutes evidence; and more (all of which are summarized in Appendix 1). Scattered throughout are also observations about the native peoples he encountered. In the Kai Islands, where he spent a week in January 1857, he encountered for the first time people of Papuan ethnicity. From the Aru Islands later that month, Wallace commented in his journal that "the human inhabitants of these forests are not less interesting to me than the feathered tribes." Humans were a single species, Wallace believed, that clearly consisted of several well-marked "varieties" as well as intermediate and intermixed forms. What was the origin of human races? His view was influenced by Sir William Lawrence's *Lectures on Physiology, Zoology and the Natural History of Man* (first published in 1819; abbreviated to *Lectures on Man*) and James Cowles Prichard's *Researches into the Physical History of Mankind* (first published in 1813), works that hinted at a transmutational origin of humans, and that he discussed and recommended to Bates in 1845. On Aru he noted in his journal that "the Malay & the Papuan appear to be as widely separated as any two human races can be . . .

It is a most interesting question & one to which I shall direct my attention in all the islands of the Archipelago I may be enabled to visit" (first Malay Journal, Linnean Society ms. 178a, entry 63).

In the longest and most important narrative section of the Species Notebook, Wallace critiqued the anti-transmutation arguments given by Charles Lyell in the *Principles of Geology* (see Appendix 2). The importance of Lyell for understanding Wallace's transmutation entries in the Species Notebook cannot be overestimated. Lyell, who by the time of Wallace's sojourn in Southeast Asia was the leading geologist in Britain, had displayed a certain openness to the idea of transmutation on his initial reading of French "transformist" notions in the late 1820s. By the time he wrote the *Principles,* however, which first appeared in three volumes between 1830 and 1833, Lyell was firmly anti-transmutationist and went to great lengths to attack the concept (see Bartholomew 1973; Corsi 1978; Rudwick 1970, 1998). Indeed, Lyell's anti-transmutation arguments in the *Principles* were widely considered to be the definitive (and damning) statement on the subject.

Of greatest interest, therefore, among the entries in the Species Notebook are certainly the twenty-five-plus narrative pages of evolutionary musings aimed at Lyell. Although these entries are undated, John Brooks (1984) reasoned that they were made in mid-1855, pointing out that in *Malay Archipelago,* immediately following material corresponding to the Species Notebook entry of June 27, 1855, Wallace wrote that he "had the misfortune to slip among some fallen trees" and hurt his ankle. He continued: "Not being careful enough at first, it became a severe inflamed ulcer, which would not heal, and kept me a prisoner in the house the whole of July and a part of August" (Wallace 1869a, 40; hereinafter cited as *Malay Archipelago*). Brooks rightly noted that it was during periods of forced inactivity that Wallace was often at his most prolific with letters and papers, so it seems quite reasonable to suppose or suggest that he used this "down time" in the summer of 1855 to (re)read and critique Lyell.

He also came up with an ambitious new project. McKinney (1966, 1972) argued that it is among the entries from the middle months of 1855 that Wallace appears to reveal his plan for a book on transmutation. The narrative section regarding Lyell commences on recto page 34, and the very next page is headed "Note for Organic law of change." Near the conclusion of this series of entries, on recto page 51, Wallace inserted an intriguing memorandum consistent with McKinney's suggestion: "? Introduce this and disprove all Lyells arguments first at the commencement of my last chapter." The reference to a "last chapter" certainly suggests a *series* of chapters—that is, a book. What does "Introduce this" refer to? It appears to refer to the very next passage, a criticism of Lyell's claim that species cannot change. In other places, too, Wallace hinted at his plan for a magnum opus on the topic of species change. The most telling of these hints is found in a letter to his friend Bates dated January 1858. Bates, still in Amazonia, had sent Wallace a congratulatory letter written in November 1856. In the letter he applauded Wallace's Sarawak Law paper, published the previous year. "I was startled at first to see you already ripe for the enunciation of the theory," Bates wrote. "You can imagine with what interest I read and studied it, and I must say that it is perfectly well done." It is worth quoting Bates more extensively:

The idea is like truth itself, so simple and obvious that those who read and understand it will be struck by its simplicity; and yet it is perfectly original. The reasoning is close and clear, and although so brief an essay, it is quite complete, embraces the whole difficulty, and anticipates and annihilates all objections. Few men will be in a condition to comprehend and appreciate the paper, but it will infallibly create for you a high and sound reputation. The theory I quite assent to, and, you know, was conceived by me also, but I profess that I could not have propounded it with so much force and completeness. Many details I could supply, in fact a great deal remains to be done to illustrate and confirm the theory: a new method of investigating and propounding zoology and botany inductively is necessitated, and new libraries will have to be written. (Wallace Correspondence Project, WCP824)

Wallace was away collecting in New Guinea when Bates's letter arrived in Macassar (Makassar). Once he returned to Macassar he had some six months of correspondence to catch up on, and he replied at last to Bates on January 4, 1858: "To persons who have not thought much on the subject I fear my paper on the succession of species will not appear so clear as it does to you. That paper is, of course, only the announcement of the theory, not its development. *I have prepared the plan and written portions of an extensive work embracing the subject in all its bearings and endeavouring to prove what in the paper I have only indicated*" (Wallace Correspondence Project, WCP366; emphasis added). Here, then, is yet another indication of the book-length treatment that Wallace was planning. McKinney (1972, 32–33) suggested that Wallace's book would have been entitled "On the Organic Law of

Change," following the convention he used for the titles of several works (including the 1855 and 1858 papers). I imagine that the book would have been structurally similar to Darwin's *Origin of Species,* with a section echoing the themes of the Sarawak Law paper on lines of evidence in support of the idea of species change, and a section echoing the Ternate essay, discussing evidence for the mechanism of species change.

That book was not to be. Wallace's most famous book-length treatment of the subject would not appear for another thirty years, and then it carried the title *Darwinism* (1889). The story of Wallace's eventual discovery of the mechanism behind species change in early 1858, sparked by his recollection of Malthus while bedridden and feverish, has been often recounted, and the events leading up to the "delicate arrangement" at the Linnean Society on July 1, 1858—the hasty reading of Wallace's essay along with Darwin's private earlier writings on the subject, arranged by Lyell and Joseph Hooker—continue to be scrutinized (see, e.g., Beddall 1988, Brackman 1980, Brooks 1984, McKinney 1972, Rachels 1986). Darwin's *Origin* was published nearly a year and a half later in November 1859, while Wallace was still in Southeast Asia. Wallace received a copy in early 1860 in Amboyna (modern Ambon), while recuperating from a collecting fiasco on Seram. He read and reread the book some five or six times, he wrote to his friend George Silk in September of that year (Wallace Correspondence Project, WCP373). It did not take long to convince him that the book was a watershed event in the history of ideas, as he expressed in a letter to Bates a few months later: "I know not how or to whom to express fully my admiration of Darwin's book . . . Mr. Darwin has created a new science and a new philosophy, and I be-lieve that never has such a complete illustration of a new branch of human knowledge been due to the labour and researches of a single man. Never have such vast masses of facts been combined into a system, and brought to bear upon the establishment of such a grand and new and simple philosophy!" (Wallace Correspondence Project, WCP374). In a subsequent letter to his brother-in-law Thomas Sims he referred to Darwin as the "Newton of Natural History" (WCP3351), and he declared to his friend George Silk that the *Origin* "will live as long as the 'Principia' of Newton," and that Darwin's name "should, in my opinion, stand above that of every philosopher of ancient or modern times. The force of admiration can no further go!!!" (WCP373).

Wallace evidently felt no need to come out with his own book-length statement on the subject, impressed as he was with Darwin's work; or perhaps he recognized how difficult it would be to do so until he was back in England and had access to libraries, and until his copious collections and notes were organized and papers written. (Wallace had commented to Darwin in a letter dated September 27, 1857, that his Sarawak Law paper was a "mere statement & illustration of the theory," which he intended to treat more extensively, "the plan of which I have arranged, & in part written, but which of course requires much [research in] libraries & collections" [Wallace Correspondence Project, WCP4080].)

Wallace's return home was a few years away, and once he got there the organization and study of his vast collection was a daunting task. He was stunningly prolific, however, publishing nearly fifty papers and commentaries in the three-year period from 1862 to 1864 alone. Several of the papers Wallace produced in the years immediately following his return were (and still are) hailed as groundbreaking: "On the Physical Geography of the Malay Archipelago" (published in 1863), "The Origin of Human Races and the Antiquity of Man Deduced from the Theory of 'Natural Selection'" (1864), and "On the Phenomena of Variation and Geographical Distribution as Illustrated by the Malayan Papilionidae" (1865), among others.

Wallace did not stop there, of course. The period of the Species Notebook is merely the first great chapter in the development of Wallace's evolutionary ideas; in an important sense, Wallace's long sojourn in the Malay Archipelago—physical and intellectual—was more beginning than end. Wallace rightly regarded his epic journey as "the central controlling incident" of his life, and he was received with laurels aplenty on his return home in the spring of 1862. The riches of his collections, notebook, and journals would serve him well for years to come as he published a steady stream of papers and book after book, beginning, as already noted, with *The Malay Archipelago* in 1869, his best-selling travel memoir that went to ten editions and has never been out of print. Beyond the travelogue, he published a host of scientific books: *Contributions to the Theory of Natural Selection* (1870); the two-volume *Geographical Distribution of Animals* (1876); *Tropical Nature, and Other Essays* (1878); *Island Life: Or, The Phenomena and Causes of Insular Faunas and Floras* (1880); *Darwinism: An Exposition of the Theory of Natural Selection, with Some of Its Applications* (1889); and *Natural Selection and Tropical Nature: Essays on Descriptive and Theoretical Biology* (1891). There were many other books and hundreds of papers on a diversity of scientific, social, and even pseudo-scientific topics, some of which

had Wallace swimming against prevailing social and scientific currents (thus diminishing his stature in the eyes of some in the scientific establishment). *The Geographical Distribution of Animals* and *Island Life,* in particular, cemented Wallace's reputation as a scientific star of the first magnitude and are recognized today as watershed works in putting zoogeography—the study of pattern and process in the geographical distribution of animals—on its modern footing.

In the wake of the excitement (and controversy) over the theory of transmutation by natural selection—what was too often called simply "Darwin's theory" or "Darwinism," a label that Wallace himself used—there was the danger that later generations would mistake Wallace's discovery of natural selection as a lucky break, a one-off coup for one who was otherwise simply a collector-explorer, even if a great one. The Species Notebook is a wonderful corrective for such a misguided view. It is nothing short of astonishing that the autodidact Wallace, a transmutationist since at least 1845, succeeded in his goal of uncovering the mechanism of species change little more than a dozen years after setting out to do so. But set out to do so he did; and what is underappreciated is the depth and breadth of Wallace's insights into the subject of transmutation, even well before the reading of the Linnean Society papers. The Species Notebook, presented in its entirety here for the first time and fully annotated, provides an unparalleled window into Wallace's thinking during what was perhaps his most creative period: a living document that puts us in touch with both the rigors of his fieldwork in the "back of beyond" that was the vast Malay Archipelago and the creative ferment of his keen and insatiably curious mind. This notebook, in short, underscores Wallace's fully independent and incisive insights into the nature of species change and his standing as an originator of modern evolutionary biology.

. . .

The brilliant Thomas Paine, self-taught, like Wallace, and influential in both the American and French Revolutions, wrote to George Washington in 1789 that having "a share in two revolutions is living to some purpose." The same can be said of Wallace, who read Paine's *Age of Reason* with gusto in 1837, just fifty years after Paine penned those words to Washington (and the very same year, as it happens, that Darwin became a transmutationist). In founding one field—modern biogeography—and cofounding another—evolutionary biology—Wallace, too, had a share in two revolutions, and surely lived to some purpose. By the time of Wallace's death in 1913 he had received many of Britain's greatest honors—medals, citations, honorary degrees, and even the Order of Merit, the greatest civilian honor bestowed by the reigning monarch, in recognition of a life of distinguished accomplishment. Musing in 1904 on who in future ages should be recognized as the "most important and significant figure of the nineteenth century," the English writer G. K. Chesterton remarked that he would "hesitate between Walt Whitman and Alfred Russel Wallace." Wallace was celebrated and feted wherever he went (not always gladly, given his retiring nature—as when he politely endured a visit to the White House at the invitation of President Grover Cleveland).

It is lamentable that while Wallace was one of the world's most famous people at the time of his death 100 years ago, he has since been all but eclipsed by his brilliant contemporary, Darwin. But honoring one of these giants need not come at the expense of the other, for both were our first guides to the evolutionary process. While certainly not all of their ideas about species change have stood the test of time in the modern view, Wallace and Darwin are equally honored as pioneers and discoverers of one of the most profound insights into nature—and ourselves—yet grasped by humanity. Darwin's monumental *On the Origin of Species* has no clear counterpart in Wallace's opus, however, since the book-length treatments Wallace produced were all post-*Origin* by many years. It is thus my hope that Wallace would approve of this presentation and interpretation of his epic Species Notebook, offered as both a window into his clarity of insight during those formative years of his "evolutionary travels" in the eastern archipelago, and as a fitting homage to "On the Organic Law of Change," the book that should have been.

## A Guide to Wallace's Travels in the Malay Archipelago

The map and chart that follow illustrate Wallace's travels in the Malay Archipelago—consisting of modern Singapore, peninsular Malaysia, Indonesia, and New Guinea—between 1854 and 1862. By his reckoning Wallace covered approximately 14,000 miles in his eight years in the archipelago and collected some 125,660 specimens, more than 83,000 of beetles alone.

## Wallace's Itinerary

| | |
|---|---|
| Singapore (first visit) | April–October 1854 |
| Borneo | November 1854–January 1856 |
| Singapore (second visit) | January–May 1856 |
| Bali | June 1856 |
| Lombock [Lombok] | June–August 1856 |
| Celebes [Sulawesi] (first visit, Macassar district) | September–December 1856 |
| Ké Islands [Kai] | January 1857 |
| Aru Islands | January–July 1857 |
| Celebes (second visit, Macassar district) | July–November 1857 |
| Timor (first visit, Kupang) | November 25–26, 1857 |
| Banda (first visit) | December 1857 |
| Amboyna [Ambon] (first visit) | December 1857–January 1858 |
| Ternate (first visit) | January and March 1858 |
| Gilolo [Halmahera] (first visit) | February 1858 |
| Dorey [Dore Baai] | April–July 1858 |
| Ternate (second visit) | August–October 1858 |
| Gilolo (second visit) | September 1858 |
| Tidore | October 1858 |
| Makian (first visit) | October 1858 |
| Kaióa [Kayoa] (first visit) | October 1858 |
| Batchian [Bacan] | October 1858–April 1859 |
| Makian (second visit) | April 1859 |
| Ternate (third visit) | April–May 1859 |
| Timor (second visit, West Timor) | May 1859 |

| | |
|---|---|
| Banda (second visit) | June 1859 |
| Celebes (third visit, NE Celebes) | June–September 1859 |
| Amboyna (second visit) | September–October 1859 |
| Ceram [Seram] (first visit, SW coast) | October–December 1859 |
| Amboyna (third visit) | December 1859–February 1860 |
| Ceram (second visit, SE and NE coasts) | February–April and June 1860 |
| Matabello [Watubela] and Goram [Gorong] Archipelago | April–May 1860 |
| Waigiou [Waigeo] | July–September 1860 |
| Waigiou to Ternate | September–November 1860 |
| Ternate (fourth visit) | November 1860–January 1861 |
| Timor (third visit, East Timor) | January–April 1861 |
| Banda (third visit) | April–May 1861 |
| Bouru [Buru] | May–June 1861 |
| Ternate (fifth visit) | June–July 1861 |
| Java | July–November 1861 |
| Sumatra | November 1861–January 1862 |
| Singapore (third visit) | January–February 1862 (Departs for England February 8; arrives there March 31, 1862) |

*Note: Wallace's travel itinerary is modified from Baker (2001). Modern spellings of place names are indicated in brackets.*

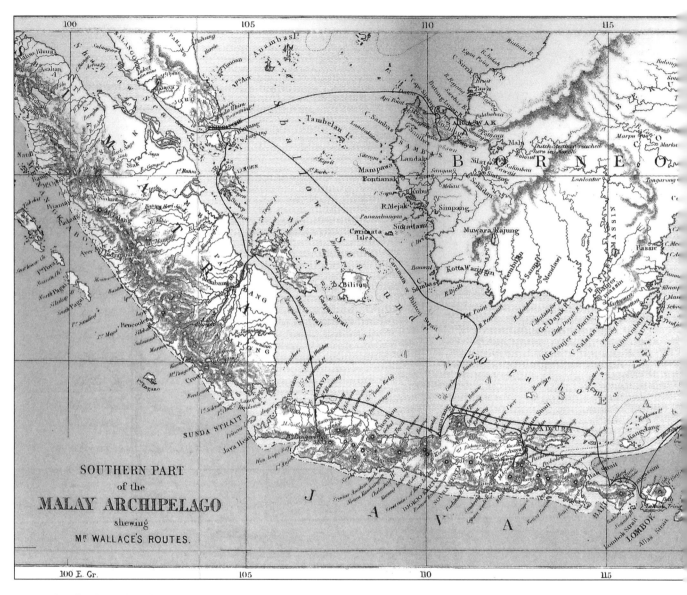

A map of Wallace's travels in the Malay Archipelago. From Alfred Russel Wallace, *The Malay Archipelago*, 4th ed. (London: Macmillan & Co., 1872). Wallace reckoned that over the course of his eight years in the region he traveled some 14,000 miles in sixty to seventy separate journeys. Note that the map shows the 100-fathom depth line in the eastern and western archipelago. The islands in the west sit on the Sunda Shelf, an extension of the continental shelf of Southeast Asia. The Sahul Shelf extends from the greater Australian landmass to include New Guinea

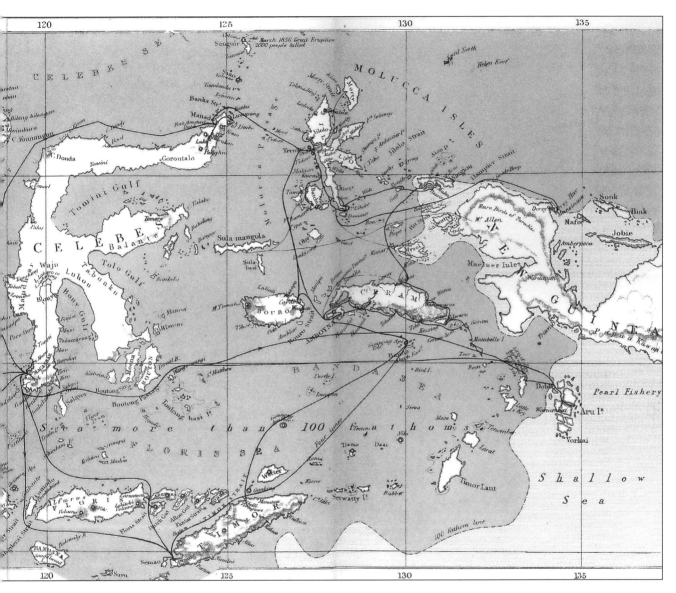

and the Aru Islands in the westernmost part of the archipelago. The eponymous Wallace Line skirts the Sunda Shelf. Wallace realized that the geological history of the archipelago shapes the distribution of its species more profoundly than environment, a key insight that resonated with the growing evolutionary vision of species origins and history.

Recto Notebook

[1]Notes

    Vertebrata

*[The inside recto cover and following endpage were left blank by ARW and are omitted]*

1. Wallace labeled the recto notebook "Notes Vertebrata." The verso notebook cover is labeled "Notes. Insects" with a prominent numeral "4."

(Note that the following two blank pages, i.e. inside recto cover and following endpage, are omitted)

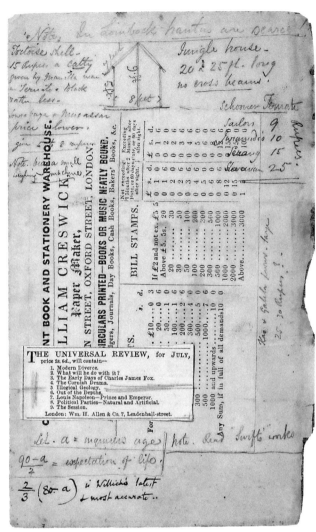

Note.   In Lombock [1]<u>hantus</u> are scarce!!

*[Sketch of house with*
*measurements below—*
*some written vertically]*                    7 feet   5 f[ee]t.
                                              9 f[ee]t  < 8 feet >

Tortoise shell—                               Jungle house—
15 Rupees a <u>catty</u>                      20 or 25 f[ee]t. long
given by Manilla man                          no <u>cross beams.</u>
in Ternate. Black
rather less.
Soura baya & Macassar                         Schooner Ternate
price lower.                                  Sailors 9      Rupees
                                                         *[Written vertically]*
_____                       *[illeg.]* 10
give 5 to 8 rupees                            Serang 15
_____                       *[illeg.]* 25
Note buy a small
weighing machine.
_____

*[The following is written vertically in right margin of page]*
Kao & Galela prows—large
25–30 Rupees?
_____

[2]*[Slip pasted in—Universal Review]*
[3]Let. a = inquirers age          | Note. Read "Swifts'" works
$\frac{90 - a}{2}$ = expectation of life.  _____

_____

$\frac{2}{3}$ (80 - a) is [4]Willich's latest
            & most accurate . .

1. *Hantu* is the Malay word for ghost. An elaborate assemblage of *hantus* are found in Malay lore and legend. Wallace's note on the scarcity of *hantus* in Lombock (Lombok)—the exclamation mark perhaps indicating mock surprise—stems from a curious exchange he recorded while visiting that island, which he relates in *Malay Archipelago*:

> A Bornean Malay who had been for many years resident here said to Manuel, "One thing is strange in this country— the scarcity of ghosts." "How so?" asked Manuel. "Why, you know," said the Malay, "that in our countries to the west- ward, if a man dies or is killed, we dare not pass near the place at night, for all sorts of noises are heard which show that ghosts are about. But here there are numbers of men killed, and their bodies lie unburied in the fields and by the roadside, and yet you can walk by them at night and never hear or see anything at all, which is not the case in our country, as you know very well." "Certainly I do," said Manuel; and so it was settled that ghosts were very scarce, if not altogether unknown in Lombock. I would observe, however, that as the evidence is purely negative we should be wanting in scientific caution if we accepted this fact as sufficiently well established. (*Malay Archipelago*, 171)

2. This slip advertises the July 1859 issue of the *Universal Review*. Wallace's interest lies in the essay "Illogical Geology," by the social philosopher Herbert Spencer (1820–1903). Spencer had a great impact on Wallace's thinking (see recto 178). This essay attacked what Spencer took as inconsistencies in the reasoning of geolo- gists.

3. Wallace's interest in life expectancy calculations was connect- ed to insurance. At the head of the third Malay Journal (LINSOC- MS178c), which was initiated March 1, 1858, there is an entry re- lating to a life annuity, including essentially the same formula as that seen here in the Species Notebook:

if a = age then $\frac{90 - a}{2}$ = expectation of life.

4. Charles Madinger Willich (d. 1867) was an actuary who pub- lished extensively on the statistics of life expectancy. It is unclear when Wallace made this entry; he here gives Willich's "latest & most accurate" formula for life expectancy, which could have come from any number of sources. Willich published the book *Popular Tables* in 1852, and several papers on life expectancy in the *Journal of the Institute of Actuaries* in the 1850s.

3.

CERAMIC COURT AT THE CRYSTAL PALACE.

THE animal is a conservative, but man is a reformer. The nest of to-day's partridge is like the
nest of the bird Noah sheltered in the ark; but
from the reed cabin or the black tents of Ham
sprang the marble palaces of Rome.

---

mient, true, and pleasing likeness is iusired; also tinted or
coloured in the highest style of Miniature Painting by the best

No 1512, OCT. 18, '56

the burning of gas, had but a very limited
existence.
The experiments were repeated on nights more
or less foggy, and the results (which, however, are
not included in the table) were in accordance with
what had been observed by the authorities already
quoted, namely, that fog does not exercise the same
power as clouds in reflecting back the heat emitted
by radiation or otherwise from the earth.

FOREIGN CORRESPONDENCE.

Bordeaux, September.

ABOUT as far away from London, in point of

These slips, advertisements for lectures and such, reflect Wallace's eclectic interests at this time.

1. Comparative anatomist and paleontologist Richard Owen's Reade Lecture "On the classification and geographical distribution of the Mammalia," given May 10, 1859.

2. Baden Powell (1796–1860) was a mathematician and an Anglican clergyman. Perhaps unusual for clerics of his time, Baden Powell was a liberal-minded advocate for uniformity and universality of natural law, leading him to accept the "development hypothesis" early on. *The Order of Nature: Considered in Reference to the Claims of Revelation,* published in 1859, argued (among other things) that natural selection made creation rational. This theme was amplified in *Essays and Reviews,* a volume published in early 1860 by Baden Powell and several other liberal theologians, joining in the debate over Darwin and Wallace's ideas.

3. The entrance to the Manchester Court, also known as the Ceramic Court, was one of the Industrial Courts of the Crystal Palace. It opened in 1857. The article is from the October 18, 1856, issue of *The Athenaeum.*

A.[lfred] R.[ussel] Wallace MSS. [manuscripts]
*[written in blue/purple ink—ball point pen?]*
<u>Addresses</u>
[1]At. Mess.ʳˢ Hamilton Gray & Co.[mpany] Singapore
    Nicol Esq.[uire]          R. Padday Esq.[uire]
    Jar*[illeg.]* Esq.[uire]
[2]J.[oseph] Barnard Davis Esq[uire] Shelton, Staffordshire

---

[3]Mrs. Wallace—Columbia Cottage
    Albion Road, Hammersmith.

---

[4]Dr. R. Bernstein
    Medicin a l'estabissement de
    Santé à Gadok
Naturalist        Biutenzorg
<u>Bird collector</u>, will exchange duplicates

---

[5]Rozenberg *[sic]*, of Darmstadt—has collection
    of Birds of Sumatra & Java—in 5 or 6 years
    will return. Dorey. June 1858.

---

*[Pasted slip with Hermann von Rosenberg signature]*

In Vol.[ume] 4. of [6]Family Herald—Leading articles
on      Old Opinions compared with new
        Reformation Revolution & Reform
        Learning Greek & Latin
        Wisdom & Knowledge
        Cruelty of Nature
        Is there a moral reforming power?
        Ireland—
        The poor—   <u>are all admirable</u>

1. Based in Singapore from 1832 to 1886, Hamilton Gray and Co. was one of several shipping agents.

2. Joseph Barnard Davis (1801–1881) of Staffordshire, was a medical doctor and anthropologist who authored many works on craniology, including *Crania Britannica* (1865) and *Thesaurus Craniorum* (1867). Davis had an extensive collection of skulls and may have been a prospective or actual buyer of primate specimens from Wallace.

3. Wallace's letters to his mother, Mary Ann Wallace (née Greenell), helped establish her places of residence. She lived with her daughter Fanny for most of the period that Wallace was in Southeast Asia, but sometime between 1853 and 1862 she evidently lived for a short time at this Hammersmith address.

4. Heinrich A. Bernstein (1828–1865) was a naturalist and physician commissioned to collect for the National Museum of Natural History in Leiden. Bernstein is described in Karl Ritter von Scherzer's *Narrative of the Circumnavigation of the Globe by the Austrian Frigate* Novara (1862):

> Dr. Bernstein is . . . at once a zealous collector, and a skilful preparer, who has already made some very beautiful collections, and who, if he stay here any length of time, will be in a position to enrich considerably the museums of natural history in Europe, with numerous rare and valuable specimens.

Wallace mentioned Bernstein twice in *The Malay Archipelago* in connection with bird collecting. In one of his letters to Stevens he wrote, with a touch of envy:

> The Dutch have just sent out a collector for the Leyden Museum to the Moluccas. He is now at Ternate, and goes to spend two years in Gilolo and Batchian, and then to N. Guinea. He will, of course (having four hunters constantly employed, and not being obliged to make his collecting pay expenses), do much more than I have been able to do; but

I think I have got the cream of it all. His name is Bernstein; he has resided long in Java, as doctor at a Sanatorium, and tells me he has already sent large collections to Leyden, including the nests and eggs of more than a hundred species of birds! Are these yet arranged and exhibited? They must form a most interesting collection. (Wallace 1861, 310)

5. Hermann von Rosenberg (1817–1888), German naturalist from Darmstadt. Rosenberg's entire career was spent in the Dutch East Indies and the Netherlands. Between 1840 and 1856 he worked as a cartographer on Sumatra and neighboring islands, and later as a cartographer and surveyor in the Moluccas and western New Guinea. Beginning in the 1860s he collected bird specimens for Hermann Schlegel at the National Museum of Natural History in Leiden. He returned to Europe in 1871, where he produced his best-known work, *Der Malayische Archipel* (1878)—like Wallace's book of the same title, this was an epic travel memoir documenting Rosenberg's thirty years in the archipelago.

Wallace crossed paths with Rosenberg in June 1858 on a steamship at Dorey, in western New Guinea, as he later recorded in *Malay Archipelago* (508): "I found a brother naturalist, a German named Rosenberg, who was draughtsman to the surveying staff." Rosenberg is mentioned in one other place in *Malay Archipelago*: of the paucity of birds in Celebes (Sulawesi), Wallace commented that "the Dutch naturalists Rosenberg and Bernstein" had made extensive collections in that region but ultimately only added eight new species to his own list of collections. (For Bernstein, see note 4, above.)

6. *The Family Herald: A Domestic Magazine of Useful Information & Amusement* (1843–1940) was a weekly paper established by James Elishama Smith in 1843. True to its subtitle, the paper featured articles, anecdotes, advertisements, and correspondence on many subjects: "Narratives, Allegories, Poetry, Philosophy, Science, Art, Customs, Manners, Historical Essays, Natural History, Inventions, Discoveries, Biography, Statistics, etc. etc."—as the "admirable" articles of volume 4 (1846–1847) that Wallace listed suggest.

[1]<u>Canaries</u> <sup>only 27 ferns.</sup> 533 sp.[ecies] of flowering plants. 310
  peculiar

---

St. Helena—61 sp[ecies] ———— 58 peculiar ————

---

? the Geological age of these islands.—
    874 flowering plants near Berlin.

---

[2]<u>Index.</u>                              N.[ew] Zealand, 750 flowering plants
Entomological notes. p.[age] 1          Madeira, under 700.
Habits of Mias .  p.[ages] 7–9          Insects of N.[orth] Africa &
                                            Europe
    "   of Mecopus .  p[age] 8                         p.[age] 177.
Mias notes on . p.[ages]                Longicorns, habits.—p[age] 136
    10. 11. 13 to. 30
Proofs of Design . p.[ages] 12. 31. 32. 53
Lyell extracts . p.[ages] 34 to 53.
Owen. Extinct reptiles . . p.[age] 54.
Name tickets for coleoptera. p.[ages] 36. 42.
Plan for arranging collection of Coleoptera 124.
On synonyms &c. . . . 126.
Eos              Batchian—  p.[age] 75
Eos              Mafor, p.[age] 107.
Von Buch on change of species . p.[age] 90.
Mites—p.[age]134
Lorius lory—habits. 160
Paradisea rubra . d[itt]o. 161
Semeroptera—p.[age] 135

---

¾ in[ch] good for        ? Put the smaller sp.[ecies] of pinned
                              coleop.[ter]<sup>a</sup>
<u>species labels</u>, for       in pairs on small squares of papered cork
<u>Coleoptera</u>,            & pin with a no 13) . or  11) pin & one
more for <u>Lepidop.[tera]</u>   locality ticket.

Paste some of the <u>local</u> tickets on cork & cut it in squares
for the purpose—each name with sq[uare] line round it ? best
½ inch ~~squares~~ by ⅜ deep.

1.  These notes pertain to proportions of endemic (unique) plant species from a host of island systems, reflecting not only Wallace's biogeographical interests but also his evolutionary interests. Note his remark regarding the geological age of the islands. He understands that island antiquity and remoteness are related to levels of endemism. The data given here indicate that the Canary Islands exhibit 58 percent endemism, while St. Helena (far more remote, but is it older or younger?) is at 95 percent endemism. This is a recurrent theme in the Species Notebook (see recto 46–49, 90, 151).

2.  As will be seen in the subsequent annotations, Wallace mined this notebook extensively for his later writings, and he prepared this index for reference. Rather than entering the index all at once when the notebook was full, he made index entries as he went, as evidenced by the different inks and pencil used.

[b]

[1]In July & August in New Guinea the Teak
trees were in fruit & frequented by <u>Paradiseas</u>
No teak in N.[ew] Guinea. They are           <u>Lesson</u>
Indian fig trees.

---

[2]Lyell —<u>1835</u>

| Mammals — | 800 | British Plants flowering | 1500 |
| Fishes — | 6000 | " Insects ——— | 12,500 |
| Birds — | 6000 | | |

---

782 species of fishes known to [3]D.[r] Bleeker
from Amboyna—in Dec.[ember] 1856.

———

906 species of d[itt]o. from Molucca sea in July
                                        <u>1856</u>.

———

? numbers of <u>large</u> species not collected.

N[ota] B.[ene] much smaller type will be better—
? have                            Specimen of Locality
room left                         Tickets to be attached
for a no.[r] [number]             to every insect in my
to every                          private collection of
specimen   *[Specimen of locality*   Coleoptera [& Lepidoptera]. To be printed
for easy   *ticket sketch and pasted]*either with lines between
reference in                      or with regular spaces
naming &c.                        so as to be easily cut.
no.[s] [numbers] a series
for <u>each local.</u>

The names to be repeated according to no. [number] of specimens.
Thus <u>Sarawak</u> will want about 3000 tickets, so
repeat the name 30 times, if to print 100 copies.
Macassar 1200, must be repeated 12 times &c.
This will make about as much as to fill an 8vo. [octavo] page
which will probably be cheaper than 1/2 size & 200 copies.

---

The lower names are best I think—printed in <u>one line</u>, with
plenty of room beneath for a <u>reference number</u>, the pin between
them. My longest name, "Singapore", is shorter than "Burlington",—most are
the length of "Glasgow"—Printed in columns—cheap
*[The following is written vertically in left margin of page]*
*[illeg.]* on stout paper.

1. Wallace spent July to September 1860 collecting in Waigiou (Waigeo), just to the west of New Guinea, with hopes of procuring more birds of paradise. This was the last Papuan territory he planned to visit, and his voyage there from Ceram (Seram) was nearly fatal, owing to sailing mishaps. At length he succeeded in procuring these coveted and secretive birds; this notebook entry reveals the key to his success: he was fortunate to have near his abode an Indian fig tree with ripening fruit, and male red birds of paradise (*Paradisea rubra*) visited the lofty tree every morning. It took perseverance, but "after several days' watching, and one or two misses . . . I brought down my bird—a male in the most magnificent plumage" (*Malay Archipelago*, 526–527). This was the first of twenty-four "fine specimens" he collected there. (See recto 161–165 for more on *P. rubra*.)

2. These figures come from the fourth (1835) edition of Lyell's *Principles of Geology* (Lyell 1835, 3: 170–171). All citations of the *Principles* in the Species Notebook are to this edition unless otherwise indicated.

Lyell cited John Curtis (1791–1862), *A Guide to an Arrangement of British Insects being a Catalogue of All the Named Species Hitherto Discovered in Great Britain and Ireland* (1829), abbreviated *Catalogue of British Insects*, for the figure of 12,500 species given for the insects, Coenraad Jacob Temminck (1778–1858) for the mammals, and the illustrious Georges Cuvier (1769–1832) for the fishes and birds. These are given by Lyell in the context of his speculations on the appearance of new species on the earth.

Sir Charles Lyell (1797–1875) studied law, but fortunately for science he came under the spell of the legendary geologist William Buckland (1784–1856) while a student at Oxford. Lyell did not long practice law, but he put its rhetorical and reasoning skills to good use in his geological writings. His *Principles of Geology*, first published in three volumes in 1830–1833 and running into ten editions in his lifetime (it has never been out of print) powerfully championed the Huttonian model of earth history first propounded by his fellow Scotsman James Hutton: an ancient earth gradually, cyclically, and inexorably modified by mundane natural processes such as volcanism, earthquakes, slow uplift and subsidence, and erosion.

Through *Principles* and other writings Lyell became the preeminent geologist in Britain; his vision of Huttonian "actualism" (now called uniformitarianism) influenced a generation of geologists—including the young Charles Darwin, who long considered himself to be a geologist first and whose reading of the first volume of the newly published *Principles* as he embarked upon his voyage round the world on HMS *Beagle* convinced him of the truth of this exciting new view of earth history. Many of Darwin's geological observations on the voyage were aimed at applying and testing Lyell's ideas; on his return the two became good friends, and later (1856) Lyell became one of just three friends in whom Darwin confided his nascent "heretical" notions of transmutation.

Prior to this, and unbeknownst to Wallace, Lyell's conviction of the impossibility of transmutation (see annotations for recto 36) was severely shaken by Wallace's paper from Sarawak (Wallace 1855c), "On the Law Which Has Regulated the Introduction of New Species." After reading the paper, Lyell opened the first of what was to become a series of private notebooks on the "species question"—notes on Wallace's paper fill the very first page (Wilson 1970). In the Species Notebook, Wallace was having a dialogue of his own with Lyell. Commencing on recto 34 is the first of many comments and critiques on the *Principles*. (See Appendix 2.)

3. Notation refers to Pieter Bleeker (1819–1878), Dutch surgeon and ichthyologist working Dutch East Indian government. (See recto 27.)

[1]In Lombok 7000 sq.[uare] y[ar]ds of ground produces
3600 cattis of Paddy = 2200 cattis white rice
$\frac{1440}{2160}$   or
4800 lbs. [pounds] of Paddy = 3000 lbs. [pounds] of rice (about)
16,000 tons of rice annually exported from Lombok
about 35 bushels to the acre.

———

Macassar.
Sept. Oct. 77° to 92° range of Therm.[ometer] in open country.

———

This [2]cabinet for Butterflies—Store a Book boxes
for Coleoptera. For butterflies 1½ in.[ch] to glass abundant.
? Amboyna gaba gaba to line them.
*[Sketch of cabinet construction with below measurements labeled]*
2 in[ches] clear under glass for Coleoptera
plate glass.
2½ inches ½ in[ch]
½ in.[ch]   ¼ in.[ch] perhaps better.

Section of a drawer, full size, for
a cheap insect cabinet: 18 in.[ch] square      or less if
all ½ inch wood with a ¼ in.[ch] ledge          glass too
                                                 expensive
at top inside to receive plate glass
with ground edges laid on lining of velvet
ribband. No cork. Drawers soft pine
Case & doors oak, or plain mahog.[any]—
                        bottom to project
                        ¼ in.[ch] each side & slide
                        in grooves easy, so as
                        to be changed if required.

———

a ½ circular thumb hole on each side to lift out
glass. Camphor in a paper ¼ in.[ch] pipe corked at ends
& pinned into drawer at top & bottom.
A 40 drawer cabinet 3'. 3" wide by 5 f[ee]t. high
allowing 8 inches for bottom ledge & 2 in.[ch] cornice.
*[The following is written vertically in left margin of page]*
? perforated card
cardboard Δ cell at top for camphor oak ends on two ends

1. Wallace arrived at Ampanam (now Ampenan), on the island of Lombock (Lombok), in June of 1856. He was impressed with the place in many respects, not least its extensive agricultural works. In *Malay Archipelago* (173–175) he related how in the interior of Lombock he beheld "one of the most wonderful systems of cultivation in the world," as he "rode through this strange garden utterly amazed" at its beautiful terracing and elaborate irrigation system. Along the road he saw "an almost uninterrupted line of horses laden with rice in bags or in the ear, on their way to the port of Ampanam." The units Wallace gives, "cattis" or "cattys" (plural) was widely used in Southeast Asia. A catty was equivalent to about 600 grams.

2. Wallace was a talented draftsman, surveyor, and artist. With an eye to practicality and economy, here and on the next page he sketches a plan for a "cheap insect cabinet," a matter of great concern in light of his ever-growing collection. Recall that one of Wallace's goals in his travels was to amass a sizable personal collection with which to investigate the species question—duplicates were sold through his agent, Samuel Stevens, to cover his expenses. He wrote to his friend and kindred spirit Henry Walter Bates in October 1847 about his "favorite subject—variations arrangement distribution &c. of species," commenting: "I begin to feel rather dissatisfied with a mere local collection; little is to be learnt by it. I sh^d like to take some one family to study thoroughly, principally with a view to the origin of species. By that means I am strongly of the opinion that some definite results might be arrived at" (Wallace Correspondence Project, WCP348). By the time of his Southeast Asian travels, he had long had his sights set even higher than "some one family" for thorough study. His concern over cabinets and collection arrangements started early, as seen in this letter to Bates written in October 1845, almost exactly two years prior to the above letter:

> I have considerable thoughts of setting up a Cabinet myself this winter—I have been thinking how to have it made the best & most economical method—How do you think

it would do to have the drawers made with a groove at the top & the glass nicely squared to slide in & fit tight, with a piece at the end to finish up the square of the box—I think it would save the expense of a separate frame for the glass. How is Mr Kirby's made? Has he one of sides made double so as to leave a space for Camphor—that is very necessary. (Wallace Correspondence Project, WCP344)

The triangular cell of perforated cardboard is designed to contain granulated camphor, a terpenoid compound derived from camphor laurel (*Cinnamomum camphora*) of Southeast Asia, which was used to repel pests that would otherwise consume the pinned specimens.

*[Pencil cabinet drawings with notes, over which entomological notes are written in ink]*

| | |
|---|---|
| frame for glass | Two of three of the |
| over *[illeg.]* | largest boxes fitted |
| tray <u>boxes</u>. | with glass thus, for |
| | <u>Papilionidae</u> |
| outside | *[Written vertically]* perforated card. |
| fitting | or with |
| frame | a <u>frame</u> divided |
| <u>best</u>. | into <u>2 halves</u> |
| has shade | Smallest d[itt]o for Lycenidae or for |
| inside | Longicorns, Cicindelidae &c. |

---

<u>Entomological Notes. Sadong River, Borneo</u>
[1]On March the 12th. 1855 I arrived at the
landing place in the Si Munjon river. Here
I found a thatch house erected by Mr. Coulson
the superintendent of the Coal Mines at the
neighbouring mountain. A rough jungle
path on logs & fallen trees leads from
this point to the hill through a swampy
forest—for about a mile & a half On the road I found a magnificent
Orchideous plant a species of [2]<u>Celogyne</u> in flower,
but saw very few insects till I arrived at
the hill, where a space had been cleared of
trees & several houses erected for Mr C.[oulson] &
the Chinese workmen. On some of these trees
in the sunshine were numbers of Curculionidae of
the genus [3]Mecopus which flew away rapidly on my
approach, and near the house I saw a specimen
of the handsome [4]<u>Mecocerus gazella</u>. One or two butterflies
of the genus <u>Danaus</u> well known to me were
the only other conspicuous insects which I

1. Wallace departed Singapore on October 16 and arrived in Kuching, Sarawak, on November 1, 1854. With his assistant Charles Allen (see next page) he spent the first several months collecting along the Sarawak River from Santubong at its mouth to the gold-mining region of Bow (Bau) and Bede, staying in a small bungalow provided to him by Sir James Brooke, who was known as the "White Rajah of Sarawak." The collecting was not ideal, as it was the rainy season. He traveled up the Sádong (or Sadang) River as the rainy season came to an end. The "landing place" was about twenty miles upriver, at the mouth of the Si Munjon (Simunjon, or Simunjan) River, a tributary of the Sádong. Here Wallace had a two-room house built, with a veranda (could this be the "Jungle House" sketched just inside the recto front cover?), and used it as his base for the next nine months. It was in this little dwelling at the foot of Santubong Mountain that Wallace wrote his landmark paper "On the Law Which Has Regulated the Introduction of New Species" (the Sarawak Law paper) in February 1855, shortly before this entry in the Species Notebook was made (Wallace 1855c). (See also verso 1.)

2. "Celogyne" (*Coelogyne*) is a genus consisting of more than 200 species of epiphytic orchids widely distributed in eastern and southeastern Asia, many of which are prized in the horticultural trade. In *Malay Archipelago* (47) Wallace describes finding this "handsome orchid," which was "very abundant, and characteristic of the district."

3. A genus of spider-legged weevil (family Curculionidae, subfamily Zygopinae). This is a type of fungus weevil, often associated with fungi on fallen trees.

4. Family Anthribidae, another type of fungus weevil, with very long antennae. Wallace mentioned in his essay collection *Contributions to the Theory of Natural Selection* that "the very rare Capnolymma stygium closely imitates" *Mecocerus gazella* (Wallace 1870, 94, 155).

noticed. Returning to the river I found that my boy [1]Charley had caught a specimen of the rare & handsome [2]Macronota <u>Diardi</u>, the second specimen I had obtained.

The next day Mr C.[oulson] returned from an excursion up another river to examine some reputed [3]coal beds, & he then invited me to occupy part of his house till I could get one put up for myself. The next day I accordingly had all my things carried up & was then in a position to commence work in earnest. The [4]wet season now seemed to be nearly over, and after the incessant rains for the last four months with the accompanying scarcity of all objects of an entomologists pursuits I looked forward with much pleasure to the next two or three months as likely to give me some real insight into the insect treasures of this interesting country. I was principally

1. "Charley" refers to Charles Allen, one of Wallace's long-time assistants in Southeast Asia. Wallace employed quite a few assistants to collect and prepare specimens, and to help with traveling, logistics, and chores. Most were short-term assistants, but Allen and a Malay boy named Ali worked with Wallace for several years.

Allen started working for Wallace at age sixteen. Wallace had high expectations for his field assistants, judging from a letter to his sister Fanny in June 1855 in which he criticizes Allen on the sloppiness of his insect pinning and bird stuffing: "This after 12 months' constant practice and constant teaching!" Wallace moans. In this letter Wallace also gives an exhaustive list of the qualities he sought in a field assistant:

> Let me know his character, as regards *neatness* and *perseverance* in doing any thing he is set about . . . whether he is he quiet or boisterous forward, or sly, talkative or silent, sensible or frivolous, delicate or strong. Ask him whether he can live on rice and salt fish for a week on an occasion— Whether he can do without wine or beer and sometimes without tea, coffee or sugar—Whether he can sleep on a board . . . likes the hottest weather in England . . . is too delicate to skin a stinking animal . . . can walk 20 miles a day . . . can work for there is sometimes as hard work in collecting as in any thing. (Wallace Correspondence Project, WCP359)

As if this were not enough, Wallace added: "Can he draw (not copy), can he speak French? Does he write a good hand? Can he make anything—Can he saw a piece of board straight?" (ibid.). On verso 3 of the Species Notebook, at the bottom of the extensive tabulation headed "Charles collected—Insects of all orders," Wallace notes: "Sept. 12 deduct 70 lost by Charles." One can detect his irritation in the underscores.

Few could meet Wallace's exacting requirements. Allen's performance improved in Wallace's eyes, however, to the point that he later made several solo collecting expeditions in New Guinea, the Sula Islands, Flores, Solor, and Borneo. Allen became danger-

ously ill in Java and had to be evacuated to Singapore at the very end of Wallace's stay in Southeast Asia. After Wallace's departure in 1862, Allen stayed behind. He married in 1883 and eventually became manager of the Perseverance Estate at Gaylang, Singapore, a sizable citronella plantation. (See Rookmaaker and Van Wyhe 2012).

2. Scarab beetle (family Scarabaeidae) named for French naturalist and explorer Pierre Diard, who collected in the Malay Archipelago in the early nineteenth century. Synonym: *Coilodera diard*.

3. Indonesia is one of the leading exporters of sub-bituminous coal, which represents the bulk of Indonesian coal production. By the time of Wallace's visit, coal mining was beginning in earnest. These are relatively recent coal deposits—late Eocene to middle Miocene. Wallace describes the geology of Borneo in *Island Life* (1880a) and his impressions of its coal seams in *Malay Archipelago* (46–47). Coal deposits are common on Borneo; undoubtedly it did not take Europeans long to discover and exploit the coal seams, as some of them are readily visible as conspicuous outcrops.

4. The word "monsoon" is typically associated with heavy rains, but in Southeast Asia monsoons are weather conditions associated with seasonal shifts in wind direction. Typically, the northwest monsoon brings rain December through March, followed by the southeast monsoon, which brings drier weather June through September. In the intermonsoon seasons, winds are light and weather is more localized.

induced to come to this river to avail myself
of the new clearings & roads formed while
working the coal as well to learn something
of the Geology of the district, & to have
an opportunity of studying the great [1]Orang-
utan (which here abounds) in its native haunts.

About 20 Chinese [2]coolies ~~wh~~ were at work
cutting timber & clearing the ground for opening
the coal levels, and as soon as I was comfortably
settled I commenced an examination of the
fallen trunks and branches. The day was a fine
one;—flies, hemiptera, & wasps were abundant &
among them a few small longicorn beetles
occasionally appeared; sometimes too a bright
green buprestis would whiz by & then settle
on some trunk or log exposed to the hottest
sun, starting off again however on the slightest
attempt to approach him. As usual coleoptera
were never abundant at any one place or time.
One by one at intervals they appeared & were
captured and it was only by several hours

1. Native to Indonesia and Malaysia, orangutans are currently found only in rainforests on the islands of Borneo and Sumatra, though fossils have been found in Java, the Thai-Malay Peninsula, Vietnam, and mainland China. There are only two surviving species, both of which are now endangered: the Bornean orangutan (*Pongo pygmaeus*) and the critically endangered Sumatran orangutan (*Pongo abelii*). The word "orangutan" comes from the Malay words *orang* (man) and *(h)utan* (forest), or "man of the forest."

2. The English word "coolie," a historical term for manual laborers or slaves from Asia, can be traced to the Urdu *qulī*, which means "(day-)laborer." It can be a contemporary racial slur or an ethnic nickname for people of Asian descent. In Malay, *kuli* is a term for "slave"; in Thai, it still retains its original meaning as manual laborer but is considered to be offensive. In Indonesia today, *kuli* is a term used to describe construction workers.

persevering search than any result could be obtained
Charley had been at a spot in the swamp
where on some fallen trees [1]buprestidae were
abundant & when in the evening we emptied
our collecting bottles the results were very
satisfactory. Of Longicorns we had ten species
mostly small but very interesting & the greater
part new to me. There were 4 species of
Buprestidae, one about an inch in length
being thickly covered with a yellow powder
as is the [2]Buprestis gigas of S.[outh] America
though it never arrives in England in
that condition as it is almost impossible
to handle it without rubbing the powders
off. [3]Besides these we had a dozen
Curculionidae, 5–6 Elaters, a Colliuris
a Taeniodera and a host of small Malacoderms
Chrysomelidae & Heteromera.

1. Metallic wood-boring beetles, more aptly known as jewel beetles; indeed these often large, cylindrical beetles, with their stunningly brilliant metallic bronze, gold, blue-green, and yellow-orange colors, are made into jewelry worldwide. The iridescence of these beetles is owed not to pigmentation but to microscopic grooves in their cuticle that act as a diffraction grating. This is why the brilliancy persists long after the insect has died: the coloration is mechanical in nature, built into the exoskeleton, rather than pigment-based. The same is true of the glowing blues and greens of the morpho and birdwing butterflies collected by Wallace.

2. *Buprestis gigas*, a New World buprestid, is now placed in the genus *Agaeocera*. This large (approximately 16mm) beetle is a brilliant metallic green. Buprestid larvae live in the heartwood of trees where they feed on the wood as they burrow.

3. This was a good collecting day, and in fact the site proved to be spectacularly productive owing to a fortuitous combination of circumstances—especially the end of the rainy season and the area's large clearings bordered by felled trees. As he related in *Malay Archipelago* (47), "During my whole twelve years' collecting in the western and eastern tropics, I never enjoyed such advantages in this respect as at the Simunjon coal-works."

Wallace's travels were essentially self-financed, in that he paid his way by selling specimens to museums and collectors back home through his talented and diligent agent, Samuel Stevens (see Stevenson 2009). Wallace paid the Chinese and Dayak laborers one cent per insect collected for him. His first lot from the Borneo interior consisted of a bounteous 719 exotic beetle specimens, all collected in the Simunjon area. Wallace sent this shipment off from Sadong on June 20, 1855, bound for London via Singapore. It arrived in London some three months later, on October 1. The British Museum (Natural History) paid £25 for 200 of the beetles (Baker 2001).

## 5

Monday.

March 19th. This was a [1]white day for me. I saw
for the first time the Orang utan or "Mias" of the
Dyaks in its native forest. I was out after insects
not more than a quarter of a mile from the
house when I heard a rustling in a tree near
& looking up saw a large red haired animal
moving slowly along hanging from the branches by
its arms. It passed in this manner from tree
to tree till it disappeared in the jungle which
was so swampy that I could not follow it.
On ɫ a tree near I found its nest or seat
formed of sticks & boughs supported on a
forked branch. Today I met with
the curious [2]Thyreoptera, apparently the same
species I had found at Singapore; as well as
some pretty small Carabidae on the ground among
rotten chips. Twelve longicorns & six elegant
Buprestidae of two species also formed part
of my captures.

1. "A white day" was a common expression in the nineteenth century, meaning a singularly memorable or fortunate experience. Darwin used it in *Voyage of the Beagle*: "Oh for Baia Blanca; it will be a white day for me, when we gain it." The phrase may have roots in ancient Greece; Plutarch wrote in *The Life of Pericles* that "those who have had a gay and festive time call it a 'white day,'" and he connects this with the symbolism of the white beans that Pericles gave to his Athenian troops to feast on during his siege of the Samians.

2. The correct genus is *Thyreopterus* (family Carabidae, ground beetles), named by Dejean in 1831. (*Thyreoptera* is an obsolete name now synonymized with the bat genus *Thyroptera*.) Wallace collected these beetles in several locales, often referring to them as "curious," as he did here for unspecified reasons. This find is reported in his *Letter from Sarawak* from 1855: "Carabidae are hardly so abundant as at Singapore, but I have some beautiful new Therates, Catascopus and Colliuris, and the curious Thyreoptera also occurs sparingly under Boleti [fungi]." (Wallace 1855a, 4804).

## 6

After some rainy days which interfered with
my operations I again had fine sport on
the 23rd when I obtained 120 coleoptera.
Many of these were of course very minute.
Small curculionidae & chrysomelidae, little cylindrical
wood-borers & several small Malacoderms—But
then there were also 3–4 new & beautiful little
longicorns, a large & handsome Colliuris of which
I had taken one specimen at Malacca &
a new buprestis several handsome Curculionidae
a fine species of Cleridae & half a dozen
Heteromera & fungicoli. .

April 21st. An immense flight of the great [1]Bat
passed over us. It continued from about 1/2 past 5
to after dark & from counting & estimation not
less than 30,000 must have passed while we
observed them. They extended apparently about
2 miles in width & came from the direction of
the E*[illeg.]* River. The afternoon had been
wet.

1. Early in Wallace's stay in Sarawak he wrote a letter to the *Literary Gazette* from his base near the Simunjon Coal Works, in which he included an account of this experience. The "immense flights of fruit-eating bats, which frequently pass over us," he wrote,

> extend as far as the eye can reach, and continue passing for hours. By counting and estimation, I calculated that at least 30,000 passed one evening while we could see them, and they continued on, some time after dark. The species is, probably, the *Pteropus edulis;* its expanded wings are near five feet across, and it flies with great ease and rapidity. Fruit seems so scarce in these jungles that it is a mystery where they find enough to supply such vast multitudes. (Wallace 1855b, 683)

Some eighteen bat species have been recorded from Borneo. The largest are the "flying foxes" of the genus *Pteropus.* The smallest may be Hardwicke's woolly bat *(Kerivoula hardwickii),* which roosts in pitcher plants. Here is Wallace's account of flying foxes, from *Tropical Nature and Other Essays* (1878, 119):

> The characteristics of the tropical bats are their great numbers and variety, their large size, and their peculiar forms or habits. In the East those which most attract the traveller's attention are the great fruit-bats, or flying-foxes as they are sometimes called, from the rusty colour of the coarse fur and the fox-like shape of the head. These creatures may sometimes be seen in immense flocks which take hours to pass by, and they often devastate the fruit plantations of the natives. They are often five feet across the expanded wings, with the body of a proportionate size; and when resting in the daytime on dead trees, hanging head downwards, the branches look as if covered with some monster fruits.

[7]

[1]Dyak's account of the habits of the "Mias".
1. States that the Mias does not attack any other
animals of the jungle nor is attacked by any. No
animal is strong enough. The only animal with which
he ever fights is the Alligator. When the Mias finds
little fruit in the jungle he goes to the banks of the
river to eat certain fruits that grow close to the
water & also the young shoots of the palms. The alligator
then lying on the bank sometimes tries to seize
him, but the Mias then jumps upon him & beats him
with his hands & feet and kills him. This man
had once in his life seen such a fight, & believes
that whenever such occurs the Mias is always the victor.

---

[2]Mrs. Somerville says—
Insects in New Holland very few.—wrong.
N.[orth] America fewer species than Europe in same
     Lat.[itude]
Asia few species in proportion to its size— because of elevated
                                              plateaux &
                                              want of wood

Europe more species than tropical Africa—no doubt false—
Caffraria the African & Indian Is.[lands] equal as to species?
Central & tropical America far richest, (because
far most collected in. A[lfred] W.[allace])

1.  The term Dyak, or Dayak, refers generally to the native peoples of Borneo, primarily of the interior, a group that includes many distinct Austronesian-derived ethic subgroups. In his notebooks Wallace typically refers to the orangutan by its Dayak name (*mias*) rather than the Malay-derived *Orang-utan*.

   This Dayak account of the *Mias* is reported in *Malay Archipelago* (71), where it is followed by a second, similar account. There, too, Wallace identifies his sources as "old Dyak chiefs, who had lived all their lives in the places where the animal is most abundant." (Dayak accounts of the *Mias* are continued on recto 9.)

2.  "Mrs. Somerville" refers to Scottish scientist and writer Mary Fairfax Somerville (1780–1872), who made significant contributions to mathematics, astronomy, and the popularization of science through several best-selling books. Somerville College at Oxford is named in her honor. Wallace's criticisms pertain to claims that Somerville makes in the chapter "Distribution of Insects" in her 1848 book *Physical Geography*—the first textbook of physical geography written in English.

   Here we can imagine Wallace poring over literature bearing on the abundance and distribution of insect species.

[8]

From [1]Knighton's Tropical sketches.
"Nowhere is vegetation more rich & luxuriant,
nowhere is woman more delicately moulded, more
finely formed than in Malacca & the Eastern
Archipelago." first line true, rest absurd;—
the women are absolutely ugly. A[lfred] W.[allace]—

Coleoptera No. [number] 398. [2]Mecopus sp.[ecies]
Runs about on dry timber with bark, in the hottest
sunshine. ♀ deposits eggs in crevices. One day
I saw a large long legged ♂ ? standing over
a ♀ his fore legs completely embracing her
& with his antennae rapidly touching her back &
head. This continued some minutes when
on a little disturbance they flew away. .

---

[3]Beetles infinitely more numerous in temperate regions
of N.[orthern] Hemisphere than in equatorial countries. ? ! oh!

---

Insects different on two sides of Col de Tende in Alps

---

In Brazil the quantity of insects is so great in the
woods that their noise is heard in a ship at anchor
some distance from the shore!        *[illeg.]* Cicadas &
                                     Locustas.—only

1. Refers to William Knighton's *Tropical Sketches, or, Reminiscences of an Indian Journalist* (1855). Wallace is clearly not impressed with the women of the region.

2. *Mecopus* is the genus of spider-legged weevils mentioned on recto 1. Wallace's reference to the "large long-legged ♂" indicates the inspiration for the common name of these weevils: their spindly stilt-like legs. The number "398" represents a collection serial number from this early period, when Wallace attempted to assign such numbers to beetles and other species he wanted to single out. Wallace kept running "species registers" separate from his journals and notebooks; for examples, see his 1858 "Birds Register" and "Insect Register" (WCP4767) online at the Wallace Correspondence Project website.

3. Continuation of the critique of Somerville's *Physical Geography* from the bottom of recto 7. Wallace takes exception to most of these statements from Somerville.

[9]

[1]Dyak's account of Mias

An old man who knows much about
the Mias says it has no enemies, no
animal dare attack it but the
Alligator & the boa constrictor. It always
kills the alligator by main[?] strength pulling
open its jaws & ripping up its throat,
standing upon it. If the boa attacks
a mias he seizes and bites it in two.
He has seen small boas served so but
cannot say about very large ones.

The mias is very strong. There is no animal
in the jungle so strong as he.
"Kesim", Orang kaya of the Balow Dyaks
in the Si Munjon River, gave the above
account.

He says the [2]Mias chappan or Mias timpan, has
great cheeks (callosities?) the Mias rimbi
is equally large has a regular face & very long
hair. The Mias kassir is much smaller
than either of the others. .

1. This is a continuation of Dayak accounts of the *Mias*, or orang-utan, from recto 7. The *Mias* contests with an "alligator" and a "boa constrictor" are also reported in *The Malay Archipelago* (71–72). There, however, Wallace corrected himself and changed "boa" to "python" and "alligator" to "crocodile," boas and alligators being animals of the New World.

This second account is attributed to an *Orang kaya* (literally "rich man") of the Balow Dayaks, on the Simunjon River. *Orang kaya* and the lower-level nobility, the *Orang besar*, were the elite rulers of the western archipelago.

2. The *Mias chappan* of the Dayaks refers to large male orang-utans that bear cheek flanges—secondary sexual characteristics that exhibit dominance to other males and readiness to mate. Wallace published several accounts of the orangs in 1856 (see Wallace 1856b, 1856e, 1856f, and 1856g). As he described them in "On the Orang-utan or Mias of Borneo" (1856f), the *Mias chappan* is "known by its large size and by the lateral expansion of the face into fatty protuberances or ridges over the temporal muscles, which . . . are perfectly soft, smooth and flexible."

Other Dayak names for orangs are given in this entry. Wallace later described them thus:

> The Dyaks of N.-Western Borneo, however, have names for three species of Mias, although I could never find any one who could determine them with precision. All the animals with large cheek-excrescences form the *"Mias chappan,"* but they declare that females are also found of the same form . . . All Orangs of smaller size and without cheek-excrescences are called by the Dyaks *Mias Kassu*, and my small males and females are undoubtedly of this kind . . . The third kind they call the *Mias rambi*, and they say it equals the "chappan" in size, but has no cheek-excrescences and very long hair. This seems very rare, and is probably one of the large species in which the excrescences have been little or not at all developed. (Wallace 1856f)

The opportunity to observe and collect orangutans was one of the chief reasons Wallace wanted to visit Borneo. He ultimately published five papers relating to orangs, based on a study of the fifteen specimens he collected plus one other. He was able to clear up a great deal of misinformation about these great apes. From his comparative analysis Wallace concluded there was but one species, albeit one with considerable variation (including the male-specific cheek flanges). Today two species are recognized: *Pongo pygmaeus* (Bornean orangutan) and *P. abelii* (Sumatran orangutan).

## 10

[1]Mias—Times seen—

March 19th. Saw one well—moved hanging by its arms

April 5th.     "     Shot it, tried to hide in foliage.
     3)        Two bullets through body.

April 24th. saw another indistinctly through branches

April 26th.     "     shot another, broken arm—
     5)        strong as two Dyaks  half grown.

May 2nd.     "     on a high tree fired at it several times
        with small gun. It screamed & broke off
        branches with its hands & threw them down.
        seemed very angry or afraid. not got it.

May 12th.     "     howled & screamed threw down branches
     6)        shot 5 times remained dead in top of lofty
        tree.

May 16th.     "     shot another Mias near the same
     7        place—2nd shot fell with blood running
        from nostrils climbed again up a high tree
     8        leaving [2]young one on ground. shot again
        & fell dead.

[3]May 23. A large Mias between road & path to
     9        mines, a sharp hunt after him—10 shots
        a large Mias chappan

1. This page and the next summarize Wallace's orangutan sightings and hunts. It is not complete, as three more were seen on June 27 (see recto 30). Wallace's hunting of orangutans may offend modern sensibilities, but was absolutely essential to him for a number of reasons both scientific and pecuniary. As he wrote in *The Malay Archipelago*, "One of my chief objects in coming to stay at Simunjon was to see the Orang-utan . . . to study his habits, and obtain good specimens . . . of both sexes, and of the adult and young animals" (*Malay Archipelago*, 51).

The specimen shipments he sent to London from Borneo included "orang-utan skeletons, skulls and limbs," and "a cask containing skins of orang-utan." He later recorded that the British Museum bought sixteen mammals and five tortoises for £12, and someone named Franks, presumably a collector, bought thirteen mammals and three skulls for £16. To put these figures in modern terms, 1 pound sterling in 1855 was equivalent to about 70 of today's pounds, so these were lucrative sales for Wallace.

The circled numbers were added later by Wallace as he sorted his specimens for shipment back to London. In his consignments notebook, page 17, he recorded, "Long packing case contains Skeletons of Orang-utan —" (Baker 2001, 326). These are organized by sex, and Wallace noted which were to be kept and which sold:

♂ adult (nos. 2, 9, 11)—3 (keep)
♀ adult (nos. 6, 7, 12)—3 (keep)
young ♂ (3)—1 (sale)
bones of limbs of nos. 1, 4, 5—

On the same page he noted that "in square case" he packed:

2 skulls of young mias [nos. 3 & 5]—1 skull of ♀ adult [no. 4]
2 skins dried of same.

2. Number 8 is the young orang, orphaned after the death of its mother, number 7, and for some months raised by Wallace (see recto 20–27).

3. The May 23 specimen is the large *Mias chappan* the hunt for which is described on recto 13–19 of the Species Notebook (and in *Malay Archipelago* on 57–60). There Wallace reports that this specimen "is now preserved in the museum at Derby." This was Lord Derby's museum, the collections of which are now part of the World Museum in Liverpool.

[1]June 4th. An old mias killed by the Dyaks
10)  nearly killed one of them. <u>Brought to me</u>
     much decomposed—[2]Skeleton in jungle to clean.

June 18th—Adult ♂—fell with 2 shots.
11)  hung a considerable time by one hand & then
     fell on face, half buried in swamp—After
     5–10 minutes groaning & panting & while we
     are standing close to him expecting each breath
     to be his last, he suddenly by a violent effort
     raised himself up. We sprung back a yard
     or two when he got up nearly erect & seizing
     hold of a small tree began to ascend. A
     shot through the back made him fall &
     expire in a few minutes. A flattened
     bullet was found in his tongue having completely
     traversed his body from the abdomen &
     broken the 1st cervical vertebra.

12)

June 21st. An adult ♀ killed while
     eating fruit in a low tree.

13)

[3]June 24th. A large Mias on very lofty tree
     remained up <u>dead</u>.

1. These orang kills are described by date in chapter 4 of *Malay Archipelago* (61–63).

2. Here Wallace refers to an orang that Dayaks had killed. It had in fact nearly killed one of the men first, when a Dayak hunter speared it and the orang grabbed the spear and, presumably pulling the man toward itself, sank its teeth deep into the Dayak's arm. The man was saved by his companions, who promptly killed the unfortunate ape. The event is described and illustrated in *Malay Archipelago*. (See recto 17.)

This incident having occurred some distance from where Wallace was staying, he offered a reward for retrieval of the orang's body. It was brought the next day, but by then it had already begun to decompose. This is why he was led to settle for the skeleton only. After first removing the head in order to retain the skull, Wallace reported he had his men build a fence around the body, "which would soon be devoured by maggots, small lizards, and ants" (*Malay Archipelago*, 60–61).

3. The account of the hunting and killing of this adult orangutan is given on recto 28–30 and recounted in *Malay Archipelago* (62–63). The tree was very large, and he could not persuade the Chinese laborers to chop it down even with the promise of four days' wages. Forced to leave the orang to decompose in the tree, two months later he hired a pair of Malays to climb the tree and retrieve the remains.

## 12

Extracts, of supposed proofs of Design.
[1]Knights. Cyc.[lopedia] [of] Nat.[ural] His.[tory]
Bats— "But it is in the bones of the metacarpus & fingers that the adaptation of the osseus parts to the wants of the animal to its necessities is, perhaps most strongly shewn."

As if an animal could have necessities before it came into existence, or as if having come into existence it could continue to exist unless its structure enabled it to obtain food. If the bat had not wings it would of course do without them & would have no more necessity for them than any other animal. A[lfred] W[allace] . .

Had the bats large eyes, it would be brought forward as an arrangement in exact accordance with their necessities, as purely nocturnal animals. But they have very minute eyes & apparently imperfect sight & they do very well without them—

1. The *English Cyclopedia,* edited by Charles Knight, consisted of sets of two to four volumes in each of four categories: geography, biography, arts and sciences, and natural history.

Here Wallace is scornful of the apparent absurdities of "proofs of design" seen in organismal structure, in this case the bat's wing. He takes the quoted statement from Knight's *Cyclopedia*—that the finger bones most clearly show the adaptation of the animal to its "necessities"—to be nonsensical. Of course its structure enables the animal to find food. Note Wallace's final statement: if bats had large eyes, which they do not, that too would be cited as a wonderful adaptation to a nocturnal lifestyle. It is a truism that any observed structure can be seen as adaptive, but to advance these as *proofs* of design is faulty logic, since by definition *whatever* structures they have will be seen as necessary and therefore indicators of good design.

Wallace continues with critiques of "proofs of design" from Knight's *Cyclopedia* on recto 31 and 32.

## 13

[1]A Mias hunt.

One afternoon I had just come home from an
Entomologizing excursion & was preparing for a
bathe when Charley rushed in, out of breath
with running & excitement & exclaimed by
jerks—"Get the gun sir,—be quick,—such
a large mias,—oh!."—"Where" said I,
"Close by"—he can't get away". So the gun
was got out & one barrel being ready loaded
with ball I started off ~~leaves~~ calling upon
two Dyaks who happened to be in the house
at the time to ~~follow~~ accompany me & ordering Charles to
bring all the ammunition after me as quick
as possible. The animal had been seen
between our house & the coal mines which
are both situated at the foot of a hill
about half a mile from each other. Just
below the hill the jungle has been cleared for
a road about 30 feet wide, and parallel
to this about a hundred yards up the
side of the hill is a foot path to the

1. This hunt, which took place about May 23, 1855, is recounted in *Malay Archipelago* (57–60). The specimen was sold to Lord Derby, whose collection is now part of Liverpool's World Museum.

mines. [1]Along this footpath we now walked
listening for the slightest sound that might
betray the presence of the monster till we came
to the place where Charley has seen him
going down towards the road. I felt sure
he would be somewhere near as there were
Chinamen working in the road, & he would
have to come down on the ground to
cross it which he would not do in their
neighbourhood, and as he has evidently
come down from the hill he would
not probably return there, but most
likely ~~remain~~ stay till night for an opportunity
to get across into the swampy extent of
jungle which is their favorite resort.
I therefore sent the Dyaks & Charley to
search in the jungle in different directions
& returned myself a little way along the
path looking up into every tree & listening
attentively. At length I heard a very slight

1. Compare Wallace's account of the episode on this and the following page of the notebook with the printed account in *Malay Archipelago:*

> The path from our clearing to the mines led along the side of the hill a little way up its slope, and parallel with it at the foot a wide opening had been made for a road, in which several Chinamen were working, so that the animal could not escape into the swampy forest below without descending to cross the road or ascending to get around the clearings. We walked cautiously along, not making the least noise, and listening attentively for any sound which might betray the presence of the mias, stopping at intervals to gaze upward. Charley soon joined us at the place where he had seen the creature, and having taken the ammunition and put a bullet into the other barrel, we dispersed a little, feeling sure that it must be somewhere near. (58)

rustling sound over head. [1]I gazed upwards but nothing was visible. I moved about in every direction to get a full view of the tree & began to think I must have been deceived, when I again heard a louder noise & saw the leaves shaking as the animal moved off to an adjoining tree. I immediately shouted out for all of them to come up, and we all tried to get a view of him for a shot. This is not an easy matter as the Mias has a knack of selecting places with dense foliage beneath it. However one of the Dyaks soon got a view of him & calling me I saw the huge red hairy body & a huge black head looking down surprised at the disturbance. I immediately fired, & he made off rapidly towards the road moving with very little noise for so large an animal. Sending the Dyaks to follow and keep him in sight I loaded both barrels as quickly as possible & started in pursuit.

1. The episode continues; again, compare the notebook with the published account:

> After a short time I heard a very slight rustling sound overhead, but, on gazing up, could see nothing. I moved about in every direction to get a full view into every part of the tree under which I had been standing, when I again heard the same noise, but louder, and saw the leaves shaking as if caused by the motion of some heavy animal which moved off to an adjoining tree. I immediately shouted for all of them to come up and try and get a full view, so as to allow me to have a shot. This was not an easy matter, as the mias had a knack of selecting places with dense foliage beneath. Very soon, however, one of the Dyaks called me and pointed upward, and on looking I saw a great red hairy body and a huge black face gazing down from a great height, as if wanting to know what was making such a disturbance below. I instantly fired, and he made off at once, so that I could not then tell whether I had hit him. (*Malay Archipelago*, 58)

## 16

The jungle was here full of great rocks &
stones fallen down from the mountain, &
obstructed with hanging & twisted creepers.
Running & creeping & climbing among these we
came up with him on a high tree near
the road where the chinamen had discovered
him & were shouting their astonishment.
[1]Ya ya Tuan orang utan tuwan! He then turned up again towards the hill
& I got two shots at him loaded again
& got two more before he again reached the
path. Once while loading I had [a] splendid view
of him walking along a large limb of
a tree in a semi erect posture, & showing
him to be an animal of the largest size.
At the path he got on to one of the loftiest trees
in the jungle—we here saw one leg hanging
down broken by a ball & there was no doubt
he had several other wounds. He now fixed
himself in a fork among some foliage &
seemed disinclined to move. I was afraid

1. In *Malay Archipelago* (58) this is rendered as "Ya ya, Tuan; Orang-utan, Tuan." In Malay, *tuan* translates as "sir" or "mister," a form of respectful address.

he would remain there & die as they frequently
do without falling & as the tree was a very large
one & evening was coming on we could not
have cut it down that night. After some
time as he did not move I fired again but
as most of his body was hidden by branches
probably did not hit him. He however moved
off & I got another and a better shot at
him which made him leave the tree &
attempt to go up the hill. He thus got
on to some lower trees on the branches of
one of which he fixed himself in such a
position that he could hardly fall, & lay
all in a heap as if dead or dying. He
was now not more than 40 feet from the
ground, & I wanted a Dyak to go up &
cut off the branch he was upon. ~~We~~ [1]They
however objected that he was not dead & were
afraid he might come to them. We therefore
shook the adjoining trees & did all we

1. Orangutan hunting was risky business. Wounded and enraged, powerful adult orangs could be deadly. Adult males can weigh in at about 95 kilograms (209 pounds), while adult females are slightly smaller, reaching 50 kilograms or so (about 110 pounds). Orangutans have a long reach—their arms are about one and a half times the length of their legs, and a large male's arms can have a spread of as much as 2.4 meters (8 feet) in length. It was not uncommon for a large orang attacked by humans to turn the tables and severely injure or even kill its tormenters, as depicted in the illustration captioned "Orang-utan attacked by Dyaks," frontispiece in *Malay Archipelago* (volume 1 of the British edition). Wallace related this incident in *Malay Archipelago*:

> A few miles down river there is a Dyak house, and the inhabitants saw a large orang feeding on the young shoots of a palm by the river-side. On being alarmed, he retreated toward the jungle which was close by, and a number of the men, armed with spears and choppers, ran out to intercept him. The man who was in front tried to run his spear through the animal's body, but the mias seized it in his hands, and in an instant got hold of the man's arm, which he seized in his mouth, making his teeth meet in the flesh above the elbow, which he tore and lacerated in a dreadful manner. Had not the others been close behind, the man would have been more seriously injured, if not killed, as he was quite powerless, but they soon destroyed the creature with their spears and choppers. The man remained ill for a long time, and never fully recovered the use of his arm. (*Malay Archipelago,* 60)

Wallace reported that the Dyaks looked upon him as a "great benefactor in killing the mias, which destroys a great deal of their fruit" (*Malay Archipelago,* 65). Sadly, orangutans are still sometimes viewed as pests or competitors and are not infrequently killed if they encroach on oil palm plantations or destroy crops. The World Wildlife Fund lists "conflict with humans" as one of the main threats to orangutans, along with habitat destruction, hunting, and the illicit pet trade.

Wallace entitled this illustration "Orang-utan attacked by Dyaks" but it shows the injured and enraged orang attacking the hunter in self-defense. Reproduced from the frontispiece of Alfred Russel Wallace, *The Malay Archipelago*, 4th ed. (London: Macmillan & Co., 1872).

could to disturb him but without effect.
[1]At length we sent for a chinaman &
axes to cut down the tree, but while
the messenger was gone a Dyak took
courage & began to climb. When he
had got about halfway up however the
Mias got up his remaining strength &
moved off getting on to a dense mass of
branches & creepers in an adjoining tree
where he managed almost to hide
himself from sight. The tree was
luckily smaller, so when the axes came
we soon had it cut down but it was
so held up by jungle ropes & creepers
to adjoining trees that it would not fall
~~that~~ remaining inclined at a slight angle.
The mias still remained in his position
and I began to think we should not
get him as half a dozen more trees would
have to be cut before the one he was

1.  Wallace does not describe here how the Dyak began to ascend the tree, but in *Malay Archipelago* he related an ingenious mode of climbing using pegs fashioned from versatile bamboo. Several pegs, sharpened at one end, were driven into the tree with a mallet. After being satisfied that the first pegs would hold their weight, the Dyaks proceeded to install a series of pegs. Wallace reported that he "looked on with great interest, wondering how they could possibly ascend such a lofty tree by merely driving pegs in it, the failure of any one of which at a good height would certainly cause their death." Their solution was to secure the pegs with a long bark cord, fed through notches at their ends. They soon succeeded in ascending many tens of feet and retrieved the dead orang. "I was exceedingly struck," Wallace enthused, "by the ingenuity of this mode of climbing, and the admirable manner in which the peculiar properties of the bamboo were made available. The ladder itself was perfectly safe, since if any one peg were loose or faulty, and gave way, the strain would be thrown on several others above and below it. I now understand the use of the line of bamboo pegs sticking in trees, which I had often seen, and wondered for what purpose they could have been put there" (*Malay Archipelago,* 66). (See recto 63–64 for Wallace on other uses of bamboo.)

on would fall. As a last resource we set
all hands pulling at the creepers, which shook
him considerably & after a few minutes
when we least expected it down he came
with a crash like the fall of giant; &
he was a giant ! his head & body being ~~larger~~ [*illeg.*]
~~than~~ as large as a man's. He was of
the kind called by the Dyaks "Mias
Chappan" or "Mias pappan" having the skin
of the face extended to a ridge on each
side,—a kind I had long wished to obtain
& examine. [1]His legs and arms were tied
together & two men carried him home on
a pole. His outstretched arms measured
7 feet 3 inches from finger to finger. His
wounds had been fearful. Both legs
were broken, one hip joint shattered to
pieces; the root of the spine completely shattered
& two bullets flattened in his neck & jaws.
Yet he was still alive when he fell. .

1. Reading Wallace's detailed accounts of shooting these great apes can be painful for modern readers. Of course the mid-nineteenth century was a different time; collecting with a gun was necessary, and ubiquitous, among naturalists worldwide (and lethal collecting is still done, sometimes unnecessarily); besides which, specimens like these were Wallace's livelihood. Today fewer great apes may be felled by guns, but the other threats they face—especially inexorable habitat loss and fragmentation—will prove more lethal if unabated.

*Pongo abelii*, the Sumatran orangutan, is classified as critically endangered on the International Union for Conservation of Nature and Natural Resources (IUCN) Red List of Threatened Species (see www.iucnredlist.org), with an estimated 7,300 individuals remaining as of the last detailed census, taken a decade ago. *P. pygmaeus*, the Bornean orangutan, is faring a bit better: it is listed as endangered, with an estimated population between 45,000 and 69,000 occupying some 86,000 square kilometers of suitable habitat—though this, again, is based on decade-old data.

## 20

[1]Habits of a young Mias . .

This little animal was probably not above
~~10 days~~ ~~fortnight~~ a month old when I obtained
~~him~~ it as it had no teeth. Two days after
however it cut its two lower front teeth. I fed
it with rice water given out of a bottle
with a quill which after one or two trials
it sucked very well. When handled or
nursed it was always quiet & contented, but
when laid down alone it would always cry
& the first night or two was very restless..
I fitted up a box for a cradle with a
mat to lay upon which was changed &
washed every day. I soon found it
necessary to wash the Mias as well which
appeared to have a good effect. It winced
a little and made ridiculous wry faces when
the cold water was poured over its head but
enjoy~~ing~~ed the rubbing dry amazingly, and
especially having the hair of its back & head

1.  Wallace adopted this orphaned orang after shooting its mother on May 16, 1856 (see recto 10). The young orang died after about three months from an unknown illness, perhaps exacerbated by deprivation of nutrients and antibodies it could obtain only from its mother's milk. Wallace later sold the specimen to the British Museum for £6, though it cannot now be found and was perhaps lost.

"I am afraid you would call it an ugly baby," he wrote to his sister, "for it has dark brown skin and red hair." See *My Life* (Wallace 1905, 1: 343–345) regarding Wallace's letter about his "dear little duck of a darling of a little brown hairy baby" (Wallace Correspondence Project, WCP359). He also published accounts of the infant orang, such as "Some Account of an Infant 'Orang-Utan,'" in the *Annals and Magazine of Natural History* (Wallace 1856g), and "A New Kind of Baby," a popular narrative printed anonymously in the November 22, 1856, issue of *Chambers's Journal* (Wallace 1856b).

brushed afterwards—[1]When first I obtained it
it clung desperately tight with its four hands
to whatever it could lay hold of, and having
once seized my whiskers & beard I could not
get it off for some time, as it doubtless
felt quite at home being accustomed to
cling almost from birth to the long hair of
its mother. When restless it would struggle
about with its hands up to catch hold of something
& might often be seen quite contented when
some bit of rag or stick was grasped in
two—three of its hands. After a week however
laying on its back in a box the muscular
activity of the fingers was considerably diminished.
It would no longer seize every thing that was
presented to it, & would lay dozing with its
hands open & relaxed instead of tightly
clenched as at first.. The extraordinary
& constant exercise of the muscles of the
limbs & hands in the young mias, remaining
as it does many hours each day

1. Wallace's account of the young *Mias* continues to the top of recto 27. Wallace surely believed that humans evolved from an ape-like ancestor probably very like the orangutan—indeed, and he playfully hints at the relationship in his 1856 paper "On the Habits of the Orang-Utan of Borneo" (Wallace 1856e), where he refers to orangs as creatures "which at once resemble and mock the 'human form divine.'" Although he does not discuss this idea in either the Species Notebook or his journals, these orang entries are nevertheless instructive, showing how intimately Wallace observed the great apes and how he came to appreciate their humanlike sensibilities.

In this there is another parallel with Darwin, who spent several weeks in the spring of 1838 making observations on a young orangutan named Jenny at the London Zoo. Jenny was the first orangutan to be displayed at the zoo and as such was both a public and a scientific sensation. Darwin, who was a transmutationist by then and keenly interested in human-primate relationships, recorded in one of his notebooks Jenny's childlike emotional behavior (Barrett et al. 1987, 545, 551, 554), not unlike Wallace's description of the emotions of his orangutan.

with the whole weight of its body ~~resting on its~~ supported by
its four extremities when in a state of
nature, must induce a development of
the limbs which can hardly take place
in ~~a state of~~ confinement. We should
therefore expect a considerable difference
in the proportions of the limb & body in
such animals as have been brought up
artificially.

I soon found I could feed my
infant "Mias" with a spoon and make its food
rather more solid. I gave it soaked a
chewed biscuit with a little sugar or egg &
sometimes sweet potatoes. These it liked
very much & it was a never failing source
of amusement to *[illeg.]* observe the curious
changes of countenance by which it would
express its approval or dislike of what
was given it. It would lick its lips, draw
out its cheeks, & turn up its eyes with an

Illustration of a young female orangutan, based on a photograph. Reproduced from Alfred Russel Wallace, *The Malay Archipelago*, 4th ed. (London: Macmillan & Co., 1872), 41.

expression of the most supreme satisfaction
when it had a mouthful particularly to its
taste. On the other hand when its food was not
sufficiently sweet or palatable, it would
turn it about in its mouth for a moment
as if trying to extract what flavour there
was, & then let it all run out between its
lips. If the same food was continued it would
set up a scream & kick about violently
exactly like a baby in a passion.

After I~~had~~ about 3 weeks I obtained a small
Hare-lipped monkey [1](Macacas cynomolgus) which though
    young had
its first teeth, was very active, & could feed itself.
I placed it in the same box with the "Mias" &
they immediately became excellent friends not
exhibiting the least fear of each other.
The little monkey would sit upon the Mias's
stomach or even on its face, would pull open
its mouth & push its little hand in, or lay down
across its ~~belly~~ body in whatever position was most

1.  The Latin name of this species is now *Macaca fascicularis*, family Cercopithecidae. Known to the Malays as *Kra*, in English it is variously called the crab-eating macaque, cynomolgus monkey, Philippine monkey, and long-tailed macaque.

This is one of the most common and widespread macaque species, with some nine subspecies recognized. It is also highly social, which made it a good playmate for Wallace's young orangutan.

agreeable without paying the least regard to the
comfort of its companion. [1]The poor little
Mias would submit very patiently to every
insult, seeming glad to have any thing warm
about its body and sucking occasionally at
a hand or tail with the greatest satisfaction.
It was curious to observe the difference
between these two. The Mias like a young
baby, laying on its back quite helpless, rolling
lazily from one side to the other, stretching
out its four hands into the air wishing to
grasp something but unable to guide its
fingers to any particular object, &
when dissatisfied opening wide its almost toothless
mouth & expressing its wants in a most
infantile scream. The little monkey
on the other hand in constant motion, running
& jumping about wherever it pleased, examining
every thing with its fingers & seizing hold
of the smallest object with the greatest

1. In his June 1855 letter to his sister Fanny, Wallace reveals the name of his infant orang's monkey companion: "When it finds no milk is to be got, there comes another scream & I have to put in [it] back in its cradle and give it 'Toby' the little monkey, to hug, which quiets it immediately" (Wallace Correspondence Project, WCP359). There seems to be no indication, however, of whether Wallace named the little orangutan too.

precision, balancing itself on the edge of the box or running up a post & helping itself to every thing eatable that came in its way. [1]There could not be a greater contrast & the baby Mias looked more baby-like by the comparison . .

In order to give my infant a little exercise & strengthen its limbs I contrived a kind of ladder upon which I would put it to hang for a quarter of an hour at a time, but this was not much to its liking as it could not get all four of its legs into a comfortable position. It would often keep hold only by one hand or foot & scream lustily till it was released. I then endeavoured to make an artificial mother for it, by wrapping up a piece of buffalo skin into a bundle with the long woolly hair outwards and hung it up about a foot from the

1.  One imagines Wallace, watching his infant orang and its play-mate, reflecting on the relationship of the three primate species residing together: human, orang, and monkey. He surely recognized that humans and orangs are more closely related to each other than either is to the monkey; indeed, humans and orang-utans share the family Hominidae, while the monkey is a member of the Old World monkey family Cercopithecidae, a diverse primate group found from southern Europe and Africa to parts of eastern Asia. Within Hominidae, orangutans (genus *Pongo*) are basal to the clade that consists (among extant species) of goril-las (genus *Gorilla*), chimpanzees and bonobos (genus *Pan*), and humans (genus *Homo*) (Purvis 1995).

The two extant orang species, *Pongo abelii* and *P. pygmaeus*, were recently subjected to genomic sequencing and analysis by the Genome Center of Washington University in Saint Louis (Locke et al. 2011), the most detailed analysis to date. In addition to identifying a number of unique features of orangutan genome structure, comparative analysis suggested a speciation time be-tween 400,000 and 1,000,000 years ago for Sumatran and Bor-nean orangutans (compared with the 12–16 million years since the last common ancestor shared between the orang lineage and the rest of the great apes).

ground. [1]To this it would cling with great
tenacity, grasping the hair tightly with its
little fingers, & seeming very much to
enjoy its position, but it would constantly
attempt to suck (thinking no doubt
it had got a real mother back again)
and getting nothing but a mouthful of
wool would be greatly disgusted &
scream violently. One day it was
very nearly choked by the wool getting
into its throat, so this plan of giving
it exercise was also obliged to be
discontinued.

When I had had it about a month it began
to exhibit some signs of learning to run
alone. When laid upon the floor it would
push itself along by its ~~hind~~ legs or roll over
& thus make an unwieldly [sic] progression. When
in its box too it would lift itself up to
the edge as if wishing to get out. It now

1. Wallace's attempt to provide an "artificial mother," made of a roll of buffalo skin with woolly hair, for the infant orangutan is reminiscent of the maternal-separation experiments of American psychologist Harry F. Harlow (1905–1981), conducted in the 1950s. In a series of experiments (many of which would likely be deemed unethical today), Harlow investigated the mother-infant bond with rhesus monkeys, using artificial "mothers" made of various combinations of wood, cloth, and wire. The infant monkeys learned to recognize and would preferentially stay near their particular "mother," and given a choice between a model mother made of cloth versus bare wire, they would consistently cling to the cloth model, even when the wire model had food (a bottle) and the cloth model had none. Harlow's work showed the importance of physical contact (as opposed to simply feeding) between mothers and infants for the cognitive development and physical and mental health of infant primates—including human infants and children. Harlow summarized these studies in his presidential address to the American Psychological Association in August 1958 (Harlow 1958). The clinging and desperate attempts to feed on the part of Wallace's infant orang underscore the importance of maternal caregiving for these primates, too.

too began to know what it wanted & when
dirty would scream violently till taken
out & washed when it would be quite happy
and contented.

In 5 weeks cut its two upper teeth but did
not increase in size.
Taken ill gave it castor oil—Got
better & want of proper food. Ill
again fever, dropsy. Loss of appetite
torpidity—Died.

---

[1]Fire Flies 4 sp.[ecies] in Europe 3 sp.[ecies] in S.[outh]
   America !!
Spiders abound more in Europe than elsewhere !!! oh!

---

853 European fish—210 fresh water.
444 fish—Mediterranean. 216 British out of
643 sea fish in Europe.
100 species common to Britain & Italy
27 species common to Black sea & Mediterranean
Caspian sea, all peculiar. A. *[Refers to note at bottom of page]*

---

782 sp.[ecies] from Amboyna alone. [2]Dr. Bleeker.

---

[3]A. Dolphins & other marine mammals of Caspian
are identical with those of Black sea &
Mediterranean—were connected recently (geologically)
        (Lyell.)

1.  It is not clear where Wallace read these figures, but he is clearly incredulous at their woeful inaccuracy. More than sixty species of firefly (actually beetles, family Lampyridae) are found in Europe, and there are far more in the tropics of South America and Southeast Asia. Spiders, too, have their greatest species richness in the world's tropics.

2.  Pieter Bleeker (1819–1878), a Dutch naturalist and surgeon working with the colonial government of the Dutch East Indies, studied a number of species groups, especially fish of the Malay Archipelago, amassing a sizable collection over nearly thirty years. After Bleeker's return to Europe in about 1860 he published a series of ichthyological volumes entitled *Atlas ichthyologique des Indes Orientales Neerlandaises* (Amsterdam, 1862–1878)—known today as the "Fish Atlas"—published in several volumes beginning in 1862 and terminating prematurely with his death in 1878. (The last plates were published by the Smithsonian Institution in 1983!) Fittingly for a man considered to be the father of Indo-Pacific ichthyology, in 2005 the Scientific Advisory Committee for the Indo-Pacific Fish Conference created the Bleeker Award for distinguished contributions to the field.

Wallace likely read some of Bleeker's 430-plus articles on the taxonomy and distribution of fish of the Dutch East Indies, aka the Malay Archipelago. Bleeker also published articles on fish of Japan, China, Australia, British India, Guiana, southern Africa, Surinam, and Europe, and some of these papers may be the source of the other fish data Wallace notes here.

3.  The reference to the cetaceans of the Caspian, Black, and Mediterranean Seas is from volume 3 of *Principles of Geology* (Lyell 1835, 3: 66). Lyell discusses the similarity of the marine mammals of these seas as part of a longer discussion of geographical distribution. Wallace was keenly interested in geographical distribution, and his interest in this observation may have stemmed from his ongoing study of faunal similarities and differences between regions.

The Lyell reference is preceded by the letter "A," which may signal that it is a note connected to the "A" following the "Caspian Sea, all peculiar" entry above the Bleeker reference. Wallace may have taken exception to a claim (by Bleeker?) that the marine life of the Caspian is unique, citing Lyell's note that in fact there was a recent geological connection between the Caspian, Black, and Mediterranean Seas, as indicated by the fact that their marine mammals are shared.

## 28

On June 24th. I was called by a Chinaman
to see & shoot a Mias which he said was
in a tree near his house at the coal mine.
Arriving at the place we had some difficulty
in finding the animal as he had gone off
into the jungle which was very rocky &
difficult to traverse. However at last
we found him [1]up a very high tree &
saw that he was a male of the largest
size. I got a good shot at him & he
soon moved away higher up the tree. I got
two more shots when he had got to the
highest part of a very large & lofty tree,
& we could see that one arm was broken.
He here began to make a nest by breaking
of large boughs, reaching over to pull
them towards him & placing these across each
other so as to make a complete nest so thick
that when laid down upon it he was quite
invisible. Before I had reloaded my

1. Orangutans are the most arboreal of the great apes, spending nearly all of their time in trees; accordingly, they have a host of anatomical adaptations for their tree-dwelling lifestyle. In particular, their arms and legs are quite long relative to the length of their trunk, their hands have long, curved fingers and opposable thumbs, and their feet have opposable big toes. Even at rest orangutan fingers adopt a natural curl owing to the arrangement of their joints and musculature, creating the suspensory hook grip found in brachiating apes and monkeys. Also, the distal carpels of their long fingers can rest against the inside of the palm, creating a tight grip around small-diameter branches, known as a "double-locked grip" (Rose 1988, 300–301).

gun he had completed this & I then waited
some time to get a shot but not being
able to see him I fired several times
by judging where his body was. [1]After
each shot he generally moved & at length
after being quiet a considerable time he
raised himself up so that half his body
was visible & then gradually lowering himself
remained with his head above the nest.
Finding it almost impossible to make him
fall I wanted the Chinamen who had
accompanied me to cut down the tree, but
as it was late & the tree was very large nothing
would induce them to attempt it. As it
was now getting dusk I could do nothing
more so fired two parting shots with great
care, but not the slightest motion was
perceptible so we concluded the Mias was
dead & were obliged to leave him there.
Next morning early I came to the

1. Like other great apes, orangutans construct daily sleeping nests from branches and foliage. As Wallace described on the previous page, recto 28, orangutans build their nests by pulling branches together. They initially select stout branches to make a foundation, and then they create a "mattress" over this with smaller leafy branches that are interwoven into the foundation branches (Didik et al. 2009). This unfortunate individual, gravely wounded by Wallace and his assistants, likely constructed its nest defensively and soon expired within it.

spot and found our conjecture was correct
as he still remained in exactly the same
position. [1]I even offered 4 Chinamen a
full days wages to cut the tree down, which
they could have done in two hours, but
after looking at it & trying it for a
considerable time they determined that
it was very big & very hard & would
not attempt it, so I was obliged to
leave this fine animal to rot on
the top of a tree hoping to get his
scull & perhaps his skeleton when it
fell.

June 27—

Charles found 3 young Mias's.
I went after him & after a considerable
chase shot one who remained dead in
a forked branch on a very high tree—
Saw him pull the small branches towards
him to get a firm hold in passing
from tree to tree.

1. In *Malay Archipelago* (62–63) Wallace gives more information about this episode. Had he doubled his offer to the Chinese laborers "they would probably have accepted it," he wrote, "as it would not have been more than two or three hours' work." The "streetwise" Wallace decided it was best not to offer more: "Had I been on a short visit only I would have done so; but as I was a resident, and intended remaining several months longer, it would not have answered to begin paying too exorbitantly, or I should have got nothing done in future at a lower rate."

In the *Malay Archipelago* narrative he also conveyed the macabre image of the result of leaving the orang's body to the tropical elements and organisms: "For some weeks after a cloud of flies could be seen all day, hovering over the body of the dead mias." A few months later he hired a pair of Malays to climb the tree to retrieve the remains for him. (See also recto 11, note 3.)

## 31

[1]Proofs of Design

Three scars on cocoa nut, one of them soft for
the exit of embryo which is immediately below it
"which without this wise contrivance would be
unable to pierce the hard case in which it is
confined." Cyc.[lopedia] [of] Nat.[ural] Hist.[ory] Vol.[ume]
2. p.[age] 55.

---

Is not this absurd? To impute to the supreme
Being a degree of intelligence only equal [to] that
of the stupidest human beings. What should
we think, if as a proof of the superior wisdom
of some philosopher, it was pointed out that
in building a house he had made a door to it,
or in contriving a box had furnished it with
a lid!—Yet this is the kind and degree of
design imputed to the Deity as a proof
of his infinite wisdom. Could the lowest
savage have a more degrading idea of his
God.

1. Wallace is quite caustic in his criticism of this passage from Knight's *Cyclopedia*. His point is a good one—it seems absurd to see benevolent design in even the most obvious necessities. We would not esteem the house builder as especially wise for including a door, while we might deem him incompetent if it was omitted. The last point gets at the heart of the matter for Wallace: pointing to such features as proofs of the creator's infinite wisdom is faint praise indeed for the creator—indeed, it is degrading. This highlights Wallace's religious sensibility: far from a personal god, Wallace's creator, insofar as he believes in one, is more deistical in nature, acting through natural laws.

## 32

Skeleton of Birds

[1]"To give greater freedom of action to the <u>bill</u> it was necessary, as the bones of the back have hardly any motion, that the neck should be long and flexible. In Mammifers the number of neck bones is 7 the Giraffe has no more & the Elephant no less. But in Birds the deficiency of motion in the back is made up by a free grant of cervical vertebrae according to the wants which the peculiar habits of particular birds require. The raven has 12. The Cock 13. The ostrich 18 the swan 23." (Cyc.[lopedia] [of] Nat.[ural] Hist.[ory] 1.461)

Here are several gratuitous statements & inferences. The writer seems to have been behind the scenes at the creation & to have been well acquainted with the motives of the creator. A humbler mortal may suppose that the same power which enabled the elephant & giraffe, the whale

1. Wallace further quotes the *Cyclopedia* article, taking exception to the author's putative knowledge of "needs," "wants," and what is "necessary" to effect what. Dismissing these as gratuitous statements, this is the rub for Wallace—how could anyone know what could or could not be done by an omnipotent creator, or what was necessary or not?

and the camel, to perform all their functions
with 7 neck bones, could also have ~~enabled~~ formed
birds to ~~have~~ perform theirs with the same number, with still
further modifications of form & structure.
The writer however places a limit to the
power of the creator in this direction. He says
this could not have been done, "it was necessary
to give them more."

Again how can any man venture to
say that, the cause of the different number
of cervical vertebrae in different birds is that "their
peculiar habits required it." Is it not
just possible that [1]some totally different
causes absolutely hidden from us determined
the form & structure of animals, & that
their wants and habits resulted from
that structure? We are like children looking
at a complicated machine of the reasons of
whose construction they are ignorant, and like
them we constantly impute as cause what is really
effect in our vain attempts to explain what we will
not confess that we cannot understand—see p.[age] 53

*[The following sentence is written vertically in left margin
 of page]*
[2]It is not easy to see how a bird could have <u>habits</u> & <u>wants</u>
 before it had neck bones, as it must
have had if the number of these bones depended on its wants
 & habits.

1. Wallace's reference to "some totally different causes absolutely hidden from us" clearly indicates his preference for a natural explanation.

Is it not possible that an omnipotent creator could fashion various species to their diverse niches while making do with the same number of vertebrae? An analogy might be the pentadactyl limb, the bones of which are modified for so many ends (swimming, flying, running, burrowing, brachiating, etc.), with the same set of bones in evidence. And so, Wallace suggests, rather than simply declare that "their particular habits require it," consider instead the possibility that "some totally different causes absolutely hidden from us determined the form & structure of animals, and that their wants and habits resulted from that structure." The number of vertebrae vary for reasons that are unclear as yet, but rather than simply declare them designed as such, look for the cause. In the interesting passage that follows, Wallace likens the author of the *Cyclopedia* article and others similarly quick to reach misguided conclusions to "children looking at a complicated machine of the reasons of whose construction they are ignorant, and like them we constantly impute as cause what is really effect in our vain attempts to explain what we will not confess that we cannot understand."

2. The comment written along the left-hand margin was likely provoked by the author's final statement in the *Cyclopedia* passage with which Wallace has been arguing: "The articulation is so contrived as to produce the greatest mobility, and that the contrivance is complete is proved by the ability of a bird to touch every point of its body with its bill."

Wallace puts his finger on the logical difficulties of such natural-theology explanations of structure and function. Which came first, the "wants and habits" or the neck bones? This is not such a difficulty for a natural theologian, perhaps, who would simply reference the archetypal "bird" as having certain "habits and wants" (preening, for example, to keep feathers clean of parasites) and in turn required a special contrivance to meet these needs, owing to the rigidity of the backbone.

## 34

[1]Notes from Lyell's "Principles"—
For near two centuries a dispute was carried on
as to whether fossils were real, or mere accidental
resemblances, freaks of nature, or special
creations of stones to imitate the remains of
animals. Men would not admit these fossils
to have been due to secondary causes, to the
ordinary Course of nature.

Later when Buffon declared "that the mountains &
valleys of the Earth were due to secondary causes,
& that the same causes would in time destroy
them & again produce new ones," he was obliged
to recant his opinions as contrary to scripture,
though they are now universally admitted to
have been correct—

---

In B.[ritish] Ass.[ociation] 1854. Prof.[essor] [2]Ramsay, shews
    that glaciers
existed in England in Permian period. Will not
this account for diminution of animal life at
that period—? which Prof[essor] Forbes accounted for
by Polarity! .

see on p.[age] 92

1. Wallace traveled with the fourth edition of the *Principles of Geology* while he was in the Malay Archipelago, published in 1835 in four volumes. Part of Lyell's successful rhetorical strategy in the *Principles* involved establishing a narrative of the history of geology that aimed to show the vast superiority of his vision for uniformity (what in his day was called actualism) over what he portrayed as the erroneous and superstitious conceptions of earth history held by his benighted predecessors. Naturally his portrayal was an oversimplification, but it is true enough that a persistent Neoplatonic strain of thought held sway through much of the Middle Ages that interpreted fossils as anything but organic remains of once-living organisms. In volume 1 of this edition of *Principles* Lyell offered a summary of the "controversy as to the real nature of fossil organic remains," summarizing the views of a host of French and Italian naturalists of the sixteenth and seventeenth centuries. Some, like the Frenchman Palissy writing in 1580, are praised for taking the minority view that fossils were organic remains. He paints French savant Georges-Louis Leclerc, comte de Buffon, as something of a martyr, a Galileo figure "forced to recant," as Wallace puts it. Buffon offered a theory of the earth in his 1778 *Les époques de la nature,* but ran afoul of ecclesiastical authorities with quasi-evolutionary speculations in his magisterial *Histoire naturelle, générale et particulière* (ultimately published in thirty-six volumes between 1749 and 1788). For rhetorical effect Lyell quotes the "grand principle" in *Histoire naturelle* (1769) that Buffon was made to renounce: "that the present mountains and valleys of the earth are due to secondary causes, and that the same causes will in time destroy all the continents, hills, and valleys, and reproduce others like them." This is, of course, Lyell's own grand principle as well: the primacy of natural law, and the steady and inexorable alteration of the face of the earth.

Wallace's concern, too, is with natural law: with the rejection in earlier times of the idea that fossils could be attributed to "secondary causes, to the ordinary course of nature," and Buffon's trouble with the church over "secondary causes." This is of a piece with some of his critiques of Knight's *Cyclopedia;* he is convinced that earth's history, and that of the species upon it, is one of slow change over time as a result of natural law.

2. Andrew Crombie Ramsay (1814–1891), a noted Scottish geologist, read two papers at the British Association for the Advancement of Science meeting in Liverpool in 1854. He startled his audience with the conclusion that glaciers existed in Britain during the late Paleozoic, and that these were responsible for the breccias and boulders of the Malvern and Abberley hills. Ramsay's papers were entitled "On the Former Probable Existence of Paleozoic Glaciers," and "On the Thickness of the Ice of the Ancient Glaciers of North Wales, and Other Points Bearing on the Glaciation of the Country." The papers were published the following year in the *Reports of the British Association for 1854,* where Wallace read them.

Wallace is noting here that Ramsay's late-Paleozoic glaciers would explain—*naturally*—the supposed "diminution of life" of that period cited by the recently deceased naturalist Edward Forbes (1815–1854) in support of his "polarity theory." This quasi-mystical theory held that species abundance changed over time according to a fixed hourglass-shaped pattern, from great abundance in the most distant past, diminishing to its nadir sometime in the Paleozoic, and steadily increasing since. (Presumably it would reach its zenith at some point in the near future and then the process would begin again.) Wallace disagreed with Forbes's theory so strongly that it prompted him to write his 1855 Sarawak Law paper in response.

## 35

[1]Note <u>for Organic law of</u> change.
We must at the outset endeavour to ascertain if
the present condition of the organic world, is now
undergoing any changes—of what nature & to what
amount, & we must in the first place assume
that the regular course of nature from <sup>early</sup> Geological
Epochs to the present time has produced the present
state of things & still continues to act in
still further changing it. While the
inorganic world has been strictly shown to be the
result of a series of changes from the earliest
periods produced by causes still acting, it
would be most unphilosophical to conclude
without the strongest evidence that the organic
world so intimately connected with it, had
been subject to other laws which have now ceased
to act, & that the extinctions & ~~renewals~~ <sup>production</sup> of species
and genera had at some late period suddenly ceased.
The change is so perfectly gradual from the latest Geological
to the modern epoch, that we cannot help believing
the present condition of the Earth & its inhabitants

1. American historian H. Lewis McKinney (1966, 1972) first suggested that this entry marked the beginning of a series of notes by Wallace for an intended book on evolution (also see recto 51).

to be the natural result of its immediately
preceding state modified by causes which have
always been & still continue in action.

---

[1]Lyell says vol.[ume] 1. p.[age] 226 that those Naturalists
are inconsistent who connect the phenomen[on]
of the early vegetation (ferns monocotyledon &c)
with a progressive advance ~~of~~ vegetation from those as a
low form, to dicotyledons later as a higher
form, & at the same time bring
them forward as proofs of a higher
temperature in cold regions where they are
now found. If the first is the
cause of their presence, the 2nd. cannot
be an inference from it, as whatever
was the climate no other form of
plants could have appeared—

2

[1.] Eurycephalus    [2.] E.[urycephalus] maxillosus. Oliv.[ier]    [3.] maxillosus. Oliv.[ier]
maxillosus. Oliv.[ier]    see p.[age] 42. for another form of label.

| maxillosus. Oliv.[ier] | E.[urycephalus] maxillosus | E.[urycephalus] |
| (Cerambyx max.[illosu][s]) | (Cerambyx pt.*[?]*) Oliv.[ier] | cardinalis |
| | | White |

3

[4.] The specific name <u>alone</u> best. author in cat[alogue]

---

1. This commences a series of critiques of Lyell's arguments against the idea of successive development of life on earth, made in his landmark *Principles of Geology*. As noted, the edition that Wallace consulted was the fourth edition, but most of Lyell's anti-transmutationist arguments were articulated beginning with the first edition. While Lyell went to great lengths to undermine the idea that the record of life in earth history is one of successive change from simpler to more complex forms, many naturalists of the day were convinced that fossils showed such change over time, though most would have understood this not as an evolutionary progression but as successively more complex creations.

The "Naturalists" mentioned here are likely botanist John Lindley (1799–1865) and geologist William Hutton (1798–1860), authors of *Fossil Flora of Great Britain* (1831–1835), a work quoted by Lyell. Naturalists pointed to the predominance of relatively simpler plants like ferns and lycopods way down in the fossil strata of England, for example, followed in the temporal sequence by dicots in more recent strata, and they saw this as evidence for progressive change over time. They also argued that those early ferns and other simple plants showed that the earth had been hotter at high latitudes in the past.

Curiously, as a young man Lyell himself was rather accepting of the idea of successive development (Bartholomew 1973, Corsi 1978), but the implications for humankind apparently so offended his religious convictions that he changed his mind and dedicated much space in the *Principles* to refuting transmutation. As the foremost exponent of transmutationism, the distinguished French naturalist Jean-Baptiste Lamarck (1744–1829) was his primary target, but Lyell also sought to undermine *any* suggestion of a progression of life forms in earth history, as it could be too easily interpreted in a transmutational context. Applying his principle of "uniformity of state" to the history of life, Lyell rejected the notion that simple forms were found on the earth earliest, followed by progressively more complex forms. He also rejected the idea that the earth was significantly hotter in the past—that, too, would violate his concept of uniformity of state. In the pages that follow we will see Wallace attacking many of Lyell's arguments against transmutation.

On this page Wallace is simply noting Lyell's argument that naturalists can't have it both ways: the predominance of relatively simpler plants such as ferns in early earth history cannot be attributed to the first step in an unfolding plan, on the one hand, and also used to argue that the early earth was hotter than at present at high latitudes. Why not? Because if you subscribe to the "unfolding plan" idea, the early earth could not have had anything *but* those simpler plants, almost by definition. As Wallace puts it here, if the cause of the preponderance of those plants in early earth history is that they represent an incipient stage in a progressive sequence, no matter what the climate no other form of plant could have occurred then. Wallace is not disagreeing with this, though unlike Lyell, he believes there is indeed a progressive series of species through time.

2. Here Wallace was experimenting with designs for specimen labels. On recto 42, as he notes here, there are two more label designs, with "best form" written beside one. The species he uses here, *Eurycephalus maxillosus*, is a long-horned beetle, family Cerambycidae.

In his 1855 "Letter from Sarawak" printed in the *Zoologist* (Wallace 1855a, 4807), Wallace reported that "numbers of the handsome red *Eurycephalus maxillosus* are here constantly flying about and crawling on the timber" in a cleared area with piles of downed trees near the Simunjon Coal Works.

3. Wallace decides that it is best to leave the author of the species name off of the collecting label, no doubt for economy of space, and simply provide such information later in the collection catalog.

## 37

[1]"Some of the more ancient Saurians approximated
more nearly in their organization to the types of
living Mammalia than do any of our existing
reptiles"—which? just what I want. Lyell vol.[ume]1. p.[age] 231.
Lyell says the Didelphys of the Oolite is fatal
to the theory of progressive development."—
Not so if low organized mammalia branched
out of low reptiles, fishes—All that is
required for the progression is that some
reptiles should appear before Mammalia &
birds or even that they should appear together. In the same manner reptiles
should not appear before fishes but
it matters not how soon after them.
As a general rule let Naturalists
determine that one class of animals is higher
organized than another, & all that the development
theory requires is that some specimens of the
lower organized group should appear earlier
than any of the group of higher organization

        A[lfred] W.[allace]

Now vertebrated animals are universally

1. Wallace continues his analysis of Lyell, whose point about the saurians is intended to further undermine the idea of progressive development: if some ancient reptiles more closely resemble living mammals than any existing reptile, the idea of progressive development of animal forms is untenable. Wallace's interest in this has more to do with his branching, treelike vision for species change over time.

The note inserted here—"which? just what I want"—related to Wallace's following argument concerning relative branching order. Wallace disagrees with Lyell that the occurrence of extinct fossil opossum (*Didelphys* or *Didelphis*) relatives in the Oolite (a type of Jurassic limestone composed of tiny spherical granules) is "fatal to the theory of progressive development."

As Lyell put it in the first edition of *Principles* (1830): "The occurrence of one individual of the higher classes of mammalia . . . in these ancient strata, is as fatal to the theory of successive development, as if several hundreds had been discovered" (Lyell 1830, 150). Not so, says Wallace, pointing out that there should be an order only to each group's *first* appearance, as groups branch from other groups. *Some* reptiles should appear before mammals, *some* fishes should appear before reptiles, and so on. Note Wallace's comment, "which? just what I want," referring to ancient reptiles more closely resembling living mammals, in some cases, than living reptiles do. Why does he want these examples? No doubt as material for his planned book. He sees mammals as having branched out of the reptile line; other reptiles, in the meantime, continued to evolve ("develop") and are accordingly highly modified from their early reptilian relatives. Those early reptiles would have more in common with modern mammals, which they gave rise to, if mammals have changed less from their early forebears than have reptiles. Wallace's main point is that different branching lineages can co-occur. The only requirement of the development view is that no representative of a group should appear earlier than the earliest member of the group from which it is supposed to have branched.

Wallace refers to the "development theory," transmutation, but more than mere transmutation, these passages show that he clearly grasps the concept of groups branching from groups. If "low organized mammals *branched out of* low reptiles," those early mammals could and should coexist with those early reptiles, and even with later-evolved more complex reptiles. Lyell thinks that the progressive development view would have groups replacing groups in a strict sequence, the "lowest" form of each group following the most "advanced" form of an earlier group.

The fossils in question, by the way, were jaws found near Oxford, in the Lower Jurassic Stonesfield Slate (actually a limestone). These fossils were famous as the first known Mesozoic mammal fossils, attracting the attention of French naturalist Georges Cuvier and the "English Cuvier," Richard Owen. These naturalists identified them as a type of marsupial, but they are now classed with triconodonts and cousins, a very early mammalian group related to the ancestors of today's mammals.

allowed to be the highest form of animal life.
They appear Geologically after the Mollusca &
Radiata. Fishes are universally declared
the lowest of the vertebrata. They appear
first. Reptiles are universally considered
as higher than Fishes but lower than
Birds & Mammalia. They are found
next in succession. The Marsupiata
are generally allowed to be one of the
lowest forms of the Mammalia & [1]it
is one of them which appears first,
while, the highest form of the Quadrumana
appears considerably later. [2]Thus not
one fact contradicts the progression.
The supposed contradictions all arise from
considering it necessary that the highest forms
of one group should appear before the lowest
of the next succeeding, not ~~succeed~~ considering
that each group goes on progressing after
other groups have branched from it.
They then go on in parallel or diverging

1. The marsupial referred to here is one of the Oolite *Didelphys* (*Didelphis*) relatives mentioned on the previous page. That these fossils are not marsupials but an even earlier mammalian group does not alter Wallace's point: the "lowest" mammalian forms appear first in the fossil record, the "highest" later.

2. To Wallace, this progression can have only an evolutionary explanation. Any apparent contradictions stem, he argues, from a misreading of the nature of this progression. He reiterates the point made on the previous page about groups branching from ancestral groups. Note his clear articulation of evolutionary change as a branching, diverging process: groups go on "progressing" [evolving] after other groups branch from them, the branches all continuing to "go on in parallel or diverging series."

Since branching groups continue to coexist and "progress" together, in principle a parental form and the daughter lineages that branched from it may continue to progress together, to the point where they attain maximal species diversity concurrently. It is possible, in other words, for "lower" or more primitive forms to be as common as later-branching "higher" relatives in certain geological periods. This is contrary to the strict successional evolutionary model that Lyell criticizes. The latter model, derived from Lamarck, saw evolutionary change as the successive replacement of lower forms with higher forms.

series & may obtain their max[imu]^m together.

---

[1]Lyell, says that varieties of some species may
differ more than other species do from each
other without shaking our confidence in the
reality of species. But why should we have
that confidence? Is it not a mere preposesson *[sic]*
or prejudice like that in favour of the
~~immutability~~ stability of the earth which he has so ably
argued against?—In fact what positive
evidence have we that species only vary
within certain limits? Let us suppose
that every variety of the Dog but one was
to become extinct & that one say the
spaniel, to be gradual spread over the
whole world, subjected to every variety of
climate & food, & domesticated by
every variety of the human race.
Have we any reason for supposing that in
the course of ages a new series of varieties

1. Using dogs as an example, Lyell argued that although species in general vary only within certain limits, it is possible for varieties of some species (such as dogs) to differ to an even greater degree than might be found between individuals of two distinct species: "The degree of possible discordance between varieties of the same species may, in certain cases, exceed the utmost disparity which can arise between two individuals of many distinct species" (1835, 2: 435).

Lyell's strategy here was to undermine the idea that the boundary between species and varieties is fluid. Transmutationists might rightly ask why varieties of a given species should differ to a far greater degree even than individuals of two clearly different species. This makes little sense under special creation, which says species are immutable, and is far more suggestive of the idea that varieties might grade into species.

Wallace takes this up: why indeed should we have confidence that species are "real" (i.e., immutable)? It is interesting that Wallace, with irony, likens belief in this to the former belief in a static earth, which Lyell argued so eloquently to overthrow. Darwin invoked much the same image in chapter 4 of *On the Origin of Species* (hereafter referred to as *Origin of Species*), where he writes that his "doctrine of natural selection . . . is open to the same objections which were at first urged against Sir Charles Lyell's noble views on 'the modern changes of the earth'" (Darwin 1859, 95). There is irony there, too, in that Lyell steadfastly rejected the efficacy of natural selection to bring about Lyellian slow, steady change in organisms. Picking up on Lyell's dog-variety example, Wallace conducts a thought experiment of his own.

quite distinct from any now existing
would not be developed,—& then should
the same process be repeated & one
of these varieties farthest removed from
the original, again be spread over the
earth & be subjected to the same
variety of conditions, [1]does it not seem
probable that again new varieties
would be produced, & have we any
evidence to show that at length
a ~~stop~~ check would be placed on any further
change & ever after the species remain
perfectly invariable under any circumstances whatever. Those who
advocate variation within definite
limits must suppose so, though the
only ground for their opinion is that
the varieties which have been produced
under the influence of man have certain
limits.—[2]But we have no proof how the varieties of dogs
were produced. All varieties we know of are produced at birth
the offspring differing from the parent. This offspring propagates
its kind. Who can declare that it shall not produce a variety
which process continued at intervals will account for all
the facts.

1. Variations, and so varieties, are inexhaustible, Wallace argues, and if it is possible to generate varieties from varieties from varieties, who is to say how far this can go?

Lyell went to great lengths in the *Principles* to attack the idea that species can change beyond certain limits, but clearly Wallace did not buy his argument. We do not have any evidence, he implies, that species might change to a certain degree in producing varieties and then "remain perfectly invariable under any circumstances whatever."

2. The kind of variety he mentions here, "produced at birth the offspring differing from the parent," sounds like a "sport" or novel variant produced by chance and then propagated for its novelty. Wallace rhetorically asks, cannot this variant itself produce further variants?

Note that Wallace does not have an understanding of varieties produced by selection at this point; indeed, perhaps because of Lyell's insistence that domestic varieties only vary within limits, Wallace sidestepped them in his opening to the Ternate essay (1858) by suggesting that domestic varieties cannot teach us anything about species in a state of nature. Being unnatural, they cannot long survive in nature anyway, he argued. This contrasts with Darwin's early vision of domestic varieties and artificial selection as a microcosm of the natural selection process (Darwin 1909). It is perhaps surprising that Wallace did not make more of domestic varieties, as he clearly appreciated that they can represent an example of transmutation, as we see on the following page.

## 41

[1]Lyell suppose some species must vary more on account of the variations of temperature &c. to which they are subject. But in the most equable climate of the tropics, ~~the~~ numberless varieties exist linking together the most closely allied species.

[2]In a few lines Lyell passes over the varieties of the Dog & says there is <u>no transmutation</u>—Is not the change of one original animal to two such different animals as the Greyhound & the bulldog a transmutation?—Is there more essential difference between the ass the giraffe & the zebra than between these two varieties of dogs. Do the carrion crow & the rook differ more essentially in specific characters than the Polish hen & the Dorking fowl. [3]And is there any other reason why these are not distinct species, than that we believe them to have been derived from a common stock, in the one case while we do not believe it

1. The idea that varying temperature and other environmental conditions somehow produce variations in species—shot down here by Wallace, who had considerably more experience with the tropics than Lyell—was widely accepted at the time. Darwin depended on environmental variation, in the *Origin*, to yield the ubiquitous variations he thought necessary as raw material for natural selection. It is curious that this rather Lamarckian idea was so readily embraced by Lyell, Darwin, and others who otherwise rejected Lamarck.

2. Here Wallace clearly "gets" the concept that domestic varieties are analogous with allied species in nature, even underscoring that the change producing domestic varieties is indeed transmutation.

   Note that here Wallace seems to define species in terms of similarity or difference in "essential" or "specific" characters. He sees constancy of character as a defining feature of species, and later, in his important paper on Malayan swallowtail butterflies, he describes his species concept more fully: "When the difference between two forms inhabiting separate areas seems quite constant, when it can be defined in words, and when it is not confined to a single peculiarity only, I have considered such forms to be species." Later in this paper he gives what is essentially the modern species concept based on reproductive competence: "Species are merely those strongly marked races or local forms which when in contact do not intermix, and when inhabiting distinct areas are generally believed to have had a separate origin, and to be incapable of producing a fertile hybrid offspring." Wallace goes on to point out that this is an operationally useless definition, however, owing to the difficulty of performing the hybridization test in the vast majority of cases (Wallace 1865a, 4, 12).

3. Putting his finger on the crux of the issue, Wallace asks rhetorically why such apparently divergent domestic varieties are not recognized as distinct species when they differ from each other to as great a degree or more than allied species such as the carrion crow and rook, which are recognized as such. He thinks the difference lies in people's recognition that the domestic varieties have derived from a "common stock," while they reject this concept for natural species, seen as specially created. Wallace highlights this inconsistency in his "Note on the Theory of Permanent and Geographical Varieties," published in the *Zoologist*:

> Now the generally adopted opinion is that species are absolute independent creations, which during their whole existence never vary from one to another, while varieties are not independent creations, but are or have been produced by ordinary generation from a parent species . . . If there is no other character [to distinguish them], that fact is one of the strongest arguments against the independent creation of species, for why should a special act of creation be required to call into existence an organism differing only in degree from another which has been produced by existing laws? (Wallace 1858b, 5888)

Although elsewhere Wallace appreciates the ability, or lack thereof, to interbreed as the defining feature of species boundaries, here he overlooks that point in suggesting that domestic varieties should be considered distinct species. Seeing domestic varieties and natural species as wholly unrelated phenomena, Lyell and others rejected the concept that they are in fact analogous. Wallace disagrees, and clearly sees the parallel—"common stock" for domestic varieties equals "common ancestor" for allied natural species.

in the other. In neither case have we
positive knowledge on this point.

[1]The varieties of the <u>Primrose</u> adduced
by <u>Lyell</u> is complete proof of the
transmutation of species. It only shows
the impossibility of convincing, a person
against his will. Where an instance
of the transmutation is produced, he
turns round & says "You see they
are not species they are only varieties."

———————

[2]"The alteration of the habits has reached
a point beyond which no ulterior
modification is possible, however
indefinite the lapse of ages during
which the new circumstances operate"
Mr Lyell must be very foresighted
to know this.                    [3]maxillosus. <u>Best form</u>.
maxillosus                  This form gives the <u>shortest</u> label, which
(Cerambyx) Oliv.[ier]   is important where there are many small
                                   species & mostly single specimens.

———————————————

*[The following two sentences are written vertically in left margin
    of page]*
Qy.*[Query?]* is not the sp.[ecies] name <u>alone</u>, sufficient & best
    for a cabinet, leaving authority & synonymes *[sic]*
to be had by reference to the catalogue. I am inclined to think
    so.—

1. In *Principles,* Lyell related an experiment by the respected clergyman and horticulturist William Herbert (1778–1847), who reported raising from the seed of one cowslip a remarkable array of individuals that Herbert considered to be distinct varieties: "a primrose, a cowslip, oxlips of the usual and other colours, a black polyanthus, a hose-in-hose cowslip, and a natural primrose bearing its flower on a polyanthus stalk"; Herbert concluded, "I therefore consider all these to be only local varieties depending on soil and situation" (Lyell 1835, 2: 446–448). Darwin's Cambridge mentor, the well-known botanist John Stevens Henslow (1796–1861), confirmed the experiment. Wallace sees the results as remarkable evidence of transmutation, but Lyell was dismissive: since the nature of the results depends on the species, he holds, they are "part of the specific character." That they consistently yield the same results indicates "certain fixed and invariable relations between the physiological peculiarities of the plant," Lyell writes, concluding that "they afford no ground for questioning the instability of species, but rather the contrary." You can detect a derisive tone in Wallace's remark here—for Lyell is dismissing what Wallace takes to be the real significance, namely, that while Herbert's experiment reveals just how malleable and intrinsically variable species are, Lyell turns this around and says that the *consistency* in their variability is tantamount to being invariable.

2. Another mildly scornful comment by Wallace—the quoted passage comes from volume 2 of *Principles* (1835, 2: 452), in a section discussing the "extent of change in species." It comes in a passage in which Lyell argues that individuals can "alter" in habits, form, or organization to a limited degree under changing environmental conditions, but with further change in environment, however slight, "all modification ceases, and the individual perishes." He then uses as an example what happens to wild animals hunted by humans: "The persecuted race soon becomes more cautious, watchful, and cunning; new instincts seem often to be developed, and to become hereditary in the first two or three generations." They then hit a wall, Lyell claims; as the skill of the hunters increases, the hunted animals cannot change further: "No further variation can take place, no new qualities are elicited by the increasing dangers." This immediately precedes the passage quoted by Wallace, who clearly feels it is absurd for Lyell to make a claim about what is or is not possible regarding the continued varying of species after a "lapse of ages."

3. Continuation of specimen label designs from recto 36. Wallace ultimately decided that simplest was best. The note "specific name alone best," back on 36, was likely written at the same time as the marginal notes here.

## 43

[1]Lyell—says, "It has been shewn that
a short period of time is sufficient to
bring about all the change possible in a
species." But this is altogether begging the
question. We can only produce a certain
change in a limited time, but it by
no means follows that other changes cannot
& have not been brought about requiring long
periods. ~~Does Mr Lyell think that~~
Does Mr Lyell think that from any
one race of dogs the greyhound
bulldog & spaniel can be produced
in a short space of time—?

————

[2]"Some new sense or organ produced in
individuals descending from a common
stock, would lead us to believe time might
bring about any metamorphosis."
This would be taking a leap with a very <u>vengeance</u>
we should have to get out of our class or
order into another passing through many
thousand species. If this is required to
prove a change from one species to another
it can never be proved.

1. The critique of Lyell continues. Here Wallace takes exception to a statement made in the concluding paragraphs of volume 2. The full passage reads: ". . . a short period of time is generally sufficient to effect nearly the whole change which an alteration of external circumstances can bring about in the habits of a species" (Lyell 1835, 2: 464). Wallace's point here echoes that on the previous page. How can Lyell presume to know what may or may not be brought about over long periods of time, given that our activity has been limited in time? He finds Lyell's argument specious.

Perhaps having second thoughts on the very idea that humans have effected change in domesticated species in only a limited time, in the final statement here Wallace steps back and asks rhetorically if a short span of time is adequate for producing, from any given dog breed, a new breed as marked as the greyhound, bulldog, or spaniel. On reflection, he implies, a long period of time indeed was necessary to produce these dramatically different dog breeds.

2. Lyell makes this statement in the *Principles* (2: 414) only to immediately dismiss it as a "gratuitous assumption." Here Wallace points out that this is an unreasonable test of transmutation, since altogether new senses or organs would be expected only after much divergence and evolutionary change.

## 44

Lyell
[1]"If we are to infer that some one of the wild grasses
has been transformed into the common wheat."
because we cannot find the cultivated wheat in a state of nature."

———————

[2]"The miserable grass <u>Aegelops ovata</u> was sown year after
year till it became wheat in no respect different
from the common hard wheat of the South of
France." Report of Horticultural Soc.[iety] 1854.

———————

Of course it will be said when this fact
is established, "The two plants are varieties
of each other". So when a plant of one
<u>Genus</u> is transformed into a plant of
another <u>genus</u>, it is still not satisfactory.

———————————————————

[3]Many of Lamarck's views are quite
untenable & it is easy to controvert them
but not so the simple question of a species
being produced in time from a closely allied
distinct species which however may of course
continue to exist as long or longer than its
offshoot.

1. Following an account of Egyptian mummies (and their apparent nontransformation compared with living species), Lyell discusses "Seeds and plants from Egyptian tombs" and then "Native country of the common wheat." This was all treated in one section in the first *Principles* edition, but in the fourth edition Lyell broke his extended argument with Lamarck down into finer units, though little new material was added.

It is the latter section that this quote comes from. The context is Lamarck's claim that "the botanist cannot point out a country where the common wheat grows wild." Lyell rhetorically dismisses this by saying that if we conclude that wild grasses transmutated into wheat just because we cannot find wheat in nature, we might as well apply that reasoning to any domesticated plant or animal—the domestic dog, or camel. (Lamarck—and Darwin— would agree that we should. So would Wallace, at the time the Species Notebook was kept.)

2. The work that Wallace quotes is probably an abstract of a paper by French botanist Esprit Fabre, which was published in translation in the *Journal of the Royal Agricultural Society of England* in 1854 (15: 167–180). In this paper Fabre makes a case for having "transformed" through successive generations of breeding the ovate goat grass *Aegilops ovata* (now known as *A. geniculata*) into the cultivated wheat *Triticum sativum*. This is in fact not the case; the common wheat (now *T. aestivum*) arose via hybridization of three species, among them another goat grass, *A. tauschii*.

Wallace sounds mildly irritated anticipating the rationalization of anti-transmutationists, who he imagines would reject Fabre's claim out of hand: even when a genus seems to be transformed into another, this would be dismissed by declaring that the two are merely varieties of the same species after all.

Fabre's experiments generated much interest, and even Darwin's mentor, John Stevens Henslow of Cambridge, attempted to replicate his experiments. Henslow worked with *A. squarrosa*, a goat grass of southern Europe, and after three years of effort reported mixed results at the British Association for the Advancement of Science meeting in Cheltenham (Henslow 1856). The goat grass was altered in several respects, but he was not satisfied that he succeeded in transforming it into true wheat.

3. Wallace's language here echoes his Sarawak Law paper (Wallace 1855c), that *"every species has come into existence coincident both in space and time with a pre-existing closely allied species"* (italics are his), and also recto 37–38 of this notebook. It was not so long before this notebook entry that Wallace had penned those electrifying words in Sarawak.

## 45

[1]Changes which we bring about artificially in
short periods may have a tendency to revert
to the parent stock though this in animals is not proved. This is considered a
grand test of a variety. But when the
Change has been produced by nature during a
long series of generations, as gradual as the
changes of Geology, it by no means follows
that it may not be permanent & thus
true species be produced.

---

[2]Lyell occupies much space in shewing how the
species which are <u>common</u> to different & distant
countries, might have been carried from one to
the other by a variety of accidents. But this has
never been felt to be a difficulty. The matter
of wonder has always been that in distant
countries of similar climate so many
should be <u>different</u>. This he gets over by
special creations of the species each in one spot
as they were wanted.

This is no doubt <u>a very</u> easy way of getting

1. A standard argument against transmutation was the supposed "reversion" of domesticated plants and animals in a state of nature. As a "grand test of a variety," in principle the tendency to revert back to the parental form would have been taken as a defining feature of a variety versus true species, which would be stable. Taking the long view of change—over geological time—Wallace maintains that permanent change may be produced. He addressed the problem of reversion in the opening paragraphs of the Ternate essay (1858f).

2. In the third *Principles* volume of 1835 Lyell addressed the many and varied ways that species become geographically dispersed. Though Wallace comments here that "this has never been felt to be a difficulty," in fact it was, from the point of view of special creation. The unique species on remote islands either were created there or somehow descended from ancient colonists. Lyell discusses at great length the means of dispersal, but it is important to note that he is talking about species moving far and wide, establishing populations in far-flung locales. In this section he does not address endemics to remote islands, but it is safe to say that he would consider these to have been specially created there.

Darwin would not have agreed that "this has never been felt to be a difficulty." He devoted a great deal of energy in the 1850s to conducting experiments aimed at showing the likelihood of the long-distance dispersal of plants over the open ocean (see Barrett et al. 1987). He did this in response to skepticism of friends like Joseph Hooker over the possibility that plants and animals could survive long transoceanic trips. Wallace and Darwin were very much in agreement, however, that invoking special creation to explain the striking differences in species composition of different locales, especially those with similar climates, was dodging the issue. In an 1858 paper about the Aru Islands, Wallace made this point forcefully in comparing the species of Borneo and New Guinea and then New Guinea and Australia. All three are geographically fairly close to one another. The first two are similar in size, terrain, and climate and yet they differ dramatically in their fauna, while the latter two differ dramatically in size, terrain, and climate, and yet their fauna is extremely similar.

over it, but [1]just as philosophical as to
say that fossils of existing species are remains
of real animals while those which are not
like any species now existing are special
creations & not fossil animals at all.

———

[2]In a small group of islands not very
distant from the main land, like the
Galapagos, we find animals & plants different
from those of any other country but resembling
those of the nearest land. If they are special
creations why should they resemble those of the nearest
land? Does not that fact point to an origin
from that land. [3]Again in these islands we find
species peculiar to each island, & not one of
them containing all the species found in the
others as would be the case had one been
peopled with new creations & the others left
to become peopled by winds currents &c. from it.
Here we must suppose special creations in
each island of peculiar species though the
islands are all exactly similar in structure

1. Wallace again pens a criticism that strikes at the very heart of what he takes to be Lyell's hypocrisy—or at least inconsistency—over the matter of species change. The readiness with which Lyell is willing to invoke special creation to get around the biogeographical patterns that present a serious difficulty for anti-transmutationists is reminiscent of the readiness with which naturalists of old disputed whether fossils were actually the organic remains of once-living creatures, as a convenient end run around the question of their origin and fate. Wallace was almost certainly aware that Lyell himself dismissed such authors as adherents of a hidebound and misguided reading of nature.

2. Wallace reads as plain as day the significance of the relationship of island flora and fauna to species of the nearest mainland. The Galapagos observation likely comes from the second (1845) edition of Darwin's *Journal of Researches* (later *Voyage of the Beagle*). There Darwin coyly wrote: "Seeing this gradation and diversity of structure in one small, intimately related group of birds, one might really fancy that from an original paucity of birds in this archipelago, one species had been taken and modified for different ends" (Darwin 1845, 380). He intimated that such change was a "fancy," but we well know that by then this is precisely what he thought had happened. For Wallace, too, entries like these reveal him to be a committed transmutationist even as he uses more circumspect language in published works such as the Sarawak Law paper.

3. The question of the "peopling" of remote islands had profound significance for the species question. Another example of Wallace's musing on island colonization is found deep in his Insect, Bird and Mammal Register (1855–1860): "In the small islands where my men were lost & in most others I have touched at, the Tummuli of Megapodius are found. These birds cannot fly 1/2 a mile. How could they reach these islands. Would their eggs be washed out in storms & float across the sea uninjured?" (WCP4766, 90). This undated entry was probably made in the summer of 1860, not long after the debacle over the marooning of his men as he struggled to travel from Goram and Seram to Waigiou (recounted in *Malay Archipelago*, chapter 25), a distance of perhaps only fifty miles that took him more than forty days of sailing in dangerous conditions. Despite the mishaps along the way, Wallace collected a host of birds and insects on the many small islands they stopped at as they attempted to hopscotch their way to Waigiou. In the entry from his collections register we see him musing on the remains ("Tummuli") of *Megapodius* birds, which construct enormous mounds in which to incubate their eggs (see recto 66). How indeed did these poor fliers make it to the remote islands? Such questions of dispersal and colonization are at the heart of any understanding of island biota.

soil & climate & some of them within sight
of each other. <sup>a work of supererogation one would suppose as they must inevitably</sup> ... It may be said it is a
<sup>in time become peopled from each other, & contrary to what takes</sup>
<sup>place elsewhere. Ireland is peopled from England.</sup>. It may be said it is a
mystery which we cannot explain, but do
we not thus make ¹unnecessary mysteries &
difficulties by supposing special creations contrary
to the present course of nature. For we
~~should~~ must conclude the course of nature in peopling
islands in the ocean to be uniform & that
all islands distant from others should ᶰᵒʷ be
stocked with animals & plants equally
peculiar. But we know this not to be the
case. Volcanic islands recently produced &
coral islands far in the ocean contain stragglers
from the nearest land & no others, nothing
peculiar!—Now we can hardly suppose that
islands would be left for ages to become
stocked in this manner, & then the new &
peculiar creations be introduced ʲᵘˢᵗ ʷʰᵉⁿ ᵗʰᵉʸ ʷᵉʳᵉ ⁿᵒᵗ ʷᵃⁿᵗᵉᵈ.

   According to Mr.
Lyells own arguments they would hardly be
able to hold their own against the previous
occupiers of the soil & there would have

1. This is an insightful passage. Wallace sees that Lyell is essentially arm waving: constructing a system not unlike the epicycle-ridden solar system model constructed by pre-Galilean astronomers to preserve, in retrospect, an untenable starting assumption. He points out that naturalists make unnecessary mysteries and difficulties through an insistence on special creation. His approach is coolly rational—recently produced islands have received chance colonist "stragglers" for ages and do not have endemic species. Old remote islands were populated in exactly this same way over longer ages, and those *immutable* species should be represented still on those islands. Where are they? Instead, those islands seem to be repopulated with "new & peculiar" creations, as he puts it on the next page. Why should this be so?

to be a [1]special extermination of them to make
room for the new & peculiar species. We
must therefore suppose that such islands as
St. Helena & the Galapagos were stocked
with their peculiar species immediately on
their being raised from the ocean, & they
would then have a chance of keeping out
the new comers which might be thrown
accidentally on their shores. [2]This supposition
will certainly explain the present condition
of those islands but it has the disadvantage
of being contrary to the present order of nature,
for none of the islands which we have any
reason to believe have been formed, since
a very late geological era, are inhabited
by such peculiar species. They generally have
not one species peculiar to themselves.
*[Written vertically in left margin]* this must be proved.

On the other hand islands which are thus peculiarly
inhabited, appear to be of a considerable antiquity.
A long succession of generations appears therefore
to ^have been requisite, to produce those peculiar

1. Wallace's ingenious argument from the previous page continues. Using Lyell's own ideas to argue against him, here Wallace points out that the descendants of the stragglers should become well adapted to their new home and therefore these "previous occupiers of the soil" should easily outcompete the come-lately specially created new species. With irony Wallace suggests a "special extermination" to make way for the specially created new species.

2. Recent islands lack endemic species, while those "of considerable antiquity" are, in contrast, inhabited by endemics. Wallace notes that he must prove this, presumably by combing through naturalists' accounts of the flora and fauna of a great diversity of islands—young and old, large and small, distant from the mainland and close by. Perhaps this analysis would have had a dedicated chapter in Wallace's planned book.

## 49

[1]productions found no where else but allied to
those of the nearest land. The change like
every other change in nature was ^no doubt gradual, &
the supposition that other species were successively
produced closely allied to those previously existing
& that while this was going on, the original
or some of the first formed species died out,
exactly accords with the facts as we find them
& the process of peopling new islands at the
present day.

---

[2]Lyell talks of the "balance of species being preserved
by plants insects, & mammalia & birds all adapted
to the purpose." This phrase is utterly without
meaning. Some species are very rare & others very
abundant. Where is the balance? Some species
exclude all others in particular tracts. Where is the
balance. When the locust devastates vast
regions, & causes the death of animals & man
what is the meaning of saying the balance is
preserved.—The sugar ants in the West Indies

1. Wallace has a clear vision for how isolation plus time leads to divergence. Note his commitment to gradualism, and how his language here reflects his Sarawak Law (see annotations to recto 44).

2. Lyell discusses checks and balances on plant and animal populations in volume 3 of *Principles*, in a section entitled "Equilibrium in the Number of Species, How Preserved" (1835, 3: 108). This section of *Principles* opens with a quote from the celebrated Swiss botanist Augustin Pyramus de Candolle (1778–1841). In his influential *Géographie botanique* (1820) Candolle presented one of the early expressions of the struggle for existence: "All the plants of a given country . . . are at war with one another." The epic struggle of individuals in burgeoning populations, as expressed by Malthus, was to inspire both Wallace and Darwin. Lyell sees this in the natural-theology tradition, however: plants and animals are adapted to serve the purpose of preserving a harmonious balance in nature.

Darwin's thinking was resonant with Wallace's: he opened his discussion of the struggle for existence with reference to the very same statement by de Candolle, also taken from Lyell. Just as Darwin showed that this sense of balance and harmony is illusory, so the observant Wallace comments here that this notion is meaningless. In fact, he finds Lyell's own subsequent descriptions of the terrible destructive potential of exploding aphid, sugar ant, and locust populations rather inconsistent with his claim of "balance." These examples are found in *Principles* (1835, 3: 115–116).

Wallace asks an incisive question here: "Where is the balance?" in these cases of species excluding one another, or of populations growing to sizes enormously destructive to plants, people, and other animals. The natural theologian would likely say that sometimes things go awry, disturbing the balance. But if such population fluctuations are common, they become more the rule than the exception. What is "normal" is therefore flux and periodic destructive potential—hardly a harmonious vision of nature.

## 50

[1]the locusts which Mr Lyell says have destroyed
800,000 men are instances of the balance of species.
To human apprehension this is no balance but a struggle in
which one often exterminates another. When animals or
plants
become extinct where is the balance.

*[The following sentence is written vertically in left margin
    of page]*
If any state can be imagined proving a want of balance then a
    balance may perhaps be admitted
but what state is that?

---

[2]Lyell—Shews the great change of species that
must result from Geological Changes.
Thus if a chain of lofty mountains were
gradually to rise in the Sahara, It might
be fertilized by rivers & rains & covered with
alluvial soil from floods & bursting of lakes &c.
"Then" he says "the Animals & plants of Northern
Africa would disappear, & the region would gradually
become fitted for the reception of a population of
species perfectly dissimilar in their forms habits &
organization." But have we not reason to
believe they would be modified forms of the
previously existing Northern African species. The
climate might then more resemble that of the
W.[est] Indies, but we know the productions would not
resemble them. It would be an extraordinary
thing if while the modification of the surface
took by natural causes now in operation

1. An interesting example of Lyell's missing the point that Wallace sees is found in *Principles* (1835, 3: 115), where Lyell describes the reproductive potential of the gamma moth (*Autographa gamma*, Noctuidae), a serious crop pest: "Reaumur observes that the female moth lays about four hundred eggs; so that if twenty caterpillars were distributed in a garden, and all lived through the winter and became moths in the succeeding May, the eggs laid by these, if all fertile, would produce more than three million moths." Lyell goes on to note that "a modern writer . . . justly observes that, did not Providence put causes in operation to keep them in due bounds, the caterpillars of this moth alone, leaving out of consideration the two thousand other British species, would soon destroy more than half of our vegetation" (ibid.). It is curious that Lyell at once agrees with the "just observation" of the modern writer, even while essentially acknowledging that there were 2,000 other vegetation-destroying species that Providence apparently did not restrain.

Wallace sees clearly the struggle for existence, and that exploding insect populations are simply part and parcel of a natural process.

2. Wallace has connected the dots and sees the reality of species change over time, and that this is intimately linked with geological and climatic change.

Wallace disagrees with Lyell that the new forms in a future altered landscape of North Africa would be "perfectly dissimilar in their forms, habits, and organization"; on the contrary, he thinks they will be modified forms of the current residents, descended from them. This passage from Lyell is quoted again in Wallace's paper on the natural history of the Aru Islands, in which he applies the ideas he set out in the Sarawak Law paper. In the Aru paper, however, Wallace flips Lyell's argument around. Zeroing in on the concluding part of Lyell's statement, "*perfectly dissimilar in their forms, habits, and organization,*" which he emphasizes with italics, Wallace goes on to argue: "Now this theory implies, that we shall find a general similarity in the productions of countries which resemble each other in climate and general aspect, while there shall be a complete dissimilarity between those which are totally opposed in these respects" (Wallace 1857d, 480). The point of the Aru paper was to show that this view is untenable: Borneo and New Guinea are about as similar as two islands can be, in size, terrain, and climate. Yet their faunas are spectacularly different: "Now we have seen how totally the productions of New Guinea differ from those of the Western Islands of the Archipelago, say Borneo, as the type of the rest, and as almost exactly equal in area to New Guinea" (481).

& the [1]extinction of species was the natural result
of the same causes, yet the reproduction &
introduction of new species required special acts
of creation, or some process which does not
present itself in the ordinary course of nature.

---

[2]? Introduce this and disprove all Lyells arguments first
at the commencement of my last chapter.

---

Lyell argues that one species could not change
into another by a change of external circumstances
because while the change was taking place
other species already accustomed to those circumstances
would displace them. But this must always be
on the supposition of a rapid not a gradual
change. He says if the [3]Temperature of Etna
was lowered the pine would descend & displace
the oak & the ches[t]nut & these latter would descend
& displace the olive & the vine before either
of these species could be modified by external
circumstances so as to support the changes."
This is highly improbable. Suppose the mean
temperature fell 1° in 1,000 years; would that
amount of change so enfeeble the oaks & ches[t]nuts

1.  This point, which has its start on the previous page, speaks to Lyell's inconsistency. The *Principles* is a long argument for the primacy of natural causes currently in operation shaping the earth over long periods of time. To invoke natural causes in shaping the earth and in the decline and extinction of species, but then to invoke supernatural causes in the introduction of new species, is nonsensical to Wallace.

2.  This important statement supports McKinney's (1972) claim that the numerous passages in the Species Notebook critiquing Lyell's arguments are notes for a book Wallace intended to write on the subject of transmutation. Note Wallace's reference here to his "last chapter." This passage was inserted after the main entries on this page.

3.  Lyell asks the reader to "suppose the climate of the highest part of the woody zone of Etna to be transferred to the sea-shore at the base of the mountain, no botanist would anticipate that the olive, lemon-tree, and prickly pear *(Cactus opuntia),* would be able to contend with the oak and chestnut, which would begin forthwith to descend to a lower level; or that these last would be able to stand their ground against the pine, which would also, in the space of a few years, begin to occupy a lower position" (Lyell 1835, 3: 161–162).

    Wallace out-Lyells Lyell with his commitment to gradualism. He cannot imagine that the plant populations would not or could not adapt to slowly falling average temperature.

in every stage of growth that they could be displaced
by seedling pines & at the end of the 1st
10,000 years would not they be probably as well
adapted to the temp.[erature] 1° lower ^a difference less than the
variation of mean air temp.[erature] as they were
previous to any change.? [1]A change of 20°
in the mean temperature might thus take
place the several species becoming so imperceptibly
modified that each would keep its ground,
& we ^should then have new species of alpine oaks
& ches[t]nuts, & subalpine vines & olives, though of
course they might be so modified that the
fruits of the latter might be no longer
eatable. [2]If the change took place rapidly
the exact results Lyell predicts ~~would~~ might
follow but how the same results would
follow from an excessively gradual change
it is impossible to understand.

———————

[3]Systems of Nature, compared to fragments of
dissected Map or picture or a mosaic.—
approximation of fragments shew that all gaps
have been filled up.

1. The argument continues: the oaks, chestnuts, grapevines, and olives could adapt to a change as great as 20 degrees mean temperature over time, if the temperature drop were gradual enough. Note that Wallace envisions that this adaptation would so modify these plants that their cold-adapted descendants would be different species; as such, there is no guarantee that current traits, like the tastiness of the grape and olive fruits, would persist.

2. A clear statement of Wallace's take on this change: if the temperature change were rapid, he concedes that Lyell might have a point and the plants might not have enough time to adapt. In that case they would be displaced by higher-elevation, cold-adapted species. But he finds it "impossible to understand" how Lyell could come to this same conclusion in the case of very slow temperature change. Again, Wallace is more Lyellian than Lyell.

3. A "system of nature" was an approach to organizing or arranging species and other taxa into a comprehensible scheme. Linnaeus's classification in his *Systema Naturae* had become the leading system by Wallace's day. Wallace may have an ecological meaning in mind—something like ecological niches, in modern terms. Camerini (1993) and Moore (1997) have noted the significance of this map imagery for Wallace, who was a trained surveyor. As Camerini put it:

> The map, as a pictorial metaphor, served as a unifying framework for disparate bits of information about insect, bird, and mammalian forms at their respective locations. The conceptual gap (between affinities of closely related forms on different islands and the lack of affinity in species on geographically proximate islands) was closed by putting the pieces of the puzzle together as a map. Given that by this time Wallace believed in common descent, the map metaphor served as a means of combining evolutionary origins and distribution patterns in a single representation. (Camerini 1993, 721)

## 53

[1]Lyell supposes 1 sp.[ecies] to die & be created per an.[num]
in the whole earth—& then shows then that in a
country the size of Europe—not more than 1 conspicuous
species could be expected to be produced—1000 years.

---

(see p.[age] 33)
The [2]Hornbills of Africa feed on reptiles, insects, such
as grasshoppers lizards &c. & even small mammals
whereas those of India eat only fruit. Yet both
have exactly the same general structure & forms of bill often
feet tail wings & stomach!. Here is the
most palpable proof that the structure of
Birds is not varied in accordance with their
habits; but that on the other hand they are
necessitated to adopt certain habits in order
to obtain a subsistence in accordance with
the peculiar circumstances by which they are
surrounded. The Trogons also of Africa & India
eat only insects those of S.[outh] America in a great
measure live upon fruits which they capture on the wing as they do insects.
   There is no difference
in their structure but being in different countries &
surrounded by different circumstances they are led
to adopt different habits.

*[The following is written vertically in left margin of page]*
so kingfishers—species of the same genus are inland or
   aquatic feeders.

1. This is a reference to a section entitled "Speculations on the Appearance of New Species" in the third *Principles* volume (Lyell 1835, 3: 172–173). Lyell was careful not to give the impression that he endorses any possibility of species change per se, but rather addressed the introduction and extinction of species.

2. This entry relates to the issue of form and function of animals discussed on recto 33, but also relates to questions of instinct versus habit. In terms of changing function without changing form, hornbills offer a lesson. The diets of the hornbills of Africa and India vary, "Yet both have exactly the same general structure & forms of bill . . . feet tail wings & stomach!" To Wallace, this is evidence that the structure of birds does not correlate absolutely with habit or habitat; rather, "they are necessitated to adopt certain habits in order to obtain a subsistence in accordance with the peculiar circumstances by which they are surrounded." Trogons, too, vary in this manner, he notes, and inland versus coastal kingfishers. These examples speak to the flexibility and adaptability of habit, underscoring that these birds' feeding behavior is not, as by instinct, tied to form.

Wallace's idea parallels that of Darwin in citing the woodpeckers of the plains of La Plata (Darwin 1859, 184) and other species with "habits and structure not in accordance," and they make essentially the same point: structure or form is not simply correlated with mode of lifestyle, as natural theology would suggest, and variance in behavior despite having the same structure is proof of this. In other respects, however, this variation means different things for Darwin and Wallace. In discussing habits and structure "not in accordance," Darwin argues that these are evidence of *transitional* habits, and that form may follow. Wallace, on the other hand, cites examples of variable habit in the context of the primacy of experience and environment over instinct. (This is a recurrent theme in the Species Notebook: see recto 112, 116, 118–119, 155, 166–173, 177; verso 8).

## 54

[1]Owen says—([Proceedings of the] Geolog.[ical] Soc.[iety]
　　May 16, 1855.) Most of the
extinct reptiles exemplify the law of the prevalence
of a more generalized structure as compared with the
more specialized structures of existing species.
The [2]Labyrinthidonts combined Sauroid with [3]Batrachian
　　characters
[4]Rhynchosaurus, Sauroid with [5]Chelonian Characters.
The [6]Icthyosaurus had modifications borrowed from the
class of fishes & the [7]Pterodactyl thus borrowed from
the type of Birds & Bats. The [8]Dicynodonts, had
resemblances to Lizards Crocodiles & [9]Tryonix.

---

[10]The above is what might be expected, if there has been
a constant change of species by the modifications of their
various organs, producing a complicated many branching
series. Those nearer the base must exhibit to some
degree a combination of those characters, which
in a higher developed condition are characteristic
each of one group of animals which have
since come into existence.

---

In Europe 503 Birds 100 common to N.[orth] America
　" Britain 277

---

[11]Birds can scarcely be called European which just enter it from Asia.
An Asiatic or African bird should extend its range over a consider-
　　able
part of S.[outh] or E.[ast] Europe to be considered as forming a true
　　part
of the European Fauna. ? How many species are thus European.

1. The specific paper or papers by Richard Owen that Wallace refers to here is unclear. Owen had only one brief memoir reported in the *Proceedings* for the May 16, 1855, meeting of the Geological Society of London, but this was on the cranium of the feline *Dicynodon tigriceps*. Wallace may have read an account of another report by Owen, such as were commonly abstracted in the *Penny Cyclopedia* and similar general science publications, perhaps erroneously reporting it from the May 16 meeting.

2. Early term for a group of Late Paleozoic and Early Mesozoic amphibians. The name, which derives from maze-like patterns on the teeth, was coined in 1850 by German zoologist Hermann Burmeister. Richard Owen coined the subclass name Labyrinthodontia in 1860.

3. Batrachia is an obsolete name for the taxonomic order of frogs and toads; now known as Anura.

4. A Middle Triassic lizard relative, family Rhynchosauridae; Owen named the type genus and species in 1842.

5. Turtles, tortoises, and their relatives.

6. Ichthyosaurs constitute an Early Jurassic reptilian group of dolphinlike marine carnivores. The name, coined by British geologists Henry De La Beche and William Conybeare in 1821, means "fish-lizard." The ichthyosaur fossils they described were found by the remarkable self-taught British fossil hunter Mary Anning (1799–1847), of Lyme Regis, who was famous even in her lifetime for her finds. Several fine ichthyosaur and plesiosaur specimens found by Anning can be seen in the Fossil Marine Reptiles gallery in London's Natural History Museum.

7. A pterosaur, or "winged-lizard," of the genus *Pterodactylus*, named by Cuvier in 1809. This group was common in the late Jurassic.

8. A group of mammal-like reptiles of the order Therapsida, most diverse and widespread in the late Paleozoic. The type genus *Dicynodon* ("two dog teeth") was named by Owen in 1845, and he later (1859) coined the order Dicynodontia.

9. Now spelled *Trionyx*, a genus of soft-shell turtles of the family Trionychidae, found in parts of Africa and the Middle East.

10. Wallace's meaning here is clear: older, extinct species have a more generalized structure than living species, often seeming to combine characteristics of different but related living groups. Note that he invokes a treelike branching and rebranching process of evolutionary change, and he rightly notes that ancestral forms should be intermediate between living forms—common ancestors of divergent lineages should exhibit elements of each of those lineages. Wallace further suggests that characters are in a "higher developed condition" in the more recent, descendant, forms.

11. This entry touches on the practical difficulties of comparative biogeography: what does it mean to speak of a "European" or "North American" fauna? To be "from" a locale, for what proportion of its life cycle or calendar does a bird have to be found there? Many temperate-zone nesting passerine (perching or song) birds migrate long distances, spending the winter in tropical Africa, South America, or Asia. Species ranges are often fluid and rarely map onto physiographic boundaries, let alone artificial human political boundaries. What is "Europe" geographically? Wallace, like all biogeographers, was forced to settle on definitions and boundaries in his pursuit of pattern and process in geographical distribution. The best-known fruit of his biogeographical labors is *The Geographical Distribution of Animals* (1876), the two-volume treatise that cemented Wallace's standing as the greatest biogeographer of his century.

## 55

[1]March. 1856. Singapore. [2]Bee Eater (<u>Merops</u>)
This like a swallow but slower. very graceful
circles round & settles on sticks & twigs & posts.
Seizes insects on the wing & rests to swallow them.
cleans its bill against the perch. Chirps
or twitters during flight.

At Singapore & Malacca migratory—appears in
November, leaves in March—April. _____

---

[3]<u>Strickland</u> Introduction to <u>Dodo</u>—observations
on laws of distribution, representation &c. &c.

---

[4]<u>Tropidorynchus</u> sp..[ecies] Common in vicinity of
Ampanam in the Island of Lombock but
said not to occur in <u>Baly</u> except on the side next [to] Lombock.
Frequents thick
foliage searches twigs & flowers for insects flies with an undulating
motion,
alternatively opening & closing its wings, called
"Quah-caich" by the natives from its loud note
frequently and rapidly repeated, has a tongue 4 times split
& frequently fibrous.

---

[5]There should be reason to believe that a bird lives
constantly & breeds in a country, or migrates to it
regularly each year to constitute part of the <u>fauna</u>.
Occasional visitors even when not uncommon, can not be so
cons[idere]d.

1. Wallace arrived back in Singapore on January 31, 1856, after more than a year in Borneo. He departed for Bali a few months later, in mid-May, so at the time of this notebook entry he was deep into preparations for the next leg of his travels. Wallace's young Malayan assistant Ali accompanied him, while he parted company with the less-than-competent Charles Allen (see annotations for recto 2), who elected to remain at the mission in Sarawak. (Charles would later assist him again with collecting, but the multitalented Ali became his main assistant for the remainder of his travels.)

2. Bee-eaters, family Meropidae, are passerine birds widely distributed in the Old World. These birds, which hunt from a perch like the New World Tyrannidae, specialize in catching bees and wasps (as their name implies), which they often "disarm" by bashing against a branch.

The particular species is not given, but based on the migration behavior Wallace describes it is likely *Merops philippinus*, the blue-tailed bee-eater. This bird begins arriving in Singapore and peninsular Malaysia late in the year, while the other common *Merops* of Singapore, the blue-throated bee-eater *Merops viridis*, is a breeding resident between March and September, after which it migrates through the Indonesian islands.

3. Reference to H. E. Strickland and A. G. Melville's book *The Dodo and Its Kindred; Or the History, Affinities, and Osteology of the Dodo, Solitaire, and Other Extinct Birds of the Islands Mauritius, Rodriguez, and Bourbon* (1848). Lyell discussed the extinction of the dodo in *Principles* (1835, 3: 133–134). (For more on Strickland see recto 76 and 77.)

4. This entry was made on Lombock (Lombok; another is found on recto [b-1]). Wallace arrived at Buleleng, in Bali, from Singapore on June 13, 1856, after a passage of twenty days aboard the steamer *Kembang Djepoon*. Two days later he left for neighboring Lombock, sailing across the narrow strait that marks the south-ern extent of what would become known as Wallace's Line, arriving at Ampenan, Lombock, on June 17.

*Tropidorhynchus* honeysuckers—also called friarbirds for the pale collarlike ring around their neck—are described in chapter 27 of *The Malay Archipelago*, which treats the natural history of the Moluccas, to the north. There Wallace had found a species of oriole apparently mimicking a species of *Tropidorhynchus*. In his first Malay Journal (LINSOC-MS178a, entries 5 and 14) Wallace described *Tropidorhynchus* as "loud screaming," and gave its onomatopoeic name as "Quaich-quaich."

Wallace likely wrote this entry on Lombock, where he first encountered these birds, but as he had come from Buleleng on the north-central coast of Bali and he had not visited the east side, along the Lombok Strait, he can only note that the bird was *said* not to occur in Bali except on the side adjacent to Lombock.

5. Musings such as this on distribution and what it means to be native to an area are likely prompted by Wallace's realization that the fauna of Bali and that of Lombock differ greatly.

## 56

[1]Dr. Johnson said
"Marriage not natural to man. In savage state, man &
his wife have dissentions & part. When a man sees another
woman that pleases him better he will leave the first."
Incorrect—The indians of the Amazon, the Dyaks of
Borneo never leave their wives, nor the Papuans the woman always submits.

---

[2]At Singapore lighthouse.
Stones of 660 lbs. [pounds] carried by 4 Chinamen
up an inclination of 15° to height of 20 feet
= 165 per man.
largest stone carried by 4 men 990 lbs. [pounds] =
247 lbs. [pounds] per man.
4 men raised 3918 lbs. [pounds] 20 feet in 4 hours
= 326 lbs. [pounds] raised 1 foot per minute for
each man, & working 9 hours a day.

————

a European can do nearly double, but
these chinamen were working by day
& not by contract work.
J.G. Thompson. Government Surveyor
at Singapore.

1. Wallace refers to the celebrated Samuel Johnson (1709–1784), English poet, essayist, literary critic, and lexicographer.

Johnson touches on marriage in many places in his voluminous writings. His famous 1759 tale *Rasselas* includes a debate on the pros and cons of marriage, ending with the pros outweighing the cons, but in Boswell's *Life of Johnson* we see a Johnson who is decidedly down on marriage. That is where we find the passage Wallace refers to: asked if the state of marriage was "natural to man," Johnson answered, "Sir, it is so far from being natural for a man and a woman to live in a state of marriage, that we find all the motives which they have for remaining in that connexion, and the restraints which civilized society imposes to prevent separation, are hardly sufficient to keep them together." An objection is raised, but Johnson maintains that the couple would never agree on anything, and concludes with the comment that caught Wallace's eye: "Besides, Sir, a savage man and a savage woman meet by chance: and when the man sees another woman that pleases him better he will leave the first" (Boswell 1900, 497).

This notebook entry does not reflect an interest in marriage per se, but rather Wallace's ethnographic interests. Wallace recorded a great deal of information about the cultures, languages, and physical characteristics of the various native peoples he was in contact with, and so could state with authority here that married Dayaks, Amazonian Indians, and Papuans never separate over dissension—the wife always submits. However, this may be a misreading of Johnson: his assertion is that when a man sees another woman who catches his fancy he will leave the first, and this is given as a further example of why marriage is not natural to humans. This is a different matter than separating in disagreement. Wallace's note about the wife submitting would seem to refer to the case of dissension.

2. Wallace is likely referring to the Horsburgh Light, named for the Scottish hydrographer James Horsburgh (1762–1836), who worked for the Honourable East India Company, and designed by British surveyor and civil engineer John Turnbull Thomson (1821–1884) (not "J. G. Thompson," as Wallace has written here).

This lighthouse, completed in 1851, is located at the eastern entrance to the Singapore Strait on a rocky island called Pedra Blanca (white rock) (Hall-Jones 1995). Wallace may have read an account of the lighthouse's construction in the *Journal of the Indian Archipelago and Eastern Asia*, informally known as "Logan's Journal" for its founder, James Richardson Logan. Volume 6 of the *Journal* (1852) featured a detailed article on the erection of the Horsburgh Lighthouse, including accounts of how the Chinese laborers lifted rocks as heavy as 7 tons using a cross-stretcher device and other tools, as well as their manner of working, wages, and the value of their labor in comparison with that of European workers. This entry reflects Wallace's interest in all things sociocultural and ethnographic during his travels.

## 57

[1]From Memoir of Emperor of China (Khang) quoted
by M.[onsier] Huc "L'Empoire Chinois" Tono. [Volume] 2.
p.[age] 359

---

"On the 1st day of the 6th moon I was walking in
some fields where rice had been sown, to be ready
for the harvest in the 9th moon. I observed by chance
a stalk of rice which was already in ear. It was
higher than all the rest and was ripe enough to
be gathered. I ordered it to be brought to me.
The grain was very fine and well grown, which
gave me the idea to keep it for a trial & see if
the following year it would preserve its precosity *[sic]*.
It did so. All the stalks which came from it
showed ear before the usual time, and were ripe
in the 6th. moon. Each year has multiplied the
produce of the preceeding *[sic]*, & for 30 years it is this
rice which has been served at my table. The
grain is elongate & of a reddish colour, but it
has a sweet smell and very pleasant taste.
It is called Yu-mi "Imperial rice" because
it was first cultivated in my gardens. It is
the only sort which can ripen north of the

1. This entry refers to *L'Empire Chinois: Faisant suite à l'ouvrage intitulé Souvenirs d'un voyage dans la Tartarie et le Thibet* (1854), by French missionary and traveler Evariste Régis Huc. Huc's book recounts the author's travels in China, Tibet, and Mongolia in the 1840s. Wallace may have read the 1855 English translation (New York: Harper & Brothers), though he gives the French title here.

Wallace copied out Huc's account of the Emperor Khang-hi's attention to a stalk of rice that was clearly superior to others growing in the same field, leading to a new and prized variety. What caught Wallace's attention here was not the question of whether it illustrated the principle of selection in agricultural improvement, but rather a potential example of the appearance of a new variety. Another such example is found on recto 59. Although Wallace later (e.g., in his Ternate essay of 1858) argued against the idea that domestication provides a useful model of transmutational species change, in the Species Notebook he cited dog varieties as evidence for transmutation (see recto 39–40). This discussion is continued on the next page.

great wall where the winter ends late &
begins very early; but in the southern provinces
where the climate is milder & the land more
fertile, two harvests a year may be easily
obtained, and it is for me a ~~great~~ *[illeg.]*
sweet ~~thought~~ reflection to have procured this advantage
for my people."

This kind of rice flourishes in Mantchouria [Manchuria] where
no other kind will grow. .

[1]Now if some of these grains had been carried
away by birds & had thus been propagated
naturally in a country where the other
kinds were not found, the differences
of form & habit would be held to
constitute a specific difference. It would
be considered a species peculiar to that
country. A[lfred] W.[allace]

---

[2]Gallinaceae & Pigeons only, among landbirds <u>walk</u>,
Raptores & Passeres all <u>hop</u>—? is this rule universal.
<u>No!</u> *[Written in left margin]*
Psittacidae <u>walk</u>; should therefore form a group alone.
<u>Secretarius</u> among Raptores walk & I think many Eagles.
? Caprimulgidae—Picidae? <u>Menuridae</u>? <u>Eupetes</u>? ask Ruffia*[?]*
                Psitta.—

1. Here we see Wallace's interest in this example: he sees its implications for the evolution of new species and varieties. If the new variant rice got established far from its paddy of origin, naturalists would have been struck by the differences in its form and habit from other rice and described it as a new species. This speaks to the arbitrariness that sometimes plays a role in the naming or recognition of species and varieties, while also underscoring that new varieties (and thus species) are derived from preexisting, closely related ones.

Wallace's point is reminiscent of Darwin's imagined labors of the young naturalist in the *Origin of Species*, tasked with describing some large and widely distributed taxon (Darwin 1859, 50–51). The naturalist is at first struck by differences and names many species, but as the scope of the investigation broadens, more and more intermediate variants are presented, confusing the issue. The naturalist is forced to acknowledge that the distinction between species and varieties is not stark nor hard and fast, something that underscores Darwin's point that species are merely "well marked" varieties and indeed represent a stage in the divergence of varieties. Wallace and Darwin were clearly thinking very much alike on this point.

2. These notes on bird ambulation continue at the bottom of recto 59; see recto 108 for a related entry.

Raptores—an obsolete term for birds of prey, referring to the taxonomic orders Falconiformes (hawks, falcons, eagles) and Strigiformes (owls).

Passeres—refers to the order Passeriformes, the song and perching birds.

Psittacidae—parrot family.

Secretarius—secretary bird (*Sagittarius serpentarius*, family Sagittariidae), a large terrestrial bird of prey endemic to Africa.

Caprimulgidae—goatsuckers, nightjars, and relatives: mostly crepuscular or nocturnal insect-eating birds with small feet, long pointed wings, and a wide gape.

Picidae—woodpecker family.

Menuridae—the lyrebird family; ground-dwelling Australian birds of the genus *Menura*, renowned for their mimicking ability and spectacular courtship displays.

*Eupetes*—the monotypic *Eupetes macrocerus*, the Malaysian rail-babbler (family Eupetidae).

The final comments on the bottom of recto 59 suggest that Wallace was making notes on distinguishing characteristics of different bird groups. Information on walking, running, and hopping in birds is given in his 1856 paper "Attempts at a Natural Arrangement of Birds," published in the *Annals and Magazine of Natural History* in September of that year (Wallace 1856a).

## 59

[1]Vinegar Polype—found in the Yellow Sea
Put in a jar of fresh water with a few glasses
of brandy, converts it into well flavoured strong
vinegar in a few days. Water being added
as the vinegar is taken out the supply is
kept up of the same strength .. Mr. Huc
had one in use for a year!

---

[2]New forms, miscalled species, are always starting up
in every Botanic Garden. In the garden of Berlin
Link states that <u>Zizyphora dasyantha</u> after many years
changed to another form which might be called
<u>Z.[izyphora] intermedia. "Lindley. Intro.[duction to] Bot.[any]"</u>
        Macassa—1 species.
In Lombock there are said to be no Squirrels, in
Baly 2–3 species—In Sumbawa many.

---

[3]The great <u>Pteropi</u> in Ceram eat holes through the thick
husk of the cocoanut to get at the fruit. ?

---

[4]<u>Crows</u>? ?Colius walk like parrots. On what muscles
&c. does this difference of action depend.. Good character
for a primary division. .
<u>Rhipidura</u> walks & <u>runs</u>.

1. Huc caused quite a sensation with his account of a vinegar-producing jellyfish, "Tsou-no-dze," and this became much recounted in an assortment of professional and amateur scientific journals.

2. German botanist Heinrich Friedrich Link (1767–1851), director of the Berlin Botanical Garden, reported a change over many years in the mint *Zizyphora dasyantha*, which had become altered to such a degree that Link suggested that it might be now a different species. This was later dubbed *Zizyphora intermedia*. Wallace may have read this account in John Lindley's *Introduction to Botany* (1832; see book 7, chapter 2, 521), or in a reprint such as that in the *Horticultural Register* for 1836, which published an excerpt from Lindley's chapter titled "Irregular Metamorphosis of Plants." Lindley was a noted botanist and prolific author, and his *Introduction to Botany* went into many editions. (Lindley's *Elements of Botany* was Wallace's very first botanical book purchase, as a teenager.)

Darwin, too, picked up on the very same report on Link's *Zizyphora*, but rather than in Lindley, Darwin read of this apparent example of transmutation in Bronn's *Handbuch einer Geschichte der Natur* (1842–1843, 2: 85). He cited the example in his draft chapter "Variation under Nature" in the *Natural Selection* manuscript (Stauffer 1975, 127): "There is a marvellous account . . . by Link of Zizyphora intermedia from Z. dasyantha." The example was not included in the *Origin of Species*, however.

3. The large fruit bats known as the flying foxes, *Pteropus* (family Pteropodidae). The Ceram (Seram) fruit bat, or Seram flying fox (*Pteropus ocularis*), is endemic to the island of Seram. (See recto 6 for Wallace's account of an immense *Pteropus* flight in Sarawak.)

4. Continuation of notes from the bottom of recto 58. See note 2 for that page.

## 60

The [1]Tucutucu (Ctenomys Brasiliensis) a burrowing
animal is often found blind. This blindness though
common cannot be a very serious evil, yet it
appears strange that any animal should possess
an organ frequently subject to be injured.
Lamarck would have been delighted with this fact,
had he known it when speculating (probably with
more truth than usual with him) on the gradually
acquired blindness of the asphalax a gnawer living
under ground & of the Proteus, a reptile living
in dark caverns filled with water, in both of
which animals the eye is in an almost rudimentary
state & is covered by a tendinous membrane &
skin. In the common mole the eye is extraordinarily
small but perfect, though many anatomists doubt
whether it is connected with the true optic nerve;
in the Tucutucu which I believe never comes to
the surface of the ground the eye is rather larger but
often rendered blind & useless. No doubt Lamarck
would have said that the tucutucu is now
passing into the state of the Proteus & Asphalax.

Darwin. p.[age] 52.

1. Here Wallace loosely quotes Darwin's discussion of the tuco-tuco of South America (*Ctenomys brasiliensis*, family Ctenomyidae), a blind subterranean rodent, in relation to the *"Asphalax"* (now *Spalax*, the blind mole rats, family Spalacidae, of eastern Europe) and the snakelike olm, *Proteus anguinus* (family Proteidae), a unique blind, cave-dwelling amphibian found in southern Europe.

The passage comes from the second (1845) edition of Darwin's *Journal of Researches* (later *Voyage of the Beagle*). Darwin's circumspect mention of the imagined delight that Lamarck would have taken in these species, as examples of a transitional series, no doubt interested Wallace. Darwin did not mention Lamarck in the first (1839) edition of his *Journal of Researches;* he quietly slipped this into the second edition—having by then been a transmutationist himself for seven or eight years, unbeknownst to Wallace.

Although this is the only explicit reference to Darwin's *Journal of Researches* in the Species Notebook (the Galapagos references on recto 46 and 48 likely come from Darwin too), Wallace had counted Darwin's travel memoir among the works that fired his scientific imagination and inspired his desire to travel to the tropics. The other notable works were Humboldt's *Personal Narrative of Travels to the New Continent during the Years 1799 to 1804* (1818–1821, in seven volumes), Chambers's *Vestiges of the Natural History of Creation* (1844), and Lyell's *Principles of Geology* (1830–1833, three volumes).

## 61

The eyes of nocturnal a darkness loving animal
are either very large & sensitive a small or
imperfect — In the first case we may well
believe the nocturnal habit to be the result
of the organisation, the full light of day being
painful to the large & highly organised visual
organs as in Owls, goatsuckers, cats &c..
In the second case the smallness or imperfection
of the visual organs, seems to be the result
of the habits which other circumstances
have bestowed upon the animal, the eyes becoming
imperfect for want of use as in all other
organs; as in the Proteus, Asphalax, & the
blind insects, found in caverns. The small
eyes of the bat, so different from other creature
of voluntary nocturnal habits, may be accounted
for by supposing them to be too weak to be
used in full day — AW.

[1]The eyes of nocturnal or darkness loving animals
are either very large & sensitive or small &
imperfect. In the first case we may well
believe the nocturnal habit to be the result
of the organisation, the full light of day being
painful to the large & highly organised visual
organs as in Owls, goatsuckers, cats &c..
In the second case the smallness or imperfection
of the visual organs, seems to be the result
of the habits which other circumstances
have bestowed upon the animal, the eyes becoming
imperfect for want of use as in all other
organs; as in the Proteus, Asphalax, & the
blind insects, found in caverns. The small
eyes of the bat, so different from other creatures
of voluntary nocturnal habits, may be accounted
for by supposing them to be too weak to be
used in full day. A[lfred] W[allace].

1. In reference to the passage from Darwin on the previous page, here Wallace speculates on the tendency of the eyes of nocturnal, cave, or subterranean animals to be either very large or very small. Note that Wallace is grappling with an evolutionary chicken-and-egg question only solvable, in modern terms, in a phylogenetic context. He muses that in some cases (animals with large and sensitive eyes) the nocturnal habit stems *from* their "organization" (bright light being supposedly painful to creatures with large eyes, deterring them from becoming diurnally active), while in the case of those with small, poorly developed eyes this "organization" results from habits arising from other circumstances, with their disuse leading to the diminution of the organ. In other words, in one case habit is a result of organization, and in others, organization is a result of habit. Although he is thinking along transmutational lines, he does not seem to entertain the idea that visual organs might increase or decrease in size or sensitivity relative to ancestors in the context of moving into new habitats or environments.

Darwin, for his part, addressed only the latter scenario: that eyes become "imperfect for want of use" is essentially how he approached the topic in the *Origin of Species*. Both disuse and natural selection play a role in the reduction of eyes' size in environments where they are not needed, he believed: "This state of the eyes is probably due to gradual reduction from disuse, but aided perhaps by natural selection" (Darwin 1859, 137).

Recall, incidentally, that Wallace commented on bats' eyes on recto 12 of the Species Notebook, where he concluded that "they have very minute eyes & apparently imperfect sight & they do very well without them."

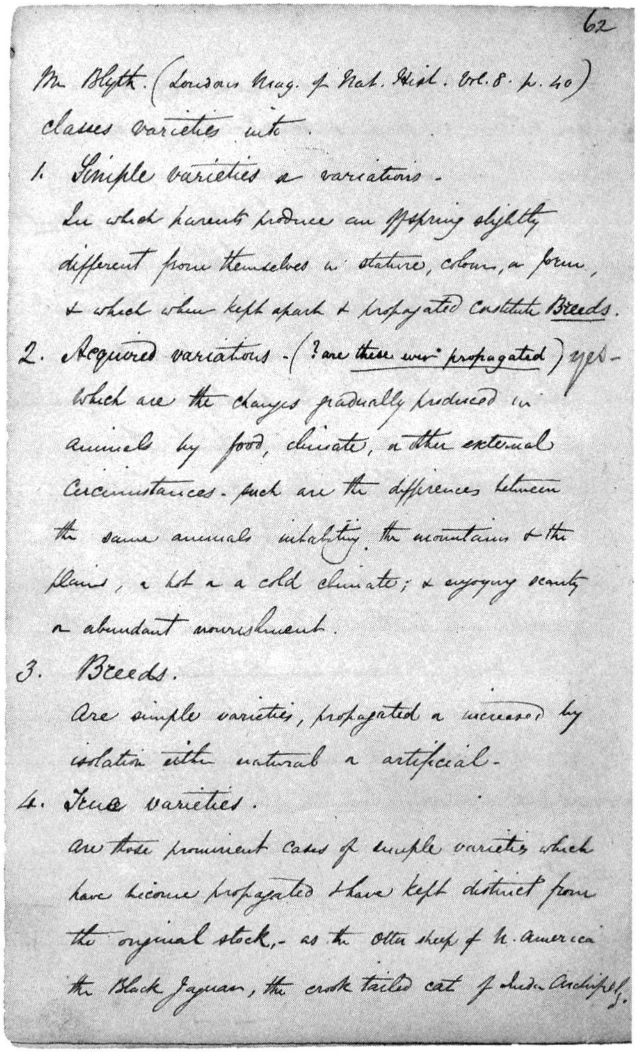

[62]

[The left column contains the handwritten manuscript version of the text transcribed in the right column.]

[62]

[1]Mr. Blyth (London Mag.[azine] of Nat.[ural] Hist.[ory]
Vol.[ume] 8. p.[age] 40)
classes varieties into.

1. Simple varieties or variations.
   In which parents produce an offspring slightly
   different from themselves in stature, colour, or form,
   & which when kept apart & propagated constitute <u>Breeds</u>.
2. Acquired variations (? <u>are these ever propagated</u>) Yes—
   Which are the changes gradually produced in
   animals by food, climate, or other external
   circumstances such are the differences between
   the same animals inhabiting the mountains & the
   plains, a hot or a cold climate; & enjoying scanty
   or abundant nourishment.
3. Breeds.
   Are simple varieties, propagated or increased by
   isolation either natural or artificial.
4. True varieties.
   Are those prominent cases of simple varieties which
   have become propagated & have kept distinct from
   the original stock,—as the otter sheep of N.[orth] America
   the Black Jaguar, the crook tailed cat of Indian Archipelago.

1. The reference to "Mr. Blyth" is further indication of Wallace's interest in the nature of species and varieties at this time. English zoologist Edward Blyth (1810–1873) resided in the West Bengal city of Calcutta (Kolkata) where he was curator of the Museum of the Royal Asiatic Society. Blyth was keenly interested in species and varieties. His 1835 paper on the subject was an attempt to clarify different categories of variation. He even had a notion of selection, though he envisioned its action as serving to preserve stasis rather than as an agent of change:

> Still, however, it may not be impertinent to remark here, that, as in the brute creation, by a wise provision, the typical characters of a species are, in a state of nature, preserved by those individuals chiefly propagating, whose organisation is the most perfect, and which, consequently, by their superior energy and physical powers, are enabled to vanquish and drive away the weak and sickly, so in the human race degeneration is, in great measure, prevented by the innate and natural preference which is always given to the most comely. (Blyth 1835, 48–49)

Blyth does not appear to have been in contact with Wallace at this time, but he was in contact with Darwin and had written in glowing terms to him about Wallace's Sarawak Law paper, recognizing it as a tour de force: "What think you of Wallace's paper in the Ann. M. N. H.? Good! Upon the whole!" (Darwin Correspondence Project, letter 1792). Here in the Species Notebook Wallace does not comment on Blyth's classification of varieties, but in brief notes he questions whether Blyth's second category, individuals with "acquired variations," are ever propagated, and later penciled in an emphatic "yes —."

The question of the permanence of varieties was integral to the transmutation process, and it is interesting to consider that these naturalists came to very similar ideas as to how varieties might remain distinct long enough to become permanent and perhaps even new species. The distinctions between simple varieties or variations, acquired variations, breeds, and "true varieties" get at the heart of the question of what constitutes a species and a variety, and the nature of variation. Blyth had put his finger on the importance of isolation in enabling a breed to expand in population while maintaining its integrity as a breed.

This view is not far from the modern one: varieties remain distinct to the degree that they don't intercross, and so isolation from other varieties (or the parental form), by preventing intercrossing, plays a key role in facilitating divergence. Note that three of the four types of variety discussed by Blyth and recorded here by Wallace are defined explicitly terms of physical separation: those being "kept apart & propagated," "propagated or increased by isolation either natural or artificial," and "kept distinct from the original stock." (Wallace's reading of Leopold von Buch was in part focused on this same issue—see recto 90.)

## 63

[1]Bamboo—Rafts are made of it in Java to convey goods down the rapid streams. The outriggers of the Canoes of Java & the East are formed of two large bamboos. In Lombock, the houses are entirely constructed of Bamboo (except only a few of the main posts) & instead of rattan, split bamboo is used to bind the several parts together.

———

In Macassa drinking cups of piece of bamb[oo] fixed into top of half coacoa/nut shell as a stand.

———

Natives of Ké often came off to our Prow in rafts made of half a dozen bamboos lashed together.

———

[2]Prof.[essor] Owen in his lecture on Orang utans &c. at B.[ritish] Ass.[ociation]
says varieties of colour in man have been produced by climate. Instances Jews of Syria black, of N.[orthern] countries of Europe some with fair hair light complexions & blue eyes. But there are also Armenian Jews at Singapore equally fair,—& it has little to do with the question as the Jewish features & form remain, & they not colour are the grand characteristics of race. Jews are always pointed to as exhibiting stability of physical characters.

1. During his residence in Borneo, Wallace had sent a letter on the uses of bamboo to Sir William Jackson Hooker, father of Sir Joseph Dalton Hooker and his predecessor as director of the Royal Botanic Garden, Kew. This was extracted in *Hooker's Journal of Botany and Kew Miscellany* (Wallace 1856d). The notes here were made later in his travels, and in some cases (the Java reference) may not have been based on first-hand knowledge. Wallace first visited Lombock (Lombok) in June 1856; Macassar (or Makassar, modern Ujung Pandang) in southern Sulawesi, September 1856; the Ké (Kai) Islands in January 1857; and he did not make it to Java until July of 1861.

Bamboo observations continue on the lower portion of the next page.

2. Richard Owen (1804–1892) was expert in the comparative anatomy of the great apes, having in his care some of the earliest available and most complete skeletons at the Hunterian Museum in London and, later, the British Museum (to which he moved in 1856). As early as 1830 he read a paper before the Zoological Society of London entitled "On the Anatomy of the Orang Utan," and he wrote later papers on the cranium of orangutans and the comparative anatomy of orangs and chimps. Wallace may be referring here to Owen's address at the Liverpool meeting of the British Association in 1854, published in abbreviated form as "On the Anthropoid Apes" and in longer form as "On the Anthropoid Apes, and Their Relations to Man"; both appeared in 1855.

Owen compared humans and anthropoid apes in quite a few papers, and these often included a consideration of human variation in such traits as stature, hair texture and color, and skin color. Owen cited the Jewish Diaspora as evidence for effects of climate on skin color: starting from the assumption that modern Jews worldwide are direct descendants of Hebrews of the Holy Land, he considered the dark complexion of those found in more tropical latitudes and the fair complexion of others at northerly latitudes to be a result of climate (minimizing the possibility of conversion of local people):

For 1800 years [the Jewish people have] been dispersed into different latitudes and climates, and they have preserved themselves most distinct from any intermixture with the other races of mankind. There are some Jews still lingering in the valleys of the Jordan, having been oppressed by the successive conquerors of Syria for ages,—a low race of people, and described by trustworthy travellers as being as black as any of the Ethiopian races. Others of the Jewish people, participating in European civilization, and dwelling in the northern nations, show instances of the light complexion, blue eyes, and light hair of the Scandinavian families. The condition of the Hebrews, since their dispersion, has not been such as to admit of much admixture by the proselytism of household slaves. We see, then, how to account for the differences in colour, without having to refer them to original or specific distinctions. (Owen 1855b, 35–36)

Wallace dismisses this by pointing out that Armenian Jews in Singapore remain fair. His point that stability of features and form define a human "race" is of a piece with his interest in species and varieties in nature.

## 64

Mr. Joseph Carter of [1]Ampanam has himself
seen in the interior of the Coti River in
Borneo a man with a <u>tail</u>, about 4 or 5
inches long. Had examined it, but does
not know if there is a tribe of the
same structure.          July 1856—
          ? gammon[?]

---

[2]Mr. C.[arter] has seen & examined the penis
with—wire inserted—some have one & some
two pieces at r[igh]t angles inserted through
the glans. It is not taken out during
coitus but then becomes buried by the
enlargement of the glans. In the
ordinary state it projects ⅛ inch
on each side.

---

[3]<u>Bamboo</u>—In Ternate excellent chairs, very
strong are made at ¼ guilder each.
Arm chairs ½ guilder & bamboo sofas 1g.[uilder]
[Sketch]

Parrots perch & feeding boxes
made of a single piece of
<u>Bamboo</u> & a <u>cross stick</u>.
Ternate, Papua.

1. Wallace traveled from Bali to Ampanam (now Ampenan), on the west coast of Lombock, in June 1856. From there he planned to obtain passage to Macassar, in southern Celebes (Sulawesi). Ampenan is now the port district of the city of Mataram, capital of the Lombock province of West Nusa Tenggara. Already a bustling trade center when Wallace visited, the city now has a population of more than 400,000. It was on this island that Wallace first noticed that he had crossed some invisible faunal boundary: "Birds were plentiful and very interesting, and I now saw for the first time many Australian forms that are quite absent from the islands westward" (*Malay Archipelago*, 165).

Joseph Carter is identified in Wallace's journals (and *Malay Archipelago*) as "one of the Bandars or licensed traders who offered me the use of his house & every assistance in my Natural History researches during my stay" (first Malay Journal, entry 3). Note that Wallace's interest in Carter's curious account of a tailed man relates to his interest in variation and varieties; he is wondering here if there is a whole tribe with the same structure. (A related observation is noted in an entry on recto 91 of the Species Notebook.) Wallace undoubtedly has in mind the idea that such a tribe, were it to exist, might represent a link with other primates. Both Wallace and Darwin realized that our vestigial "tail," the coccyx, is evidence of our descent from progenitors with tails. This point was made by the then-anonymous Robert Chambers in his *Vestiges of the Natural History of Creation* (1845, 147): "Man has, again, no tail; but . . . between the fifth and seventh week of the embryo a tail does exist, and in the mature subject the bones of this caudal appendage are found in an underdeveloped state in the os coccygis . . . the tail of the human being is shrunk up in the bony mass at the bottom of the back."

2. Wallace's Victorian sensibilities may have prevented his including Carter's further account of male genital piercing, termed *ampallang*, in *The Malay Archipelago*. *Ampallang* was commonly practiced by the Dayaks of Borneo, and to some extent it is still practiced.

3. The reference to Ternate suggests this entry was made no earlier than January 1858, the time of his earliest visit to that island.

## 65

[1]Malay races . . characteristics.
Colour reddish brown of various shades.
Hair black straight, on body & beard scanty or none.
Stature low or medium, form robust, breasts very
much developed feet small, thick, short, hands small
& rather delicate. Face broad & rather flat, eyes oblique
distinctly but slightly. Nose small with no prominent ridge
straight, nostrils broad [wings inclining rather downward] cheek bones
   rather prominent but
less so than in Chinese, mouth large but lips not thick [or
   prominent].

In many cases principally among Rajahs & Priests a
taller stature [yellower skin] finer and slightly aquiline nose with a
tendency to beard is observed—Due probably to Arab
Hindoo or European mixture, former principally.
Among lower classes, curly or wavy hair, frizzly beard
darker skin & larger nose are often observable
due probably to Papuan mixture.
   Oct.[ober] 56. Makassar. A.[lfred] W.[allace]
[2]Moral characteristics . . Reserve, dissimulation, no
exhibition of feelings,—"Nil admirari", no appreciation
of the sublime or beautiful . .

1. As Wallace traveled east in the archipelago he came in contact with a greater range of ethnicities and racial types, in particular people of Papuan affinity. His eastward progression had him arrive at Macassar, Celebes (Sulawesi), in September 1856, then the Aru and Ké (Kai) Islands in December and January 1857, back to Macassar that July, on to Banda and Amboyna (Ambon) in November of that year, and finally arriving at Ternate on January 8, 1858. Accordingly, a greater attention to ethnology is evident, though this was certainly not a new interest.

Since becoming a transmutationist in the mid-1840s, Wallace was as keen on the question of human origins as he was on the question of the origin of species generally. As he put it in his second Malay Journal (LINSOC-MS178b, entry 71): "The human inhabitants of these forests are not less interesting to me than the feathered tribes." McKinney (1972, 88–89) suggested that the large number of ethnological notes made in the Species Notebook between November 1, 1857, and late February 1858, immediately before the Ternate essay was written, indicates that Wallace was thinking about human racial diversity just before discovering the principle of natural selection. In that process he had recalled Thomas Robert Malthus's famous *Essay on Population*, which he had read years before, and its accounts of intertribal strife and struggle (see Moore 1997).

Lists of physical characteristics like these provided raw material for the extensive ethnological discussions in *Malay Archipelago*; the particular observations given here can be found in the last chapter, "On the Races of Man in the Malay Archipelago."

2. Wallace also observed behavioral and what he termed "moral" characteristics of the native peoples. Although generally quite a sympathetic observer who never doubts the intrinsic intelligence and personal dignity of the natives, his pronouncements not infrequently have an air of Eurocentric paternalism—which is perhaps inescapable, given his imperial culture.

His claim that the Malays had "no appreciation of the sublime or beautiful" is of course of the time. Their aesthetic sense was likely quite different from that of a contemporary European, but that is something very different from having *no* appreciation of the sublime or beautiful. Wallace's claim may have been based on the relative lack of ornament and personal adornment among the Malays compared with the Papuans. Or it could have stemmed from the fact that while he was agog with the beauty of *Ornithoptera* butterflies or birds of paradise, his native companions seemed unmoved. In *Malay Archipelago* he even muses on the irony that such exquisite beauty, such as the birds of paradise, is wasted on the benighted natives of the archipelago:

> The remote island in which I found myself situated, in an almost unvisited sea, far from the tracks of merchant-fleets and navies; the wild luxuriant tropical forest, which stretched far away on every side; the rude uncultured savages who gathered round me—all had their influence in determining the emotions with which I gazed upon this "thing of beauty." I thought of the long ages of the past, during which the successive generations of this little creature had run their course—year by year being born, and living and dying amid these dark and gloomy woods, with no intelligent eye to gaze upon their loveliness—to all appearances such a wanton waste of beauty. (*Malay Archipelago*, 448)

## 66

[1]Egg of <u>Aru</u> Megapodius. 3¾ in.[ch] x 2¼ in[ch]
of rusty ochre red, surface harsh.

---

Papuan races use bow & arrow, Malays not.
query is this a universal difference?. If so, [2]good
proof of diversity of origin.
(Dyaks of Borneo totally ignorant of Bow—
Malays d[itt]o.. Javaneese, used as a game
by chiefs &c. ? introduced from India . .
Natives of Ké & Aru use bows . .

---

Chinese or Tartar races? do they use bows.

---

Africans—do all use bows?

---

[3]Alfurers of Minahassa,—have native names
for bow.—
Papuans have flat forehead, <u>projecting brows</u>
large thick nose, apex <u>bent down</u> & wings of
nostrils <u>inclining upwards</u> from apex, hair on bodies &
considerable beard.

*[The following sentence is written vertically in right margin of page]*
Papuans have the lips thick
projecting, & <u>sharply cut</u>.

Malays have, rounded forehead, <u>flattish brows</u>
small nose, <u>rounded apex</u> & nostrils inclining
<u>downwards</u> from apex, scarcely any hair on bodies &
no beard.

Papuans tall, rather slender limbs. Malays short stout
thick limbed.

*[The following is written vertically in left margin of page]*
<u>Noted at Aru.</u>

1. *Megapodius* (family Megapodiidae) are mound builders or scrubfowl, chickenlike mound-building birds of the forests. Some dozen or more species are found from northern Australia and the western Pacific to Southeast Asia. Wallace first encountered these birds on Lombock, the westernmost extent of the Austro-Pacific realm. The Latin name refers to their large feet, which are put to good use scratching together massive mounds of decaying vegetation. Their eggs are buried in these mounds, incubated by the decay-generated heat. The mounds built by these birds can be enormous: "The immense mounds of earth and leaves formed by them are scattered all over the forest. These mounds are generally from 5 to 8 feet high, and from 15 to 30 feet in diameter," Wallace wrote in his paper on the natural history of the Aru Islands (Wallace 1857d).

2. This passage underscores Wallace's interest in human origins. In his chapter "On the Races of Man in the Malay Archipelago," he delineates the distribution of the Malay-affinity peoples from the Papuan/Polynesian-affinity peoples but does not find that the boundary coincident with that of the "animal productions," as he puts it, along his now-eponymous line delineating the Indo-Asian from the Australian halves of the archipelago. The human line lies to the east of the zoological line, something Wallace says is handily explained by penetration of the "superior" seafaring Malay races into the former territory of the "inferior" Papuan races. Wallace believed, incidentally, that the Asiatic groups (including the Malays) had their origin in continental Asia and bore no relationship at all with the "Papuan" groups, which he took to be of Pacific origin "from lands which now exist or have recently existed in the Pacific Ocean" (*Malay Archipelago*, 30). In fact these groups, too, have an Asian origin in some sense, and the Polynesian groups are relatively recent arrivals, having moved through the Pacific islands only since the second century or so BCE. It would take another century for anthropologists to realize that all human groups have an African origin.

Wallace must have eventually found sufficient variation in the knowledge and use of the bow and arrow that he decided it could not be cited as a consistent difference between the Malay and Papuan/Polynesian groups, and depended instead on physical and mental/behavioral characteristics as well as language in his discussion of the subject in *Malay Archipelago*.

3. Minahassa [Minahasa in *Malay Archipelago*] is a region of the northernmost arm of Sulawesi, called Celebes in Wallace's day. "Alfurers" (Alfuros) are the indigenous people of the island of Halmahera (Wallace's Gilolo), not Sulawesi, but Wallace may have encountered Alfuros in Menado, the largest town and trading center in Minahasa. Alternatively, he may be using the term "Alfuro" here as a label for what he took to be a certain racial type, as when in *Malay Archipelago* he states that "in Ceram, the Alfuros of Papuan race are the predominant type" (370).

"Bow" is one of the 117 words Wallace lists in thirty-three languages in the appendix to *Malay Archipelago*. There Wallace reports the word in two languages from Halmahera: *pusi* in the Gani language, and *Ngámi* in Galela.

## 67

[1]Plan to <u>stop the further increase</u> of <u>Synonyms</u>.
Let 3 periodicals be appointed in each principal
Country of Europe & in the United States, in which
alone <sup>after a fixed date</sup> New species can be described <sup>so as to be</sup>
    adopted
by Naturalists. For example, let the Proceedings
of the Linnaean Zoological & Entomological Soc.[ietie]<sup>s</sup>
respectively be the medium for making known
New species of Plants, Animals (Insects excep.[te]<sup>d</sup>)
and Insects <sup>described in England</sup> & let the directors of all the
public Museums & all the chief Naturalists
of Europe &c. declare their determination to
recognize no names of species described in
other places unless repeated here also. Let
the Proceedings of all the appointed Societies
be regularly published say ~~quarterly~~ <sup>monthly</sup> in sheets
& mutually exchanged, by which means the
whole body of Naturalists would become
immediately aware of all descriptions of
New species & all hunting through the
Proceedings of ~~all~~ Scientific Societies & ~~all~~
Periodicals become unnecessary.

*[The following two sentences are written vertically in*
*left margin of page]*
To make sure of not having more synonyms each Society
    should have certain N.[ew]S.[pecies] meeting in the year so
arranged as to come <u>in rotation</u>. Every person could then be
    certain whether his species had been previously published.

1.  This plan shows yet another aspect of Wallace's creativity. We have seen him as a keen observer and critic of scientific ideas; here, he offers a remedy for a thorny problem of scientific practice. It may be the earliest example of "Wallace as reformer"— later, he suggests remedies to many problems, real and perceived, scientific and social (see Berry 2002, Smith and Beccaloni 2008).

This remedy is typically practical. As a collector whose professional activities revolved around selling hard-won zoological specimens, the specific identity of those specimens was of central importance to Wallace; as a philosophical naturalist it was no less important, since confused taxonomy made it difficult to accurately map and understand, as Wallace was striving to do, the geographical distribution of species and varieties. In this light we might understand Wallace's characteristically creative plan to reduce confusing synonyms as a plea.

Wallace never put this plan forward, but he did address the problem of synonyms and nomenclature in various writings (reviewed in Costa 2013), such as in his review of David Sharp's book *The Object and Method of Zoological Nomenclature* (Wallace 1874). Wallace argued for the adoption of a "law of priority" in assigning species names, where the first name to be published is applied—an approach practiced today. The "law of priority" was discussed as early as 1842, when a specially appointed committee on zoological nomenclature submitted its report to the British Association for the Advancement of Science—the lead author was Charles Darwin. This report evolved into the rules of nomenclature used today. Among zoological disciplines, the International Commission on Zoological Nomenclature (ICZN), founded in 1895, acts as advisor and arbiter for the scientific community regarding the correct use of scientific names (see iczn. org).

Even today there are nomenclatural disputes, but there were considerably more in Wallace's day, before an internationally recognized code was adopted. In an 1858 letter to the Entomological Society of London, Wallace weighed in on what he called a "novel and most erroneous as well as inconvenient interpretation of the law of priority," regarding a case where erroneous names erect-ed by Linnaeus were being given priority (Wallace 1858a). A few years later, in a short paper regarding *Iphias* butterflies, he wrote: "I presume that the proper application of the law of priority is to determine among conflicting names still in use, and thus establish a uniform nomenclature. To apply it to rake up obsolete names, and thus create synonyms and produce the confused nomenclature it was intended to abolish, is an abuse which ought not to be tolerated" (Wallace 1863a).

Yet another example from that year is found in a note regarding Wallace's naming of a species of mynah bird, genus *Gracula*. Wallace had named a new species *Gracula pectorialis*, and to his consternation the zoologist George Robert Gray proposed that this name was synonymous with *Gracula anais*, a name that had priority (having been given previously by the French naturalist René Lesson—see recto 105). Wallace charged that Lesson's description was based on a partial specimen, with missing parts added from other species. "Now, I contend," wrote Wallace, "that this is not a case for the application of the law of priority, and would inevitably lead to further confusion" (Wallace 1863d).

## 68

[1]Plan to obviate the necessity for quoting any Synonyms <u>for the future</u>.

For this purpose it is necessary that a complete & authorized catalogue, should be published in all branches of Natural History, (in groups of moderate extent) giving all the synonyms under which each species has ever been described or figured since the establishment of the binomial nomenclature, with full references;, at the same time determining, by ~~the law of priority alone,~~ authority the true & standard specific name to be henceforward used by all naturalists without quotation of Synonyms. In order that every Naturalist may use this Synonymical Cat.[alogue] it must be prepared & corrected by Committees of Naturalists in every country of Europe, the true Synonyms being <u>finally</u> determined by comparisons of the original specimens in all doubtful cases & the location of those spec[i]m[en]s indicated; & it should be published in the very cheapest possible form

1. The proliferation of synonyms meant that in order to avoid ambiguity as to precisely what species they were writing about, authors would often parenthetically list synonyms after giving the species name they had adopted. This can become quite cumbersome each time a species is discussed in a scientific paper. Wallace's practical solution is to develop a universally recognized reference catalog—he terms it a "Synonymical Catalogue"—of accepted species names. Like his proposal for an authoritative body for nomenclatural issues, this proposal, too, has in effect been realized. The International Commission on Zoological Nomenclature maintains and publishes the *Official Lists and Indexes of Names and Works in Zoology*, and there are now rules and codes for all phyla (e.g., the International Rules of Botanical Nomenclature, International Code of Botanical Nomenclature, International Code of Zoological Nomenclature, International Code of Nomenclature of Bacteria, the International Association for Plant Taxonomy's Nomenclature Committee for Fungi, and so on).

so that its expense may not be burthensome.

This [1]Catalogue being published, uniformity
& simplicity of nomenclature will reign ~~in~~
among Naturalists. In all Catalogues
Lists, Synopses &c. & in all exchanges
of specimens & communications among
naturalists one specific name only
need be used,—every one being supposed
to have a copy of the Catalogue in the
department he studies & all collections to be named by it. The expense
of all future Catalogues & systematic
works will thus be much diminished
a great portion of their space being now
occupied by references to the synonyms.
Uniformity in the naming of collections will
be introduced & thus a fertile source of
error & perplexity removed, & all
those numerous "aliases" which are a disgrace
to Nat[ural] History will be kept out of sight,
& only referred to for purposes of study.
Feb.[ruary] 1857.

1. We see on this page that Wallace's "Synonymical Catalogue" proposal is dated February 1857, when he was collecting in the Aru Islands. It is understandable that questions of nomenclature would be very much on his mind at this time: early the following year he would publish an account of his insect collecting in Aru, having tallied a remarkable 1,364 species in "about four months' clear collecting" (Wallace 1858d).

## 70

[1]Formation of <u>a complete library</u> of Nat.[ural] Hist.[ory]

That such does not exist is discreditable to
Naturalists. It is proposed that the chief
Nat.[ural] Hist.[ory] Societies (Linnaean, Zoological
& Entomological) should, while keeping
their Libraries distinct, have them under
one roof in adjoining rooms & under
the care of one Librarian. Members
of all the Societies to have free use of
all in the Library, duplicates only to
be taken out, except ~~by members of the~~
~~Soc[iety] to wh[ich] the work belongs and then only~~
for ~~very~~ short periods & on leaving a <u>deposit</u>
of ~~its~~ the <u>value</u> of each work. Saving of Expense in rooms
& Librarian to be spent on Books, each
adding works in its own department. To
such a joint library many expensive works
would be given by foreign governments wh[ich]
could not be afforded to ea[ch] of the three.

1. Yet another creative Wallace proposal. This idea of resource sharing among the three leading natural history societies of London is sensible; a common library would at once provide naturalists with ready access to a greater range of resources and save on administrative expenses by requiring only one library staff rather than three. It is not clear if Wallace ever advanced his proposal, but he would have been delighted with the invaluable resources available to naturalists today, such as the Biodiversity Heritage Library (biodiversitylibrary.org) and Hathi Trust Digital Library (hathitrust.org). See discussion in Costa (2013).

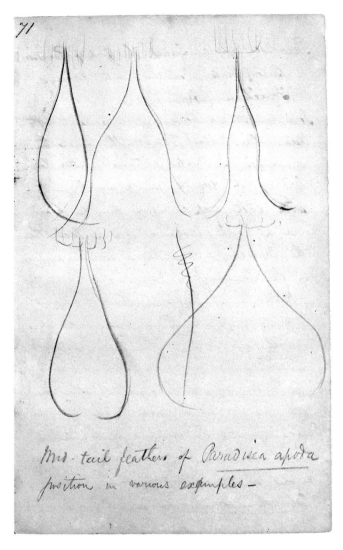

71

Mid-tail feathers of Paradisea apoda
position in various examples —

[1] *[Sketches of tail feathers]*
Mid-tail feathers of <u>Paradisea apoda</u>
position in various examples.

1. These simple sketches were intended to represent variation in the long, wire-like central tail feathers of the stunning greater bird of paradise, *Paradisea apoda*. It was this fabled bird that drew Wallace to the farthest eastern reaches of the vast archipelago, and he succeeded at last in finding specimens in Aru. (See following page and verso 46).

Wallace arrived in the Aru Islands in early January 1857, at the trading post of Dobbo (now Dobo) on the small island of Wamma (Wamar). He and his young assistants collected on Wamma for a few weeks, but, anxious to collect in the interior of the islands, Wallace soon sought to move on. To his consternation, however, the locals would not budge, owing to pirate activity in the area. "The Aru natives are of course dreadfully alarmed," he recorded on February 18 in his first Malay Journal (LINSOC-MS178a, entry 65), "as the pirates attack their villages burn murder & carry away their wives & children for slaves, & I shall probably have difficulty in getting a boat or men to venture to take me anywhere, especially as the chiefs are made responsible for my safety which will furnish them an excellent excuse for refusing to stir." Indeed, the next entry, for February 23, reported that a trader's boat was attacked in the area, with nearly the entire cargo lost and the crew wounded.

At last, on March 13, 1857, Wallace and his native guides left Dobbo for the island of Maykor (Maikoor), establishing a base at "Wanumbai" in the interior. The site must have been dismal indeed judging from Wallace's first impression, considering he was no stranger to arduous conditions and less-than-palatial accommodations: "We reached a house or rather small shed of the most miserable description which our steersman the Orang-kaya [head man] of Wamma said was the place we were to stay at" (second Malay Journal, LINSOC-MS178b, entry 69). He likely would have rejected the site, but the head man fell ill. Resigned, he got to work. "One of my first objects," Wallace related in *Malay Archipelago* (446), "was to inquire for the people who are accustomed to shoot the paradise birds." He was soon rewarded for his perseverance.

## 72

[1]Paradisea apoda. (Linn.[aeus]), "The Bird of Paradise";
    "Burong Mati" (Dead bird) of the Malays;
    "Fanéhan" (Arru.)
Is a very strong & active bird, frequents lofty
trees flying about constantly among the
branches. The males continually utter their
loud cry of wawk, wawk, wawk;—
wōk, wōk, wōk; which can be heard
at a great distance & is in fact the
loudest of all animal sounds
in the forests of Aru.

Natives say it makes its nest of leaves
laid on the top of an ants nest or some
projecting part of a tree, & has never
more than <u>one young</u>. They do not
know the eggs.. The throat when alive
is remarkably broad & swollen, showing to great
advantage the fine metallic green plumage.
The skin of this part has a lining of hard fatty
mucous similar to that of the neck of the
[2]<u>Cephalopterus ornatus</u>, & serving no doubt

1. Many of the observations recorded on this and subsequent pages provided material for a characteristically thorough paper on *Paradisaea apoda*, published in the December 1857 issue of the *Annals and Magazine of Natural History* (Wallace 1857c) and, of course, in *The Malay Archipelago* (chapter 31, 449). It is not surprising that a bird of paradise graces the title page of the book, or that an entire chapter (38) is dedicated to them; these legendary and immensely valuable birds were among Wallace's most prized finds. Wallace was in fact the first to bring living birds of paradise to England—he negotiated a deal through his agent that would secure him free passage home and a handsome payment for living birds. He arrived in London in the spring of 1862 bearing two lesser birds of paradise (*Paradisaea minor*) procured in Singapore, their breeding plumage developing just as he arrived home; both Wallace and the gorgeous birds—the latter were displayed in a roomy aviary at the Zoological Garden—quickly became the talk of the town. One of the birds survived for just over a year and a half, until late 1863, and the other lived almost exactly two years in England, dying in March 1864.

The specific epithet *apoda*, given by Linneaus in 1758, stems, by the way, from a belief supposedly held in the West, that this species was footless—the specimens that reached Europe by trade were prepared without feet, leading some, apparently, to jump to the dubious conclusion that these beautifully plumed birds were strictly aerial, never alighting and so created *sans* feet. The first bird of paradise skins are thought to have arrived in Europe in 1522, brought by the surviving crew of Ferdinand Magellan's first voyage. One of the crew, Antonio Pigafetta, recorded that the skins were a gift from the sultan of Bacan (in the Moluccas) to Holy Roman Emperor Charles V. One of these specimens was likely the model for the first color illustration of a bird of paradise, an exquisite rendering executed between 1539 and 1546 by Croatian miniaturist Julije Klović that adorned a prayer book known as *The Farnese Hours* (Mužinić, Bogdan, and Beehler 2009). No feet are in evidence in Klović's painting—nor are they seen in Conrad Gessner's illustration in his encyclopedic *Historiae Animalium* (1551–1558, 1587), or in the several bird of paradise illustrations found in Ulisse Aldrovandi's *Ornithologiae* (1599), where the birds are depicted floating among clouds and subsisting on rainwater.

Although European naturalists did not see a living specimen until 1825, by the eighteenth century specimens with feet firmly attached were known—see, for example, plate 63 of Albertus Seba's 1734 *Locupletissimi Rerum Naturalium Thesauri*, volume 1, illustrating "Avis Paradifiaca Aroëfica." Linnaeus's use of *apoda* some twenty years later was thus a nod to the earlier belief of footlessness. (See the opening paragraphs of chapter 38 in *Malay Archipelago* for Wallace's account of early European encounters with these fabled birds.)

2. *Cephalopterus ornatus*: the Amazonian umbrellabird, large passerine birds of the family Cotingidae that Wallace collected in South America.

in both cases to nourish the dense mass of
plumage which springs from it.

It is after the commencement of the East
Monsoon <sup>only</sup> that the [1]natives obtain them; this has
no doubt led to the statement of their migrating
from New Guinea, which is altogether [at Aru]
erroneous as they are found in Aru all the
year round. About March & April when the
change from the W.[est] to the E.[ast] Monsoon ~~begins~~ <sup>takes place</sup>,
the Paradiseas begin to shew their fine ~~plumes~~ <sup>feathers</sup>.
It is not however till May & June that they
arrive at full perfection. This is probably the
season of pairing,—the males ~~is~~ <sup>are</sup> then in a state
of excitement & delight in exhibiting their plumes.
For this purpose they assemble on certain lofty trees
which they prefer, ~~[illeg.]~~ & there take short flights at
intervals from branch to branch, uttering a
lower & more chirping note than usual. While
perched they keep their wings half open & quivering
their neck stretched out their head looking downwards
while the long golden downy tipped feathers which

cont.[inue]<sup>d</sup> at p.[age] 85.

1. The curious Malay name for these birds, *Burong mati*, or "dead bird"—mentioned on recto 72 and 86 and included in the title of Wallace's paper (1857c)—likely reflects the difficulty of obtaining the birds with their prized plumage in the remote reaches of the archipelago. Neill (1973) suggested that Malay traders in the western parts of the archipelago simply never saw live specimens. The difficulty was compounded by the biological realities that Wallace relates here: the coveted feathers are breeding plumage, seen only a few months annually.

Soon after returning home in 1862 Wallace read a paper at the Zoological Society of London in which he recounted the trials, tribulations, and triumphs he met with seeking birds of paradise. In his words, "Five voyages to different parts of the district they inhabit, each occupying in its preparation and execution the larger part of a year, have produced me only five species out of the thirteen known to exist in New Guinea." He lamented that "Nature seems to have taken every precaution that these, her choicest treasures, may not lose value by being too easily obtained" (Wallace 1862, 159–160). He astutely concluded his paper with a wistful eye on the likely avian riches of unattainable New Guinea:

It may be considered as certain that every species of Paradise Bird yet obtained from the natives has come from the north peninsula of New Guinea, that being the part most frequented by the Malay traders. The vast extent of country east of long. 136° is quite unknown; but there can be little doubt that it contains other and perhaps yet more wonderful forms of this beautiful group of birds. If we look round the whole circumference of the globe, we shall be unable to find a region at once so promising to the naturalist and so absolutely a "terra incognita" as this great tropical land; and it is to be hoped that our explorers and naturalists may soon be induced to direct their attention to this hitherto neglected country. (ibid., 161)

Today forty-one bird of paradise species in fourteen genera are recognized, and indeed most species are found in New Guinea. Wallace's hope was realized fifteen years after his death: in 1928, the then-twenty-three-year-old German naturalist Ernst Mayr carefully surveyed birds of paradise in the interior highlands of New Guinea. Mayr was to become one of the foremost evolutionary biologists of the century, and as a result of his bird of paradise surveys, these birds became an iconic example of geographical speciation (Mayr 1963); Wallace would have been delighted.

## 74

[1]"Goby-goby", Aru. "Burong Rajah" Malay..
Paradisea regia. Linnaeus (Cicinnurus *[sic]* regia. Viell.[ot])

———

Frequents the thick jungle on trees. Eats fruit
swallows very large stone which pass through its
stomach. This with a whirr.r.r. Is very
active on wings & legs. Often flutters its wings
very much in the manner of the S.[outh] American
Marakias. Has a wide gape. Nostrils linear,
long, close to margin of mandible. Eyes pale
olive; legs fine cobalt blue; bill orange
yellow. The middle tail feathers cross each
other near the base. When the wings are closed
the breast plumes are concealed beneath them, when
the bird is excited or pleased it opens & quivers
its wings displaying the beautiful fan shaped
green tipped plumes to great advantage.
*[Pencil sketch of tail feathers is under text]*

Pencil lines show nat[ural] position of tail feathers.

1. *Cicinnurus regia* (now *regius*), the king bird of paradise, was named by Linnaeus in 1758. Known as the *Burong rajah* to the Malays and the *Goby-Goby* to the natives of the Aru Islands, this was the first bird of paradise species Wallace collected (more accurately, it was collected by his assistant Ali, who was a crack shot), early in 1857. As Wallace related in his second Malay Journal (LINSOC-MS178b, entry 70): "On the second day after my arrival my boys returned from the jungle with a most beautiful specimen of that superb little creature the King bird of Paradise *Paradisea regia* of Linnaeus. Thus one of the great objects of my coming so far was accomplished!" His delight was complete: he rapturously declared in *Malay Archipelago* that this bird was "one of the most perfectly lovely of the many lovely productions of nature," continuing:

> I knew how few Europeans had ever beheld the perfect little organism I now gazed upon and how very imperfectly it was still known in Europe. The emotions excited in the minds of a naturalist who has long desired to see the actual thing which he has hitherto known only by description, drawing, or badly-preserved external covering, especially when that thing is of surpassing rarity and beauty, require the poetic faculty fully to express them. (448)

This is the bird that led Wallace to muse on the nature of beauty (see recto 65), and to ruminate about the irony in the fact that the arrival of those with an "intelligent eye" for their beauty—Europeans—would spell the doom of these jewels of the forest. His lament is at once an example of Eurocentric paternalism (if not racism) as well as a fine expression of the fragile ecological context in which this bird lives:

> Should civilized man ever reach these distant lands, and bring moral, intellectual, and physical light into the recesses of these virgin forests, we may be sure that he will so disturb the nicely-balanced relations of organic and inorganic nature as to cause the disappearance, and finally the extinction, of these very beings whose wonderful struc-

ture and beauty he alone is fitted to appreciate and enjoy. (*Malay Archipelago*, 448–449)

Note Wallace's light pencil sketch of the long wire-like tail feathers of the male, ending in a tight spiral disk of a shimmering emerald green. These highly modified feathers were coveted by fashionable milliners of Europe (see annotations for recto 87); although most of Wallace's specimens went to museums, scientific societies, and individual naturalists, some doubtless found their way onto hats—a form of conspicuous consumption driven by not so much an "intelligent" as an avaricious eye for their beauty (see, again, annotations for recto 87).

## 75

[1]Casuarius ~~emeu~~ . . . galeatus. Cassowary "Bidári". Aru
Has 4 to 7 eggs, lays them on the ground, has no
nest & does not cover them up nor sit on them. Eats fruits of
every kind that falls in the forest. Sleeps at
night ~~laying down~~ crouching down on belly at the roots of a tree.
Wandering about all day in search of food.

Young ones seen at Dobbo, of[ten] sit down on
their tibiae, roll on their backs jump & play
together like young kittens. When very young
have bright broad dark brown stripes on a
buff ground. This fades gradually till the full
grown bird becomes black.

[2]Eos garrula ?                                     Batchian
Red, wings & thighs green, a large oval patch between
the shoulders, the shoulder broadly & the under tail
covers bright yellow; wings with inner web & all
beneath black, first ten quills with basal 2/3 red, 11th & 12th
with small red patch. tail red with a broad term.[ina][1]
band dull green purple tinged, the 6 centre feathers with
a small green patch at base above; bill orange red
base of up.[pe]r mand.[ible] & cere dusky,—orbits bare
    black.—feet b[lac]k
tongue with a strong marg.[?] brush. Iris pale yellow.

1. Cassowaries (family Casuariidae) are ratites, related to emus, rheas, ostriches, and kin; the name derives from the Malay *kesuari*. The three extant species, found in the tropical forests of New Guinea, northeastern Australia, and nearby islands including Aru, have raised, hornlike, but soft and spongy crests called casques on their heads. This species, now *Casuarius casuarius*, was formerly called the helmeted cassowary and now goes by the name of southern or double-wattled cassowary. It is brown to jet black, with striking electric-blue and red coloration on the wattle, neck, and head. Wallace correctly surmised that the Aru Islands were once a part of greater New Guinea, and cassowaries were part of his biological evidence:

> The distribution of the animals of Arru and New Guinea proves the close connection between these countries, it being evident that, where a considerable number of animals which have no means of passing from the one to the other are common to two countries, some former communication must have existed between them . . .
>
> . . . out of the [one hundred or so] small number of land birds known from all parts of the coast of New Guinea . . . I have myself found thirty-six in Arru. This renders it highly probable that all the birds of Arru are also found in New Guinea . . . These facts, scanty as they must necessarily be in the present imperfect state of our knowledge of the zoology of New Guinea, certainly support the view I have taken of the former connection of the Arru Islands with that country." (Wallace 1858c, 166–167)

These observations were also incorporated into *Malay Archipelago*, chapter 33.

2. The species that Wallace describes here is *Lorius garrulus*, the chattering lory, a vibrant red parrot with green wings endemic to the Moluccas. *Eos* is also a genus in the lory and lorikeet group, characterized by their unique brushlike tongue, which they use to feed on nectar and soft-pulped fruits. Wallace realized that the specialized tongue of these parrots could be used to unite them in classification.

This entry was made in Bacan (Wallace's Batchian), a small island just to the southwest of Halmahera (Wallace's Gilolo). He arrived there on October 21, 1858 (see his letter from Batchian [Wallace 1859b], written one day after arrival), having spent the previous half year or more collecting in Ternate, Halmahera, and western New Guinea. With the benefit of comparison now possible, he noted of his New Guinea specimens that "*the tongue has precisely the same structure as in* Trichoglossus *and* Eos!" (Wallace 1859a, his emphasis).

Despite "the absolute identity in the external form and structure of the Lories and some of the *Trichoglossinae* [a now-obsolete subfamily name for the brush-tongued parrots], while they were supposed to belong in reality to distinct groups," he concluded that they could be combined in a common subfamily: "The *Trichoglossinae* will now form a well-marked and highly natural group, characterized by a peculiar compressed form of bill, compact glossy plumage, a brush-tipped tongue, graceful forms, and active habits. They may be called in English, Lories,—a name by which many of the species are known both in Europe and in the Indian Archipelago" (Wallace 1859a, 148).

Indeed this is the case: these formerly separated groups are united within the subfamily Psittaculinae, tribe Lorini.

Note, incidentally, that this entry was written in the fateful month that Wallace received letters from Darwin and Joseph Hooker informing him that his Ternate essay had been read, along with Darwin's writings on the subject, at the July 1, 1858, meeting of the Linnean Society. He had received these letters by the time this entry was made, having departed Ternate (which had a mail depot) for Batjan on October 9. The letters are lost, but they are mentioned in a letter Wallace wrote to his mother just before departing Ternate (see recto 118).

## 76

[1]Is not the Cetaceous group rather a modification
of mammalia to an aquatic life than a
link connecting them with fishes. In essentials
~~characters~~ they exhibit ^nearly^ all the mammalian
& scarcely any piscine characters. The
skeleton is mammalian, the highly developed
fore arm has its parts more perfect than
many terrestrial quadrupeds. In the essential
characters of generation & nourishment of the
young they shew no tendency to depart from
the mammalian type. The skin the layer
of dermal fat, the respiration are all
truly mammalian. In the seals we have
the first steps of such a modification to
the aquatic form of the feroe[?], in the
cetacea we see it carried much farther
from the pachyderm type. The very fact
of there being aquatic forms of <u>two orders</u> shows
that they are both aberrant developments of their
respective orders, not the foundation of the whole class
which both of them can not be.

1. Here commences an eight-page discussion of issues of morphological "affinity and analogy" and their bearing on transitions—*evolutionary* transitions in Wallace's mind, no doubt. He does not indicate here who or what prompted this extended discussion, but chances are good that he was reading or thinking about the writings of entomologist William Sharp Macleay (1792–1865) or geologist Hugh Strickland (1811–1853). Macleay's "quinarian system" of classification arranged taxonomic groups in overlapping sets of five circles, rather like an elaborate Venn diagram. Influential for a short time but out of favor by the early 1840s, Macleay's system sought transitional forms for every intersection of circles. Hence bats linked birds and mammals, cetaceans linked fish and mammals, and so on.

It was Strickland who dealt Macleay's system a "deathblow," as Wallace later put it, in a paper given at the 1841 meeting of the British Association for the Advancement of Science. Strickland (1841) argued against any a priori symmetrical system of classification, suggesting instead that variable degrees of relationship were to be found between taxa. He also discussed at great length how "affinities" (in modern terms, *homologies*) could be used to identify relationships and thus inform classification.

The idea that all groups of organisms can be arranged in some sequence is very old—manifested, for example, in the Aristotelian idea of a linear "great chain of being." Elements of that ladderlike way of thinking about the natural world persisted even in Wallace's day, but what he is criticizing here is the tendency of naturalists to seek links between disparate taxonomic groups and, in so doing, to fixate on outward similarities of selected members of those groups and cite them as transitional forms.

In the final sentence on this page Wallace nicely pinpoints the crux of the issue: the specializations of, say, a mammalian group like cetaceans for aquatic conditions should be seen as "aberrant" or extreme developments within their lineage, not as evidence of a link with fishes, since in all other respects they bear all the morphological features typical of the Mammalia.

Implicit in this assertion is evolutionary-tree thinking: linking or transitional forms are to be sought basally, not apically, in the evolutionary branches of two groups being compared. This idea is developed more fully in the following pages.

## 77

Notwithstanding the distinction between
[1]affinity & analogy is now generally
recognised by Naturalists & we have
no such absurdities as considering
the Hummers to connect birds with
Insects or the Bats to form a natural
transition from Mammalia to Birds
yet the <u>principles</u> of ~~difference~~ the distinction are
~~still~~ often lost sight of, & we still
have resemblances nearly as superficial
as those above mentioned alluded to
as exhibiting a natural transition.

When two groups are broadly distinguished
by certain characters which are universal
in each of them & are not found in ~~every~~ any
other group, we must in order to
establish a transition from one to the
other shew that a [2]decided diminution of these
peculiarities takes place, that the essential
characters of each groups have begun
to vanish. As long as these most

1. Strickland (1841) gave a succinct definition of these terms: Affinity "consists in those *essential* and *important* resemblances which determine the place of a species in the natural system, while [analogy] expresses those *unessential* and (so to speak) *accidental* resemblances which sometimes occur between distantly allied species without influencing their position in the system" (185; emphasis in original). Wallace was inspired by Chambers (1844) to use "affinity" in an evolutionary sense.

The concept of affinity has much in common with the term "homology," coined two years later by anatomist and paleontologist Richard Owen and defined as the "same organ in different animals under every variety of form and function" (Owen 1843, 379); see also Wood (1995).

2. A true transitional series between two groups would show a gradual reduction in the unique and defining features of each group as they approached each other. Were cetaceans truly transitional between mammals and fish, for example, they should show a diminution of defining mammalian traits, and, for that matter, some evidence of defining fish traits beyond mere external appearance. In other words, Wallace argues, the *essential characters* of each group must show evidence of vanishing.

important characters remain undiminished
no alterations of external form or habits
can be held to shew any signs of or
transition. [1]Now as the essential
characters of the group are always most
highly developed in its most perfect
or typical forms it follows that the
transition between groups can never
take place by means of such typical
forms, but always by the most imperfect
or least developedment considered with
regard to the whole. It is thus that
the Bats being a highly developed form
of mammalia can not form a transition
to any other group, the essential mammalia
characters all existing in their highest
perfection, the external form only being
altered by a modification of the fore
limbs, which, while it adapts them to flight
removes them farther from the structure

1. Here Wallace elaborates on why typical representatives of a taxon can never serve as a transitional form—by definition, being a typical representative, the morphological and physiological features that define the group are well developed. This would run counter to the point made on the previous page, that in transitional forms the essential features of the group are poorly developed.

Wallace cites bats as a good example: typical mammalian characteristics are fully developed and functional in this group, the main departure from other mammals being their modified forelimbs. They thus cannot represent any kind of transitional form, having the full complement of fundamental mammalian features (hair, mammary glands, uterine development, etc.).

of the bird's wing than in any other ~~member~~ group
of the class. [1]So the hummers can not possibly
form the point of transition by which the birds
are connected with other parts of the animal
Kingdom, because the wings, plumage & whole
internal structure, far from shewing any
departure from the typical form ~~of~~ in birds, exhibit
them all in one of their very highest developments.
So also the Insecta must not be looked at in
order to discover the connection of the Annalosa
with the Vertebrata, for they exhibit the
perfect & highest development of the annaloses
form, in a complicated structure on a totally
different plan from that of the vertebrata,
so that we can hardly find analogy much less
affinity between them;—the external skeleton
six true (articulated) limbs & horizontal jaws of the
one; can by no possibility be brought into
agreement with the internal skeleton, 4 ~~jointed~~ true
limbs & vertical jaws of the other. Yet there
is a connection & a very satisfactory one between

1. Wallace continues this train of thought with further examples. These can be seen as good examples of groups cited by some earlier authors as transitional: hummingbirds, being insectlike, as transitional between birds and insects; insects, being segmented yet with limb development, as transitional between "Annelosa" (an obsolete term for annelid worms) and vertebrates.

Ironically, recent insights from genomics and developmental genetics have revealed the fundamental underlying unity even between organisms with body plans as seemingly different as arthropods and vertebrates—there is a high degree of homology between the developmental genes responsible for their respective forms of "segmentation" and differentiation of body regions, limbs, and so on (see, for example, Carroll 2005).

the two great groups of Vertebrata & Annulosa
but it is to be seen only in the very lowest
forms of each, where all the essential
characters are in such a rudimentary state
that we can very easily conceive by a few
more modifications much less important
a perfect transition from one to the other.
[1]In the curious                the lowest
vertebrate form known, a fish without
skeleton or members & in some of the
aquatic vermiform animals we see ~~the~~ still existing
nearly the earliest types of Vertebrate &
Annulose animals & can scarcely doubt
that such a close approximation could,
amid the varied changes the world has
undergone, have been left for our examination
had not hundreds of forms ~~species~~ of
allied animals once existed ~~by~~ which by
a perfectly gradual transition from species
to species united these two grand divisions

1. By "lowest vertebrate form known, a fish without skeleton or members," Wallace likely had the cephalochordate *Amphioxus* in mind, and intended to look up the name and fill in the blank space later. Cephalochordates (commonly called lancelets) are basal chordates; the last common ancestor between the lancelet lineage and that of Chordata is thought to have lived more than 520 million years ago, in the Cambrian. The most recent fossil cephalochordate described may be the oldest yet discovered: *Yunnanozoon*, dating to the early Cambrian in what is now southern China (Chen et al. 1995).

This group, together with the unnamed "aquatic vermiform animals" Wallace mentions, provide good examples of the point he has been making over the past few pages. These groups do, in his estimation, represent basal groups ("earliest types") that reveal a transitional link between the vertebrates and annelids. Hundreds of allied species must have existed that, could we have known them all, would have revealed a gradual transition uniting these two disparate taxonomic groups.

of the animal kingdom.

[1]We may make another observation on the
kind of characters which are the most important
as shewing *[illeg.]* affinities between extensive groups.
It is not generally those on which the habits, motions & food
of the ~~individuals~~ various species depends that
are of the greatest value for the purposes of
classification, but often on ~~the other~~ characters
which seem to have little influence & to be of
little importance but which yet by their
universality, & by the very gradual manner
in which they shew any signs of change, are
strikingly characteristic of natural groups. [2]Thus
in Birds, the wing may vary in the most
extraordinary manner or even become almost
obsolete, the legs & feet & bill shew their greatest &
least development, & yet be only characters
on which to found genera or minor groups.
But the feathers occur with scarcely any variation
throughout the whole class & the horny covering of
the bill & the hard scale like skin of the feet

1. Wallace is exactly right; characters that have been adaptively modified are of little use in classifying a species, while seemingly unimportant underlying traits can reveal true relationships. Darwin made the same point in chapter 13 of the *Origin of Species* (1859, 414):

> It might have been thought (and was in ancient times thought) that those parts of the structure which determined the habits of life, and the general place of each being in the economy of nature, would be of very high importance in classification. Nothing can be more false. No one regards the external similarity of a mouse to a shrew, of a dugong to a whale, of a whale to a fish, as of any importance. These resemblances, though so intimately connected with the whole life of the being, are ranked as merely "adaptive or analogical characters"; but to the consideration of these resemblances we shall have to recur. It may even be given as a general rule, that the less any part of the organisation is concerned with special habits, the more important it becomes for classification.

2. Wallace's example is well chosen. Wings and beaks are notoriously variable in shape and size, and beaks, in particular, vary so much according to niche that the beak alone of an unknown specimen suffices to say with confidence how it procures food. But more mundane structural features, like the scaly skin of the feet, vary little and can serve as a character at a higher taxonomic level.

change not with all the variations of size
habitat & habits. [1]So in the internal structure
the form of the legs & wing bones & of the
scull, & the length of the intestines & the form
& size of the stomach vary from species to
species & from genus to genus in strict
accordance however with the changes of
external form habits & food of which they
are the immediate causes. They therefore furnish
us with no additional information than that
which we have already obtained from observation <sup>of internal</sup>
<sup>form & habits</sup>.

But we have in the sternum ~~character~~ modifications of form which
do not depend immediately on external characters.
The general form indeed exhibits to some degree
a dependence on the length <sup>of wing</sup> & strength of
flight, but there are changes of ~~form~~ <sup>proportion</sup>, apertures
sinuosities ~~&c~~ <sup>& processes</sup> which vary independently of these
& furnish us with a new series of characters
highly valuable in the determination of
doubtful affinities. There are also minute
modifications in the texture & arrangement

1. Here Wallace makes the same point by turning to anatomy of birds: wing bones and the form of the intestinal tract are highly variable, but other anatomical features, such as the shape of the sternum, are much less so and thus have taxonomic value at another level. While most birds have a keeled sternum where the wing muscles attach, the ratite birds (ostriches, emus, rheas, and their relatives) lack this structure. The name ratite derives from the Latin word for raft, *ratis*, in reference to the flatness of the sternum—a character that holds at the ordinal level, the Struthioniformes.

of the plumage, in the nature of the skin & in the form of the nostrils which, though they do not seem immediately connected with the habits of the animal ^yet perhaps for that very reason furnish excellent & very constant characters for the determination of larger groups.

[1]Now let us apply the principles here enunciated to the elucidation of some of those doubtful affinities about which Naturalists are still disagreed, namely ^those of the Cetacea to the fishes, the Ostriches to the Mammalia, the Penguins to the Reptiles and the Caprimulgidae to the Owls..

The natural group of the Marsupials & their affinities,—the lowest of mammalia.

1. Some of these putative links may seem quite odd to a modern reader. Relating cetaceans to fish is plain enough—even the ancients called whales fishes—and Caprimulgidae, the nocturnal birds called goatsuckers and nightjars, might seem owl-like with their large eyes, but penguins a linking form with reptiles, and ostriches with mammals? Such seemingly absurd connections were indeed suggested in a time when links were sought between all taxonomic groups.

Here is Macleay in his *Horae Entomologicae* (1819–1821, part 2, 264) on the penguin-reptile connection:

> The active tenant of the air and the proverbially sluggish reptile seem, at first sight, to have no quality in common . . . We require, for instance, to see some animals of an intermediate construction . . . by examining a common turtle, we may obtain the knowledge of some curious points of natural arrangement. Thus we conclude in the first place, that the birds which come the nearest to this animal in structure must be aquatic; that they ought to be covered with scales rather than with plumes; that their sternum ought to be very large, protecting all their viscera; their wings short, of no use for flight, but serving rather as fins to swim with; finally, that their legs ought to be placed so far behind as to render the bird almost incapable of walking . . . If such then be the sort of bird we are to look for, who does not see the Patagonian Penguin . . . in the above description?

And consider the rationale linking ostriches with mammals, as expressed in chapter 14 of the book that made Wallace a transmutationist, *Vestiges of the Natural History of Creation:*

> The struphionidae (birds of the ostrich type) form a link between birds and mammalia, and in them we find the wings imperfectly or not at all developed, a diaphragm and urinary sac, (organs wanting in other birds,) and feathers approaching the nature of hair. Again, the ornithorynchus belongs to a class at the bottom of the mammalia, and approximating to birds, and in it behold the bill and web-feet of that order! (Chambers 1844, 194–195)

(The bird-mammal link of the unlikely duckbill platypus, *Ornithorhynchus*, would have been well known to Chambers's readers.)

|  | [1]Birds, having |
| the plumage covering breast & belly &c. | plumage in lines belly &c. quite bare. |
| --- | --- |
| Megapodius. | King fishers, intermediate |
| Pigeons | Thrushes |
| Alcyone. breast downy—. | Trogons |
| Parrots | |
| waders . . | |

1. It is not clear what point Wallace intended to make with these lists, but they likely relate to the topic of the previous two pages, principles that could be used to resolve some of the instances of "doubtful affinity" that naturalists still disagreed over. Note at the bottom of recto 82 and top of 83 the comment about "minute modifications in the texture & arrangement of the plumage." Here Wallace seems to be exploring just that as part of his continuing inquiry into informative characters to use in bird classification, as seen, for example, in his paper "Attempts at a Natural Arrangement of Birds" (1856a).

## 85

[1]<u>Paradisea apoda</u> . . cont.[inue]$^d$ from p.[age] 73.
spring from beneath the wings are opened in a
manner which neither artist nor taxidermist
have yet ventured to represent. Instead of
hanging down on each side of the bird as $^{in}$ all
drawings & mounted specimens I have seen, they
are erected vertically *[illeg.]* inside the wing &
spread open to their full extent so that every
individual feather of the mass has full room to
display itself. They thus form two glorious fan
shaped plumes, bending gently outwards & overshadowing
the whole body head & tail of the bird, their
base a shiny mass of intense silky yellow $^{(marked\ with\ a\ stripe\ of\ deep}$
$_{blood\ red,[])}$

gradually opening $^{with\ perfect\ regularity}$ into *[illeg.]* with downy
    hair tipped
plumes $^{all\ the\ filaments\ of}$ which shake & quiver unceasingly. In
this attitude alone can the bird be seen to full
advantage & claim our admiration as the
most superb production of the *[illeg.]* animal
kingdom. Its large legs are no more a deformity
as the bird crouches upon them, its dark brown
body & wings are only a central support
to the splendour above from which more

1. This narrative is continued from recto 72–73.

Several of the passages on this and the next two pages are repeated almost verbatim in Wallace's paper "On the Great Bird of Paradise, *Paradisea apoda* . . ." (Wallace 1857c). It is a testament to Wallace's remarkable literary skills that his field notes could be so easily transcribed into a scientific paper.

Note Wallace's attention to the details of how the feathers are naturally held. Here is how these observations translated into the poetic expression of the delivered paper:

> The long, downy, golden feathers are, however, displayed in a manner which has, I believe, been hitherto quite unknown, but in which alone the bird can be seen to full advantage, and claim our admiration as the most beautiful of all the beautiful winged forms which adorn the earth. Instead of hanging down on each side of the bird, and being almost confounded with the tail (as I believe always hitherto represented, and as they are, in fact, carried during repose and flight), they are erected vertically over the back from under and behind the wing, and there opened and spread out in a fan-like mass, completely overshadowing the whole bird. The effect of this is inexpressibly beautiful. (Wallace 1857c, 412)

brilliant colours would distract our attention;
while its pale yellow head, swelling throat of
rich metallic green & golden eye, give
contrast vivacity & life to the whole.
~~In the freshly killed birds the effect can~~
~~be well imitated, the plumage opening~~
~~naturally & regularly as in life~~
                                but it
will require all the skill of the taxidermist
to call back this ^most beautiful^ attitude *[illeg.]* from
the dried specimens. [1]A. see over leaf.

[2]When the natives wish to obtain "Burong mati"
they choose one of these assembly trees & erect
among its branches a little hut of boughs in which
they can lie concealed with just room to
shoot *[illeg.]* their arrows through small openings
left for the purpose. They ascend to this
nest before daylight in the morning with a
sufficient stock of arrows some [3]sirih & tobacco
a few sago cakes & a bottle of water, a boy accompan.[yin]^g^
them who crouches at the foot of the tree.
When the birds appear which they often do

1. Wallace intended that this passage, on the necessity of "all the skill of the taxidermist to call back this most beautiful attitude . . . from the dried specimens," should be inserted at the appropriate spot on recto 88.

2. Wallace drew on these notes to relate, in chapter 31 of *Malay Archipelago*, how the Aru natives hunted birds of paradise. Ascending to their blind before dawn, they would have to bide their time for many hours at a stretch, waiting for the wary birds to make an appearance.

    *Burong mati* is Malay for "dead bird." Wallace says in *Malay Archipelago* (553) that these birds "were first described from skins preserved by the savages of New Guinea, and generally more or less imperfect. These are now all known in the Malay Archipelago as 'Burong mati,' or dead birds, indicating that the Malay traders never saw them alive." (See recto 72–73.)

3. *Sirih* is the Malay term for betel vine leaves (*Piper betle*, Piperaceae), chewed as a mild stimulant and for their medicinal properties. Sago cakes are made from meal derived from the stems of the sago palm (*Metroxylon sagu*, Arecaceae). In Wallace's day, sago cakes constituted a principal food of the natives of the Moluccas. Wallace gave a memorable account of these cakes in his fourth Malay Journal (LINSOC-178d, entry 216), his attention to detail reflective of his ethnological (and economic) interests:

> The hot cakes with the addition of a little sugar or cocoa nut powder are very agreeable. They are soft & something like corn flower cakes. When required to be kept they are dried for several days in the sun & then tied in bundles of 20 when they will keep for years uninjured. When thus dried they are very hard & taste rough like sawdust bread, but the people are used to them from infancy & little children are seen gnawing them as ours with their bread & butter. When dipped in water & then toasted they become almost like new again & with a little butter I eat them every day with my coffee. Soaked and boiled into a kind of pudding it is also very good & I thus eat it once a day to

economise our rice to get a constant supply of which is somewhat difficult in these remote regions.

This illustration of bird of paradise hunters on Aru is one of two frontispieces in the American edition of *The Malay Archipelago* and the one in the second volume of the British edition, both published in 1869. It is easy to see how "some sirih & tobacco a few sago cakes & a bottle of water" would be nice amenities while lying in wait. Reproduced from Alfred Russel Wallace, *The Malay Archipelago*, 4th ed. (London: Macmillan & Co., 1872), 443.

soon after sunrise they wait till a sufficient
number are assembled & till they are in
full enjoyment of their exercise. Then choosing
those nearest to them they often bring down
several in succession the boy securing them
as they fall & killing those which are still
alive. If the rest take alarm & fly away,
the marksman waits patiently some hours till
they again assemble and thus often stays till
night obtaining a dozen or more birds. [1]These
are skinned then next day the bones of the head
& legs being taken out, the body filled with ashes
a string tied round the wings, & a stick pushed from below till it comes out at the mouth & A peg is then put through
the nostrils by which they are the skin is hung up to dry.
The head is thus elongated & narrowed, the eyes concealed
and as the skin dries to it shrinks & assumes
a form & proportions very different from those
of the living bird. In damp weather the skin
of the head often breaks or the beak altogether comes
out which is then repaired with pitch. When dry
the skin is wrapped in a fibrous palm spathe

1. This passage underscores the great contrast between the living and preserved birds. Remember that in the mid-nineteenth century few westerners had ever laid eyes on a living bird of paradise; thus Wallace could open his 1857 paper on these birds by declaring he had "enjoyed the rare privilege of a personal acquaintance with this remarkable bird in its native haunts" (Wallace 1857c, 411). As he became more familiar with the birds and their characteristic coloration he realized that the difference between the preservation technique of the natives of Aru and his own resulted in a taxonomic error: the naming of a new variety of the red bird of paradise based on his collections. He explained the problem to G. R. Gray, who in turn communicated it to the Zoological Society:

> Mr. Wallace has offered some remarks in reference to the supposed variety of Paradisea apoda, which will be best explained by quoting his own words:—"I cannot consider the Paradisea apoda of Aru a new variety, because I believe all the specimens known have come from there. You will find, I think, the same difference of colour between my Paradisea papuana and the native skins, which arises from my care in covering up the plumes during the period of drying, which preserves their colour, while the natives bleach them by weeks of exposure to sunshine. The pale colour of the head also is from my specimens not being shrunk and smoked, as all the native ones are." Under these circumstances, I now propose that the name of Var. Wallaciana (given Proc. Z. S. 1858, p. 181) should be erased from the list. (Gray 1859a, 157–158)

In light of their almost fabled beauty and (to Europeans) rarity, it is a pity to think that they were also merely commodities, the spectacular plumes and long, modified feathers of the males, having fallen prey to the hunters' arrows, destined in large part to adorn the hats of fashionable ladies back in Europe. The hunters did their job too well, causing the price for specimens to fall dramatically: "Some years ago," Wallace noted in the 1857 *Paradisea* paper, "two dollars each were paid for these skins, but they have gradually fallen in value, till now there is scarcely any trade in them. I purchased a few in Dobbo at 6d. [duits] each." [To modern readers it may seem odd that the currency of the remote eastern archipelago was dollars; despite the fact that the guilder was long the currency of the Netherlands (until 2002), it was the Dutch traders who introduced the use of a coin called the "lion dollar" in the Dutch East Indies (and in New Amsterdam—now New York). Duits were copper penny coins used by the Dutch East India Company. The word *duit* is still the Malay and informal Indonesian term for the English "money."]

According to Swadling (1996), at the peak of the demand for bird of paradise feathers, 1905–1915, the wholesale price in London was $21.00 to $24.60 per plume, and during this period an estimated 80,000 bird of paradise skins were auctioned in London, Paris, and Amsterdam. (See also the excellent website for the exhibition *Fashioning Feathers: Dead Birds, Millinery Crafts and the Plumage Trade*, FAB Gallery, University of Alberta: fashioningfeathers.com/birds-of-paradise/.)

and hung up horizontally in the roof of the house,
where, the ends of the feathers seldom being
covered become darkened by smoke as does
more or less the whole plumage.
[1](.A. from last leaf . . In the freshly killed bird
it can be readily seen, [illeg.] even did not observation
of the living bird prove it, that this is the
natural position of the long plumes. They all
spring from an oval fold of the skin about an
inch in length situated just below the elbow
or first joint of the wing. [2]On this point they
turn as on a hinge & admit only of being
laid down closed beneath the wing or erected
in the manner described which position they
take of their own accord if the bird is held
by the legs with the head rather inclining
downwards & the whole gently shaken. In
this manner by slightly altering the position
of the body all the forms which the plumes
assume during life can be correctly imitated.)

1. This is where the passage from recto 86 should be inserted. The resulting construction, in polished and embellished language, is what we find in the delivered paper:

> In the freshly killed specimens it can be easily seen (even did not observation of the living bird prove it) that this is the natural position of the long plumes. They all spring from an oval fold of the skin, about an inch in length, situated just below the elbow or first joint of the wing. On this point they turn as on a hinge, and admit only of being laid down closed beneath the wing, or erected and expanded in the manner described, which position they take of their own accord, if the bird is held up by the legs, with the head inclining a little downwards, and the whole gently shaken. In this manner, by slightly altering the position of the body, all the forms which the plumage assumes during life can be correctly and beautifully imitated. If I am right in supposing this attitude to be now first made known in Europe, and our taxidermists succeed in properly representing it, the Bird of Paradise will, I am sure, excite afresh universal admiration, and be voted worthy of its illustrious name. (Wallace 1857c, 413)

2. Wallace describes here the deployment of the bright yellow plumes of the great bird of paradise, *P. apoda*. This is the species depicted in the frontispiece illustration of *Malay Archipelago*, "Natives of Aru Shooting the Great Bird of Paradise" (reproduced on annotations page for recto 86).

89
ad naturam

[Pencil sketch of bird head on manuscript page]

This peculiar habit of assembling to play, & enjoy
that perfect condition of health & activity which
accompanies the full development of the ~~plumage~~ feathers,
seems common to many birds as the Rupicola
the Turkeys the Argus pheasant &c, and more
especially to those adorned with remarkable
& brilliant plumage. The soaring of the lark
is probably a modification of the same
habit & as in all song birds is perhaps the exertion
of singing.

¹ad naturam
*[Pencil sketch of bird head]*
This peculiar habit of ²assembling to play, & enjoy
that perfect condition of health & activity which
accompanies the full development of the ~~plumage~~ feathers,
seems common to many birds as the Rupicola
the Turkeys the Argus pheasant &c, and more
especially to those adorned with remarkable
& brilliant plumage. The soaring of the lark
is probably a modification of the same
habit & as in all song birds is perhaps the exertion
of singing.

1. We have seen that Wallace was quite skilled at drawing, and this sketch of a *P. apoda* head is accurate.

2. Wallace's choice of terms here, describing these birds' "peculiar habit of assembling to play," is a revealing, and likely common, misinterpretation of their behavior. He may, however, have simply been relating the natives' description of the behavior. In *Malay Archipelago* he noted that the birds gather in "what the people here call their 'sácaleli,' or dancing-parties, in certain trees" (466).

The plumage itself is only part of the story of what makes these birds so interesting to ornithologists, behaviorists, and evolutionary biologists—not to mention birders of all stripes. The male courtship displays, something that could never be conveyed by dead, stuffed specimens, are nothing short of stunning.

Precisely why these birds exhibit such beautiful plumage and elaborate displays is understood in the context of sexual selection, which was first elaborated by Darwin in his *Origin of Species* (1859). Sexual selection, he wrote, "depends, not on a struggle for existence, but on a struggle between the males for possession of the females; the result is not death to the unsuccessful competitor, but few or no offspring" (88). In one form of sexual selection, now termed male-male competition, the males engage one another to secure a territory, say, or dominance, and therefore access to mates (see verso 8 for an example Wallace illustrated). In another form of sexual selection, female choice, males struggle with one another indirectly, vying for female attention. The females choose mates based on their perceived qualities of plumage and display.

Birds of paradise fall into the latter category: "Amongst birds, the contest is often of a more peaceful character," Darwin wrote. "All those who have attended to the subject, believe that there is the severest rivalry between the males of many species to attract by singing the females. The rock-thrush of Guiana, birds of Paradise, and some others, congregate; and successive males display their gorgeous plumage and perform strange antics before the females, which standing by as spectators, at last choose the most attractive partner" (Darwin 1859, 88–89).

At the time Wallace was in Aru, January to July 1857, he had not yet arrived at the principle of natural selection, and even when he did the following year he did not conceive of this "special form" of selection that depended on mating success. Initially Wallace accepted Darwin's sexual-selection concept, but it was not long before his thinking about the origin of sexually dimorphic coloration led him to change his mind about female choice. He argued that ornamentation of the kind seen in male birds of paradise was useful in various (but unobvious) ways and so evolved by natural selection and not according to female preference. See Helena Cronin's lucid discussion of Wallace and Darwin's long argument over sexual selection in her 1991 book *The Ant and the Peacock*.

Incidentally, recent phylogenetic studies of the evolution of sexual dimorphism have shown that in some groups, such as *Papilio* butterflies, Wallace's view of the primary importance of natural over sexual selection is likely correct (e.g., Kunte 2008). Papilionids are an especially interesting group for exploring such questions, as many members exhibit not only sexual dimorphism but often also sex-limited Batesian mimicry (see Joron and Mallet 1998, Kunte 2009).

## 90

[1]Pigeon . . (vinago ?) 35. Macassar.
  has the 3rd. quill emarginate *[Sketch of feather]*
ground pigeon. 48. Lomb.[ok] has all the feathers entire &
  regular . .

---

[2]On continents the individuals of one kind of
plant disperse themselves very far, and by the
difference of stations of nourishment & of soil
produce <u>varieties</u>, which at such a distance
not being crossed by other <u>varieties</u> & thus
brought back to the primitive type, become
at length permanent & distinct <u>species</u>.
Then if by chance in other directions they
meet with another <u>variety</u> equally changed
in its march, the two ~~are~~ have become very distinct
<u>species</u> & are no longer susceptible of intermixture.
  "<u>Flora of Canaries</u>"   <u>Von Buch</u>
[3]He then shows that plants on the exposed peak of
Teneriffe where they can meet & cross do not form
varieties or species, while others such as <u>Pyrethrum</u>
or <u>Cineraria</u> living in sheltered vallies & low grounds
often have closely allied species confined to one
valley or one island.

1. "Vinago" was a large genus of bright green and yellow tropical African and Asian pigeons, named by Cuvier. These are now mostly genus *Treron*, but "vinago" persists in the Spanish name for these pigeons. The number 35 may refer to a specimen number. Wallace noted in his *Letter from Macassar* (1857b; written September 27, 1856, and printed in the *Zoologist* April 1857) that he collected a new pigeon there, but does not (there) venture to guess its affinity. He collected "great green pigeons" in Lombock (commenting in *Malay Archipelago* that they were good eating), but it is not clear if he took any "ground pigeons" there.

Ever a keen observer, Wallace evidently noticed the curiously shaped feathers of his Macassar pigeon. *"Emarginate"* refers to having a "cut away" or notched edge or tip.


2. Here Wallace takes notes from an important book by Christian Leopold von Buch (1774–1853), the noted German geologist from Brandenburg whose statue now adorns the facade of the wonderful Museum für Naturkunde in Berlin. This passage is from von Buch's influential *Flora of the Canary Islands*, first published in German as *Physikalische Beschreibung der Canarischen Inseln* (1825–1831) and later in French as *Description physiques des Îsles Canaries* (1836). The passage is found on pages 132–133 in the 1825 edition, pages 147–148 in the 1836.

Wallace included this very passage in a letter to Darwin (letter no. 2627 of the Darwin Correspondence Project, tentatively dated December 1860). The entry here was thus perhaps made in 1860, in the months after Wallace had received a copy of *Origin of Species* from Darwin in February. Other contents of the letter suggest that Wallace may have been responding to Darwin's letter of May 18, 1860 (Darwin Correspondence Project, letter 2807), which mentions Patrick Mathew and others who had anticipated the theory of natural selection (Darwin correctly writes "*our* view of natural selection"; my emphasis). "My Brother," Darwin concluded, "who is [a] very sagacious man, always said you will find that some one will have been before you." Perhaps Wallace copied out the passage from von Buch in agreement, to underscore that comment and provide Darwin with another example of one who had seemingly grasped the essential point that varieties are but incipient species.

Unbeknownst to Wallace, however, Darwin was already familiar with the passage. He had the 1836 French edition of von Buch, and was apparently as struck as Wallace with von Buch's observations on the relationship between the flora of islands and that of the nearest continent. He wrote in his B Notebook in late 1838 regarding the flora of the Canaries and St. Helena: "analogous to nearest continent: poorness in exact proportion to distance." He continued two pages later: "Von Buch distinctly states that permanent varieties. become species. P. 147 . . . not being crossed with others.—Compares it to languages But how do plants cross?—= admirable discussion" (Barrett et al. 1987, 209–210).

Wallace and Darwin both appreciated the significance of von Buch's observations, namely, the origin of varieties and the role of intercrossing versus isolation in permitting these varieties to become distinct species.


3. Wallace further notes von Buch's example of plant varieties on the exposed peak of Tenerife, highest peak in the Canary Islands, where they are capable of intercrossing, versus those of the sheltered valleys—the observation being that those in valleys often have closely allied species geographically separated in other valleys or on other islands. To Wallace this underscores the role of isolation in the speciation process.

*Pyrethrum* is a the name of several Old World daisy relatives (Asteraceae) in the genus *Chrysanthemum*; *Cineraria* is a large genus also in the daisy family, but the Canary Island species formerly of this genus have been transferred to the genus *Pericallis*.

(See recto 151 for further reference to von Buch.)

## 91

The duck of Aru near Anas Radjah. Less in often
bred becomes tame & mixes with domestic breed
the influence of its peculiar colours may be constantly
seen . .
The same occurs with the Muscovy duck in S. America.
The wild cock of Macassar mixes with
the domestic breeds from which it can
hardly be distinguished.

Dufour in Hist. of Prostitution
says that inhabitants of ancient Italy are
represented on vases &c. with a tuft of hair
in place of a tail . He says it is evidently
a natural character! — now lost! —

Cocoa nuts in Ceram have three Enemies, the.
Cockatoo (Cacatua moluccanus) eats into the young fruits
when about half grown, a Pteropus also eats the
young fruits or flowers, & a large beetle (Oryctes)
eats into the very young fruits — ? Does not this
prove that we are at or near the native country
of the Cocoa nut.
In Batchian the Belideus only attacks the cocoa nut.

[91]

[1]The duck of Aru near <u>Anas Radjah</u>. Less[on] is often
bred becomes tame & mixes with ~~native~~ domestic breeds
the influence of its peculiar colours may be constantly
seen . .
The same occurs with the Muscovy duck in S.[outh] America.
The wild cock of Macassar mixes with
the domestic breeds from which it can
hardly be distinguished.

[2]Dufour in Hist[ory] of Prostitution
says that inhabitants of <u>ancient Italy</u> are
represented on vases &c. with a tuft of hair
in place of a <u>tail</u>. He says it is evidently
a <u>natural</u> character!—now lost!—

[3]<u>Cocoa nuts</u> in Ceram have <u>three</u> Enemies, the
cockatoo (Cacatua moluccanus) eats into the young fruits
when about half grown, a <u>Pteropus</u> also eats the
young fruits or flowers, & a large beetle (Oryctes)
eats into any young fruits—? Does not this
prove that we are at or near the native country
of the Cocoa nut.

In Batchian the Belideus only attacks the cocoa nut.

1.  The radjah (raja) shelduck was named *Anas radjah* in 1828 by French naturalist René Primevère Lesson (1794–1849); it is now called *Tadorna radjah*. Wallace's radjah shelduck specimens from Halmahera, Morotai, Bacan, and other locales in the Moluccas were acquired by the British Museum.

2.  Paul LaCroix [aka Pierre Dufour] published his *Histoire de la prostitution* in several volumes appearing between 1851 and 1856. The observation given here relates to notes on recto 64. Wallace said in a letter to his friend George Silk dated September 1, 1860, that *Histoire de la prostitution* was one of two books "of the highest interest" and "of most diverse characters" that he had been reading of late, recommending both highly. (The other book was his personal copy of Darwin's *Origin of Species.*) The observation from *Histoire de la prostitution* that Wallace notes here is incidental to his broader interest in the work—namely its bearing on social justice—as his letter to Silk reveals:

> If there is an English translation . . . pray get it. Every student of men and morals should read it, & if many who talk glibly of putting down the 'Social Evil' were first to devote a few days to its study they would be both much better qualified to give an opinion on the subject & much more diffident of their capacity to deal with the question. The work is truly a history & a grand one, & reveals pictures of human nature more wild & incredible than the pen of the romancist [*sic*] ever dared to delineate . . . Again I say read it. (Wallace Correspondence Project WCP373)

LaCroix's observation in *Histoire de la prostitution* regarding tailed inhabitants of ancient Italy is curious; he likely mistook representations of satyrs for ordinary humans. Satyrs bearing tails—some long, some short, but generally horse-like—were often depicted on ancient Greek and Roman vases, drinking cups, and the like.

3.  An interesting suggestion: on the basis of what he takes to be more "natural enemies" of coconuts (*Cocos nucifera*, Arecaceae) in Ceram (Seram) as compared with other locales, Wallace suggests that the species originated there. Just one species, the small flying opossum *Belideus*, in Batchian (Bacan) attacks the coconut, while three apparently do so in Seram. (Wallace described the flying opossum, *Belideus ariel*, as "a beautiful little animal, exactly like a small flying squirrel in appearance, but belonging to the marsupial order"; *Malay Archipelago*, 398).

This is reasonable evolutionary thinking, though it has its pitfalls. Pantropical in distribution today, the region of origin of the coconut palm is much debated. A problem with Wallace's suggestion is that none of the three "enemies" he mentions is restricted in its diet to coconut plants; the cockatoo, flying-fox fruit bat *Pteropus*, and rhinoceros beetle *Oryctes rhinoceros* are all generalists. A better gauge of region of origin of a plant might be the occurrence of a significant number of species specialized on that plant. (Note, however, that as species disperse, their specialists might easily disperse with them.) In the Pacific, caterpillars of several species of the microlepidopteran family Batrachedridae putatively specialize on coconut, including *Batrachedra atriloqua* and *B. mathesoni* (Greenwood 1940).

## 92

[1]Trias of Germany, *[illeg.]* lowest of Mesozoic
rocks, is conformable with Permian or
highest Palaeozoic, yet totally new forms of
animal life are introduced—many
large Saurians now first appearing.

———————

[2]This does not prove a sudden change as it
at first sight appears to do. During the
Permian period this area of sea in central
Europe received deposits from a cold
country as shown by the existence of icebergs
& Glaciers. Let us now suppose it &
a great extent of adjacent country to
be gradually raised, till the sea became
dry land, & thus to remain for such a
long Geological period that in the ordinary
course of change of species an entirely
new creation was introduced. This elevation
being accompanied by a change of climate
from a frigid to a temperate or even subtropical,
the inhabitants of the land & adjacent

1. "Trias" refers to the Triassic Period, the first period of the Mesozoic era, about 250 to 200 million years ago. The name was coined in 1834 by the German geologist Friedrich August von Alberti (1795–1878), for the three (hence "tri-") well-defined sets of strata common in central and western Europe, and especially well developed in Germany: weathered red sandstones, white limestones, and dark shales. These Triassic strata are conformable (contacting and sequentially continuous) with the oldest Paleozoic strata (Permian period), yet, as Wallace mentions here, "totally new forms of animal life are introduced." Indeed, the mass extinction that marks the Permian-Triassic boundary is the most extensive known in earth's history, with an estimated 57 percent of all families and 83 percent of all genera going extinct. The cause or causes of this greatest of all mass extinction events are debated, but it seems clear that this mass extinction occurred progressively, in up to three distinct phases (Jin et al. 2000, Sahney and Benton 2008). Still, the transition from the Permian fauna to that of the Lower Triassic is dramatic. (The geological epochs and periods are largely defined in terms of such transitions.)

2. The argument that commences on this page and runs through recto 100 is a return to considerations of the fossil record, both its imperfection and the record of faunal succession. The length and narrative structure of the arguments suggest that this was intended as material for Wallace's planned book, "On the Organic Law of Change" (see notes for recto 35).

Wallace addresses the apparent suddenness of the appearance of a new fossil fauna at the transition from the Paleozoic to Mesozoic, with the appearance of whole suites of saurians. Wallace commences an argument explaining how the dynamically changing earth, though regional subsidence and elevation, is responsible for the apparent gaps between fossil fauna of different periods. Darwin grappled with precisely the same issue, and offered much the same explanation as Wallace, in the *Origin of Species* (chapter 9). The use of the phrase "new creation" might seem curious for this convinced transmutationist, but note his reference to "the ordinary course of change of species" in this same sentence. "New creation" here can be taken simply as shorthand for the arising of a suite of new species. Wallace's casual use of the word "creation" in his 1855 Sarawak Law paper, where it appears no less than eight times, led Darwin to mistakenly dismiss him as a creationist. But he often used this word in the sense described here, both before and after his Ternate essay. In such later writings as *Darwinism* (1889), however, the word is almost always presented in the context of "special creation," which he attacks.

seas would assume a very different character.
Now let the same district sink & have
its principal source of animal bearing deposits
from the south, & we shall have the
fossil fauna of the Trias, & we have
abundant evidence <sup>in the Coal formation for example</sup> that changes to this
amount may take place without producing
any unconformity of the strata.

But it may be asked, [1]where is the evidence
of this change where are the fossil remains
of the animals that flourished during this—
long period during which the organic world
underwent such remarkable modifications.
We answer this evidence is all buried
beneath the ocean, for as a great portion
of what is now land was land then, it
is highly probable that the <sup>present</sup> ~~adjacent~~ oceans
were oceans then & received then as
now the organic remains of the adjacent
countries. Such gaps must occur wherever
the ocean which received deposits during

1. Wallace had a very Lyellian view of the dynamically changing surface of the earth. Geological history is one of slow uplift and subsidence, coupled with cycles of warm and cold climate. Gaps in the fossil record arise when uplifted fossil-bearing strata are eroded: the ocean receives "the organic remains of the adjacent countries."

Wallace rhetorically sets up his discussion that follows—"where is the evidence" for this change? He tackles the pernicious problem of an imperfect, gappy fossil record. The points he makes in these notes are echoed in later writings, such as his review of the tenth edition of *Principles*, titled "Sir Charles Lyell on Geological Climates and the Origin of Species":

Even in the places where we have access to the strata formed at any former period of the earth's history, we know that those strata will not contain anything like a consecutive record of the animals and vegetables that lived in its vicinity. As long as the surface at that spot was dry land, few or no organisms could be preserved, except at the mouth of some large river which might carry down their remains and cover them with sediment. But when the strata thus formed came to be raised up again above the waters, they would be exposed to the action of the waves and in most cases be eaten away and redistributed on the sea-bottom. Only when a long-continued subsidence had allowed strata of great thickness to be formed, and when the subsequent elevation was so rapid or the situation was so favourable that they could resist the denuding action of the sea, would fossiliferous strata ever rise above the surface. We might, therefore, anticipate, that between the formation of any given beds and others of different composition lying upon them, there would often elapse vast and unknown intervals of time, allowing for a complete change in animal and vegetable life; and this is exactly what geologists find. In the infancy of the science it was supposed that this indicated a catastrophe by which hundreds of species were destroyed, and a sudden creation of hundreds of new ones. Now it is universally admitted to prove only a vast lapse of time. (Wallace 1869c, 386–387)

any given period is now again ocean,
& looking at the large proportion of sea
to land such must often be the case.
[1]It is next to impossible therefore that we
could have a geological record without
any break, the wonder rather is that the
gaps are not still wider & still more
numerous than they are. It does
not seem to be generally remembered that
to have in any one spot a complete record
by fossil remains of two or more Geological
periods, it is necessary that the area
in question should have never risen above
the waters except for short intervals during these immense periods
of time, for whenever it was land the
registry of the animal life of the adjacent
Countries must have ceased. And not
only this but no great changes of level
or of drainage must have taken place

[The following sentence is written vertically in left margin of page]
a deep ocean would have the same effect, if no continents near
there could be no deposits.

1. Wallace makes a key point here: it is indeed impossible to have a continuous fossil record, free of gaps, and if anything it's a wonder that the record is as good as it is. This point was poetically described by Darwin too:

> I look at the natural geological record, as a history of the world imperfectly kept, and written in a changing dialect; of this history we possess the last volume alone, relating only to two or three countries. Of this volume, only here and there a short chapter has been preserved; and of each page, only here and there a few lines. Each word of the slowly-changing language, in which the history is supposed to be written, being more or less different in the interrupted succession of chapters, may represent the apparently abruptly changed forms of life, entombed in our consecutive, but widely separated formations." (Darwin 1859, 310–311)

to divert the general currents & deposits
in other directions. ~~Such~~ This stability of large
areas  for such immense periods must
be very rare if it ever has occurred,
for in every formation we have evidence
of great changes of level  having repeatedly
~~occurred~~ taken place even between the deposition of
individual beds of deposit. Interruptions
therefore in the series of organic remains
sometimes for small sometimes for immense
periods must be expected. [1]Continuity
must be the exception not the rule, &
all the evidence ~~of the~~ afforded by the
peculiar fossils of different beds, groups &
formations, shews it actually to be so.

———

We have therefore the strongest reason to believe
that from the very earliest geological periods
to the present time, the changes in the organic
as well as in the inorganic would have been
continuous & gradual in the record only of those changes
open to us, being discontinuous and incomplete.

1. Or, inversely, interruptions or gaps in the fossil record are the rule rather than the exception. Note that Wallace (and Darwin— see comments on previous page) referred to gaps in the *record* of what is actually a continuous process of ancestor-descendant relationship. Broadly, however, there is still ample evidence of continuity or a "law of succession," a term Lyell coined to describe the sequential replacement of one fossil group by another, related fossil group. This empirical observation was one of the two central pillars of Wallace's Sarawak Law paper of 1855, the other being geographical proximity of newly arising species to previously existing, closely related species.

## 96

It may be urged as very improbable that at any former epoch the whole of the existing lands should have been in exactly the same state as now & if any part was under water we should probably there find some organic remains of that period. This is admitted, & the immense areas in Asia Africa & America still unexplored for fossils leave ample room for the future discovery of strata which will fill up some of the gaps existing in Europe. [1]But at the same time it must be remembered that in distant localities we shall have great difficulty in tracing the synchronous nature of the fossils. At some future period how little resemblance will be traced between the fossil remains of Europe & Australia or even of Australia & S.[outh] Africa so much nearer to each other. On one side of a then existing continent we may have beds containing almost

1. Wallace's point here is that while fossils have been (and are being) deposited in every epoch, he suspects it will be difficult to identify correspondences between the fossil fauna of widely separated locales.

His suggestion that uncovering fossil strata in as-yet-unexplored regions will help fill gaps in the record is borne out: while different geological periods are better represented in some parts of the world than others, geologists have been able to painstakingly construct a continuous record of formations from the fragments available in different regions, and to correlate these (first using mineral and fossil analysis, later corroborated, where possible, by radiometric dating; minerals in fossil-bearing sedimentary rock have been eroded from older sources, but such formations can be bracketed in age by dating igneous formations above and below them).

Wallace's concern that the fossil faunas of different continents would be difficult to correlate has not been borne out, however. To Wallace, living at a time when continents were thought to be static, this task would have seemed impossible considering, as he points out here, how very different are faunas of continents such as Australia and Africa. Correspondences of fossil formations that do exist across continents were in fact used as an important line of evidence for continental drift, first proposed by Alfred Wegener in 1916 and confirmed in the 1960s. Ironically, the very difficulty identified by Wallace here—correspondence of fossil fauna of different continents—turned out to be one tool used by geologists in reconstructing the relative positions of continents over time. The earth does indeed dynamically change, far more so than Lyell, Wallace, or Darwin could have imagined.

exclusively Marsupials, on the other
may be found the remains of <sup>the</sup> Elephant,
Rhinocerus & hundreds of ~~Ruminants~~ <sup>Antelopes</sup>.
The Geologist of that age will hardly
consider these to mark one & the same
age.

---

[1]What is the import of the doctrine of Morphology
of plants.? The carpels are said to be
leaves the edges of which are joined together
as beautifully seen in the common pea.
The stamens are modified or peculiarly
developed petals,—these are sepals under
another form; sepals again are leaves
under various forms & changes. These
statements are supported by hundreds of
cases of monstrosity & disease in which
these changes & no others take place, &
also by numerous species in which one
or other of these parts actually exists in its

1.  The "doctrine of morphology" was a phrase then in common use, referring to fundamental affinity of the structural parts of an organism. In this case it refers to homology of such plant structures as leaves, petals, sepals, stamens, and pistils. Homology can be seen in the context of comparative anatomy across species (Richard Owen had this in mind when he defined homology in 1843 as "the same organ in different animals under every variety of form and function"), but it can also apply to differentiated structures of a species derived from common, inherited genetic pathways. This is now termed *serial homology*; such traits have similar embryological origins and developmental pathways.

The first person to articulate this concept in terms of plant structure was the German polymath Johann Wolfgang von Goethe (1749–1832), in an essay published in 1790 entitled *Metamorphosis of Plants (Versuch die Metamorphose der Pflanzen zu erklären)*. He later presented this theory in a poem of the same name. Goethe worked on plant and animal morphology for some years, and had evidently arrived at his idea of serial homology of plant parts a few years earlier, expressing it in a letter dated May 1786 to his friend Herder: "I must moreover tell you in confidence, that I am very near the whole secret of the generation and organization of plants, and that it is the simplest thing that can be imagined" (Austin 1833, 13). In the botanical world the idea that stamens and pistils could be thought of as modified leaves gained acceptance slowly, however, as reflected in this editorial note in *The Gardener's Monthly and Horticulturist* (21 [1879]: 278):

> *The Doctrine Of Morphology*
> It is almost wonderful that the doctrine which teaches that all parts of a flower are modified primary leaves, should have such universal assent, when but a comparatively few years ago it was laughed at by the most intelligent men of the day. Speaking of the theory, an editorial article in *Paxton's Magazine* for 1844, says: "There is something so monstrous, so degrading in the idea, that the mind which contemplates all things as beautiful and perfect in their creation, revolts at it."

lower or less developed form. Then again
there are the cases of retarded development
by which some ~~of~~ or other of these parts become
smaller imperfect or altogether absent.
By these laws all the countless modification
of flower & fruit can be reduced to
a common type & most of the excentricities *[sic]*
of vegetation explained & accounted for.
[1]Now what does all this beautiful law
mean, what does it teach us? Is it
a substance or a shadow, a truth or a
fallacy?—For if we are to believe that
each & every species ~~has been specially~~ is an absolute &
~~created without any~~ distinct creation independent of ~~its~~ any of
its closest allies, & has come into existence
at a different time & in a different place
from them, if those having the highest developed
parts may have had their origin before those others

1.  The following passage, which continues to recto 99, is remarkable. If "those having the highest developed parts" have had their origin—were found in the fossil record, say—"before those others of which we suppose them to be a more complete development," the doctrine of morphology would be meaningless, because simple structure should not come before complex structure.

Wallace sees that the principle of homology, in this case as embodied in the "doctrine of the morphology of plants," makes perfect sense in an evolutionary context.

of which we suppose them to be a more
complete development, then all this doctrine
of morphology is meaningless & leads to
error. [1]For if stamens & petals & carpels have
been in every case <sup>independently</sup> created as such, it is
absurd to say they are modifications or
developments of any thing else, & the
absurdity is still greater if that of which
they are said to be the development
came into existence after them. In that
case all the beautiful facts of morphology
are a delusion & a snare, as much so
as fossils would be were they really not
the remains of living beings but chance
imitations of them.

[2]The natural inference of an unprejudiced
person however would be that both are true
records of the ~~steps~~ progress of the organic
world. Nature seems to tell us that as
organs ~~may~~ are occasionally changed & modified
now, in individual plants, we may learn

1. Wallace's point is well taken—if plants bearing structures like the specialized stamens and pistils of flowering plants have been created independently of nonflowering plants, it nonsensical that these structures could be *derived from* other structures. Then, too, if such "higher" plants came into existence before "lower" plants, having their specialized structures derived from other, preexisting structures does not make sense.

Note how Wallace likens rejection of the reality of the serial homology of plant parts to rejection of the reality of an organic origin for fossils (see recto 34). (Darwin made a very similar point in his *Origin of Species* [1859, 167]—declaring that he would sooner believe, with the cosmogonists of old, that fossils have an inorganic origin than that the patterns he was uncovering did not say something about transmutation and common descent. This is another example of the resonance in thinking between Wallace and Darwin). To wave away this evidence as incompatible with an a priori worldview is to be willfully ignorant, we can almost hear Wallace thinking.

2. Indeed, Wallace all but says so in the next sentence, which is completed on the following page. Note that the evolutionary inference is a *natural* one to an *unprejudiced* person. In essence Wallace is suggesting that to refuse to use a key proffered to unlock the answer to a great mystery, otherwise inaccessible, is to remain willfully ignorant.

how the actual changes have taken
place in the species of plants. A
key is offered us to a mystery we could
otherwise never have laid open, why
should we refuse to use it?

---

[1]The lower orders of Javaneese *[sic]* however often approach
nearer to the Chineese *[sic]* physionomy *[sic]* & colour.
Javaneese—Handsomest of Malay
races, nose well formed, mouth often
very small, chin small face very oval—
probably an extensive mixture of Hindoo
race as many have all the usual
Malay characteristics fully developed.

---

[2]The flying lizards describe a curve descending
& ascending—thus *[Sketch]* the velocity
attained in the first descent
enabling them by placing the body & wings sloping
upwards to ascend a little & alight on the tree
vertically & with head erect.
    Observed at Macassar Nov.[ember] 57.
Will pass 20–30 feet.

1. Another ethnological observation; Wallace wrote that the Javanese formed one of four great "Malay races" collectively occupying the western half of the archipelago. The Javanese are found, according to *Malay Archipelago*, not only in Java proper but also in Madura, part of Sumatra, Bali, and part of Lombock. He further reported that they spoke two languages: Javanese and Kawi.

Lying along a major sea-lane trading route, Java and adjacent islands have been the sites of cultural intermixing for millennia, with southern Chinese (indigenous Taiwanese) influence from the north and Indian influence from the west.

Wallace's reference to "intermixture of the Hindoo race" is interesting, in view of the modern understanding of the relatively recent origins of the Javanese and Kawi peoples and language. Extensive interaction with traders from the ancient Tamil kingdom of Pallava beginning in the eighth century CE brought many elements of southern Indian culture, including writing, to Java and the neighboring islands. Thus Kawi has a strong Sanskrit influence. Wilhelm von Humboldt (1767–1835), philosopher, linguist, diplomat, and brother of the illustrious Alexander, argued in his final philological work, *Über die Kawi-Sprache* (*On the Kawi Language*, published posthumously 1836–1839), that Kawi is the stem language of all Malay-Polynesian languages.

2. The flying lizards are a large group of diminutive and often colorful agamid lizards of the genus *Draco*. The type species, named by Linnaeus in 1758, is *Draco volans*—the species name meaning flying or gliding (as in "volant"). These remarkable lizards glide long distances by extending winglike patagial membranes, which are supported by greatly elongated thoracic ribs.

The island of Sulawesi alone (Wallace's Celebes) is home to some half dozen flying lizard species; based on geography, the species he spotted at Macassar is likely *D. walkeri* (McGuire et al. 2007). As Wallace's sketch indicates, these lizards are capable of generating some lift from their membranous "wings," enabling them to control their approach to their destination tree and land upright on the trunk.

Incidentally, Wallace came across another "flying reptile"—actually an amphibian—during his stay in Sarawak in 1855:

> One of the most curious and interesting reptiles which I met with in Borneo was a large tree-frog, which was brought to me by one of the Chinese workmen. He assured me that he had seen it come down, in a slanting direction, from a high tree, as if it flew. On examining it, I found the toes very long, and fully webbed to their very extremity, so that when expanded they offered a surface much larger than the body . . . This is, I believe, the first instance known of a "flying frog." (*Malay Archipelago*, 49)

Wallace's flying frog. Based on specimens collected by Wallace, this species was formally described and named *Rhacophorus nigropalmatus* in 1895 by Belgian zoologist George Albert Boulenger (1858–1937). Wallace's lovely watercolor of his flying frog, executed while in Sarawak, is now in the collection of the Natural History Museum (London). Reproduced from Alfred Russel Wallace, *The Malay Archipelago*, 4th ed. (London: Macmillan & Co., 1872), 38.

same at Dorey & Amboyna.

nigrogularis G.[eorge] R.[obert] Gray

[1]Trichoglossus sp.[ecies] (4. Aru.) near cyanogrammus.
Wag.[ler]

?T.[richoglossus] ornatus B[ritish] M.[useum]

Grass green ; / head olive b[lac]k. the forehead & cheeks bright
blue

shading into the black behind, the crown & occiput dark olive

brown; nuchal collar bright greenish yellow; / breast &

lower neck orange crimson, the feathers broad scaly & narrowly

margined with dark purple & green; / rest of body beneath rich
green

mixed with roundish spots of dark purplish b[lac]k.; sides
beneath

wings red & green banded; thighs entirely green & yellow

banded as are the under tail coverts; the feathers of

the upper part of the back have a red central band

hardly seen when the plumage is accurately disposed. /

*[The following two lines are written vertically in left margin
of page]*

Amboina specimens want the b[lac]k.

spots beneath.

/. The wings have the inner web of all the feathers dusky b[lac]k.

with a large central patch of bright yellow which in

the primaries covers ¾ of the feather from its base; the

under wing coverts are crimson /. Tail dusky olive beneath

with the inner web of the 8 outer feathers broadly yellow.

/. Bill orange red, orbits bare black, irides golden red,

feet pale olive. ♂ & ♀ alike. .

/. Length 12½ in.[ch] tail 5½ in.[ch] wing 6 in.[ch] 3 first quills
equal.

[2]A most lovely bird, abundant in Aru: frequents the

Casuarina trees & also in the deep forest.

1. This is *Trichoglossus haematodus*, the rainbow lorikeet, a spectacular bird that looks like it met with an accident in a paint shop. The specimens Wallace collected in Aru were called *T. nigrogularis*, a species named by G. R. Gray in 1858 (written in pencil at the very top of the page). This name was later synonymized with *T. haematodus*, and today it is recognized as a subspecies: *T. h. nigrogularis*. Wallace's Aru specimens were obtained by the British Museum.

Note that in the field Wallace is not quite sure of the species and indicates it is "near cyanogrammus"—the green-naped lorikeet. This, too, is now recognized as a subspecies of rainbow lorikeet: *T. h. cyanogrammus*. ["Wag." is German zoologist Johann Georg Wagler (1800–1832), director of the Zoological Museum at the University of Munich. Wagler authored a treatise on parrots and relatives (*Monographia Psittacorum*, 1832).]

2. The observation of this "lovely bird" frequenting the *Casuarina* trees is mentioned in Wallace's paper on the natural history of the Aru Islands, published in December of 1857 (Wallace 1857d). *Casuarina*, commonly known as ironwood, is a group of some seventeen species in the family Casuarinaceae, which is widely distributed throughout the tropics of Australasia.

## 102

[1]For a Motto.

"He who in place of reasoning, employs authority,
assumes that those to whom he addresses himself
are incapable of forming a judgement of
their own. If they submit to this insult, may
it not be presumed they acknowledge the
justice of it?" Bentham.

---

"Speculative"

"An epithet in use among official persons, for the
Condemnation of whatever is too adverse to private
interest not to be hated, and at the same time
too manifestly true to be denied." Bentham.

---

"Theoretical"

Every man's knowledge is ~~to~~ (in its extent), proportioned to
the extent (as well as) [and] number of those general propositions,
of the truth of which (they being true) he has the persuasion
in his own mind: the extent of his theories comprises
the extent of his knowledge.

If indeed his theories are false these in proportion as they
are more extensive he is more deeply steeped in
ignorance & error.

1. These quotes are from the *Book of Fallacies* by Jeremy Bentham (1748–1832), English utilitarian philosopher, jurist, and social reformer. Bentham's advocacy for such radical notions as the rights of women, the right to divorce, individual and economic freedom, the abolition of slavery, elimination of capital punishment, and many other then-outlandish ideas would have resonated with Wallace, who, ever since attending lectures on social reform when he lived in London with his brother John as a teenager, had developed a passionate sense of social justice. The Wallace brothers spent many evenings at the local Mechanic's Institute, where all manner of lectures on social and scientific (and pseudo-scientific) topics were given. Wallace attributed his lifelong embrace of socialist ideas, and his skepticism of organized religion, to this experience (see *My Life* [Wallace 1905, volume 1, chapter 6]). Many radical reformers of Wallace's day were followers of the charismatic Robert Owen, a utopian social reformer who founded New Lanark community, in Scotland, a town built around a jointly owned cotton mill—an early version of a modern co-op. Bentham supported many of Owen's ideas, and was a shareholder in New Lanark.

Bentham was a prolific author, and the *Book of Fallacies*, a collection of his writings, was published in 1824. Wallace may have written out these passages for future use as epigraphs. It does not appear that any of these were ultimately selected, though he made liberal use of epigraphs in various works:

*On Miracles and Modern Spiritualism* (London: James Burns, 1875): five epigraphs

*Land Nationalisation* (London: Trübner and Co., 1882): six epigraphs

*Bad Times* (London: Macmillan and Co., 1885): one epigraph

*The Wonderful Century* (London: Swan Sonnenschein and Co., 1898): three epigraphs

## 103

[1]But from the mere circumstance of its being theoretical, its falsehood is no more a necessary consequence, than is falsehood a necessary consequence of speaking.

One would think, that in thinking there were something wicked or else unwise: every body feels or fancies a necessity for disclaiming it. "I am not given to speculation" "I am no friend to theories." Speculation,—theory,—what is it but thinking? Can a Man disclaim speculation & theory without disclaiming thought,—for unless they mean a little more thought than ordinary they mean nothing.

To escape from the imputation of meditating destruction to mankind, a man must disclaim every thing that puts him above the level of a beast.

---

### [2]Belief.

What is in a man's power to do, in order to believe a proposition, and all that is in his power is to keep back & stifle the evidence that is opposed to it. For when all the evidences are equally present to his observation & equally attended to, to believe or disbelieve is no longer in his power, but is the necessary result of the preponderance of evidence for or against.

1.  This is a continuation of the Bentham quote from the previous page. It is one of the "Fallacies of Confusion," taken from Bentham's *Book of Fallacies* (1824).

2. The following quote, labeled "Belief," appears in a number of collections of the writings, correspondence, and aphorisms of Jeremy Bentham that appeared after his death. For example, it is found in the section entitled "Extracts of Bentham's Commonplace Book" in *The Works of Jeremy Bentham*, edited by John Bowring (Bentham 1838–1843, 10: 146). This would have made a good epigraph for Wallace's 1875 book *On Miracles and Modern Spiritualism.*

## 104

[1]Judges recommending Criminals not to plead Guilty.
"The wicked man repenting of his wickedness, offers
what atonement is in his power: the judge, the chosen
minister of righteousness, bids him repent of
his repentance, and in place of the truth, substitute
a barefaced lie. Such is the Morality such the
holiness of an English Judge.

---

[2]The inhabitants of Floris are Alfurus, some have
frizzly hair, some wavy & some straight but all
have dark skins & papuan features.—(Capt.[ai]n Drysdale[])

---

Inhabitants of Baly have much mixture of
Hindoo. In the interior are one or two villages
of the indigenes who are Alfurus ?—(Mr. King).

---

Tomore people from E.[ast] Celebes settled at
Batchian, almost exactly resemble the
Macassar & Bugis people;—They have
all the characteristics of a Malay race.

---

1. From Bentham, on "incongruities" in English common law. This appears in Book III, chapter 15, "Rationale of Judicial Evidence," in *The Works of Jeremy Bentham* (Bentham 1838–1843, 6: 473).

2. The "Alfurus" mentioned here (and on recto 134) are the people Wallace calls the "true indigenes of Gilolo" in *Malay Archipelago* (321). The term is a broad one—also called Alfurs, Alfures, Alifuru, Alfouro, and Alfuren, the name was originally applied by the Malays to all the non-Muslim peoples of the eastern portion of the archipelago. Today it is restricted to the indigenous peoples of the islands of Halmahera, Buru, and Seram.

In May of 1859 Wallace traveled from Ternate to Coupang, on the western side of Timor. In his third Malay Journal (LINSOC-MS178c, entry 175), Wallace describes Captain Drysdale, who lived in Coupang, as "a scotchman long settled here & married to a Portuguese lady from Delli." Drysdale gave Wallace "much information about the country to the East & south of whose capabilities he thinks very highly." No doubt this included information on the inhabitants of Flores and Bali, mentioned here.

In the next entry (176) in this Malay Journal, Wallace gave more detailed notes of the peoples of that part of Indonesia: "We have therefore in the islands from Lombock to Timor the confluence of three strongly marked primary races, & it is not to be wondered at that these compound mixtures should have produced many puzzling combinations & rendered the idea of transition races highly probable to casual observers." These observations were also incorporated into his discussions on race in chapter 13 of *Malay Archipelago*.

## 105

[1]D'Urville—says in Tahiti commence ferns which are
common to the Moluccas & even to Mauritius—
Casuarina in Tahiti,—not on Sandwich—

[2]Lesson—says the Papuans are a mixed race, negro
& Malay inhabiting the N.[orth] Coast of N.[ew] Guinea,
    Aru
Waigiou &c. The alfurus & natives of the interior
of New Guinea he says are straight haired blacks
and are quite distinct. The papuans inhabit
also New Britain & New Ireland.

The bow & arrows unknown over all Australia. Less.[on]

Alfurus of N[orth] Celebes.
Neither of Malay nor Papuan race—
Lighter colour than Malays—Hair straight black
rather finer & lanker than that of Malays—
beard little, bodies smooth—
Features very European, head wide above.
nose straight, apex rather produced.—not
thickened as in Papuans.
Character mild.
? Philippine or Pacific race.

1. Jules Dumont d'Urville (1790–1842), French naval officer, explorer, and naturalist, traveled to the Pacific in two voyages spanning the 1820s: those of the *Coquille* and the *Astrolabe*. In 1837 he commanded a second voyage of the *Astrolabe* to the Antarctic, accompanied by a sister ship called the *Zélée*. Several geographical features and the French research base in Antarctica are named in his honor.

On the *Coquille* Dumont d'Urville oversaw scientific collections in botany and entomology. The *Coquille* sailed to Tahiti, then through the island groups of the South Pacific to New Guinea and the Malay Archipelago. Wallace may here be referring to Dumont d'Urville's September 1825 paper on the distribution of ferns in the *Annales des sciences naturelles* (Darwin made a note about this same paper in his C Notebook; Barrett et al. 1987, 241). More likely he found the observation in the chronicle of the *Coquille*'s 1822–1825 voyage written by René Lesson, who served as naturalist. It is evident from the further notes on this page that Wallace had consulted Lesson.

2. René Primevère Lesson (1794–1849), French surgeon and prolific naturalist best known for his ornithological work, had visited New Guinea and environs aboard the *Coquille* in 1822–1824. In 1838–1839 he published his two-volume *Voyage autour du monde entrepris par ordre du gouvernement sur la corvette "La Coquille,"* which is referenced often in the Species Notebook. Wallace seems to be relating information on the distribution of species such as ferns and *Casuarina* (see recto 101) from the South Pacific and Malay Archipelago to patterns in the distribution of human groups. This is evident by his transition from the fern observation to anthropological observations, citing Lesson. Both Dumont d'Urville and Lesson wrote extensively on the peoples of "Oceania," which to them extended from the Pacific islands west to Malaysia.

## 106

[1]Lesson's distribution of the Malay & Polynesian Races.
Hindoo.

      Malay . .      of the Indian Arch..[ipela]go & Malacca.

      Oceanic . .   Tahitians, Sandwich Is..[lan]ds &c. N.[ew] Zealand

Mongolian . .

      Mongol pelagian of the Phillippines Carolinas & Mid*[illeg.]*

Black.

      [2]Caffre Malegassee including

            Papuans of N.[ew] Guinea coast & islands.

            Tasmanians . . of Tasmania

      Alfurus. including

            Endamene [Endamênes]. . of the interior of N.[ew] Guinea &

                other large islands

            Australians . . of New Holland

---

The Malays are Hindoos & Mongol mixed & isolated..

---

The true Malays have nothing of Hindoo.
They are a Mongol race.

1. Wallace's notes from René Lesson continue.

In 1832 Dumont d'Urville published a classification of Oceania and its people into four cultural regions and two racial types; it was through this paper that the labels Melanesia, Micronesia, and Polynesia were popularized (Tcherkézoff 2003). Lesson, in contrast, in his *Voyage médical autour du monde* (1829, 153–228), divided the "Oceanians" into a three-part racial hierarchy on the basis of physical and cultural characteristics and supposed origins.

Wallace came to his own conclusions about the native peoples of the region; he summarized these in a paper read in 1864, and later expanded and published them in a celebrated paper entitled "On the Varieties of Man in the Malay Archipelago" (Wallace 1865b). This later version was largely incorporated into the final chapter of *Malay Archipelago*.

"Two very strongly contrasted races inhabit the Archipelago," Wallace declared: "the Malays, occupying almost exclusively the larger western half of it, and the Papuans, whose head-quarters are New Guinea and several of the adjacent islands" (*Malay Archipelago*, 584). Wallace considered the "Papuan" type to be one of a constellation of racial groups that collectively made up one Pacific group, all "varying forms of one great Oceanic or Polynesian race" (*Malay Archipelago*, 593).

2. Caffre—also Caffer, Kafir, Kaffir—is from the Arabic *kafir*, "infidel," as applied by Arabs to non-Muslim peoples and nations. Also refers to Bantu people (and languages) of eastern South Africa (*Oxford English Dictionary*). In modern times this is an offensive and insulting term for any black African.

Notions of the Tropics.

"There are lands" said Cleveland "in which the
eye may look bright upon groves of the palm
& the Cocoa, & where the foot may move light
as a galley over fields carpeted with flowers,
& savannahs surrounded by aromatic thickets."
                                    Scott—Pirate

~~Lorius superbus. Fraser.~~
Eos sp.          Psittacus lory ?          Mafor Island.
Back & upper tail coverts, face throat & breast, crimson red;
as well as flanks & base of thighs;— lower breast,
belly, under tail coverts, a broad band on the nape &
a narrow one just below between the shoulders, thighs, shoulder
margin & smaller under wing coverts purple blue,— the nape
collar very brilliant extended narrowly on each side & almost
meeting in a point in front of neck;— crown to the
nape black; wings rich green, inner webs of all
feathers black with the central portion of primaries & a
small spot on secondaries yellow;— tail above purple,
base of feathers green & central portion reddish, beneath
greenish yellow with a red band abt ⅓ length from base
upr mand or. red, lower m. yel — iris orange red.
(greater under wing coverts bk.)

---

[1]Notions of the Tropics.
"There are lands" said Cleveland "in which the
eye may look bright upon groves of the palm
& the cocoa, & where the foot may move light
as a galley over fields carpeted with flowers,
& savannahs surrounded by aromatic thickets."
                                    Scott—Pirate

---

[2]Lorius superbus. Fraser.
Eos sp.[ecies]          Psittacus lory ?          Mafor Island.
Back & upper tail coverts face throat & breast, crimson red;
as well as flanks & base of thighs;—lower breast,
belly, under tail coverts, a broad band on the nape &
a narrow one just below between the shoulders, thighs, shoulder
margin & smaller under wing coverts purple blue,—the nape
collar very brilliant extended narrowly on each side & almost
meeting in a point in front of neck;—crown to the
nape black; wings rich green, inner webs of all
feathers black with the central portion of primaries & a
small spot on secondaries yellow;—tail above purple,
base of feathers green & central portion reddish, beneath
greenish yellow with a red band ab[ou]t ⅓ length from base
up.[pe]r mand[ible] a red, lower m.[andible] yel[low]—iris
    orange red.
(greater under wing coverts b[lac]k.)

1. This passage from Sir Walter Scott's *The Pirate* (1831; volume 25 in the Waverly Novels series) must have struck Wallace as a nice evocation of the tropics, despite its romanticized view. Note that Wallace left off the last part of the (future) pirate Cleveland's soliloquy: "'There are lands,' said Cleveland, 'in which the eye may look bright upon groves of the palm and the cocoa, and where the foot may move light as a galley under sail, over fields carpeted with flowers, and savannahs surrounded by aromatic thickets, and where subjection is unknown, except that of the brave to the bravest, and of all to the most beautiful'" (Scott 1831, 36).

2. Wallace gives Mafor Island, now Numfor Island, as the locality for this specimen of *Lorius superbus*. He did not collect there himself, however; his easternmost point in the archipelago was Dorey Harbor (now Manokwari), just west of Mafor, where both he and the steamer he traveled on were laid up—the steamer for want of coal, and he with an ulcerated ankle. He acquired what he could through the locals. During that time he complained, "I got nothing from the natives, every specimen being taken on board the steamer, where the commonest birds and insects were bought at high prices" (from "Narrative of Search after Birds of Paradise," a paper read at the Zoological Society of London meeting of May 27, 1862; see Wallace 1862).

This specimen was eventually obtained by the British Museum—in the catalog it is listed as *Lorius cyanauchen*, with "Mafoor" given as the locality. See "List of the Birds Lately Sent by Mr. A. R. Wallace from Dorey or Dorery, New Guinea," communicated by George Robert Gray (1808–1872), keeper of birds at the British Museum, to the Zoological Society of London (Gray 1859a).

## 108

[1]Wagtails & Larks walk, *[illeg.]* moving the feet alternately, Crows & [Jack] Daws swagger in their walk." (White.)

---

[2]Migratory birds which pass the winter only with us retire Northwards in the spring. This has been supposed a difficulty; But we must consider that their breeding place is their true home which they only leave in the extreme severity of winter for our milder insular climate,—where they can still find food.

---

[3]Plains in the tropics. Why are some covered with lofty forests,—others with grasses only.? This for a long time puzzled me, but I think I have found the explanation. It depends upon the more or less rapid rise of the plain up from the ocean. Where a forest covered plain gradually increases by deposit, as in a delta &c. the seeds of trees vegetate in the mud & each spot is occupied as it is formed,—but if a rapid elevation should convert a shallow sea into a muddy plain, the

1. An observation that originates with the celebrated *A Natural History of Selborne*, first published in 1789 by English naturalist and clergyman Gilbert White (1720–1793). Much of White's work is presented as letters to his fellow naturalists Thomas Pennant and the Honorable Daines Barrington; the birds Wallace cites comes from letter 84, to Barrington, dated August 7, 1778. He opens the letter with a declaration that Wallace surely agreed with:

> Dear Sir,—A good ornithologist should be able to distinguish birds by their air as well as their colours and shape, on the ground as well as on the wing, and in the bush as well as in the hand. For, though it must not be said that every species of birds has a manner peculiar to itself, yet there is somewhat in most genera at least that at first sight discriminates them, and enables a judicious observer to pronounce upon them with some certainty. Put a bird in motion, *Et vera incessu patuit.*

Wallace may have gleaned his information from White directly, but this particular passage on wagtails, larks, crows, and daws was repeated in several of James Rennie's books on bird biology, for example *The Faculties of Birds* (1835) and *Natural History of Birds: Their Architecture, Habits, and Faculties* (1839).

Wallace likely read one of the editions of Rennie that included the several kinds of bird-related observations (in particular locomotion, migration, and nest building) that Wallace discusses and extracts on recto 112–116.

2. White discussed "birds of Winter passage" and other examples of migration in his *Natural History of Selborne*. This was a special interest, and he kept detailed notes on the arrival and departure times and nesting periods of birds. The difficulty Wallace refers to is the debate over whether birds migrate or hibernate during the winter months; White was a "migrationist" (Lyle 1978), but others held that birds hibernated in sheltered places as mammals did. Wallace takes seasonal migratory movements as a given, and seems to be concerned more with using breeding range to define the "true home" of a species than with the supposed difficulties concerning migration.

Wallace's comment may have been taken from Rennie (1839, 297–298) or a later reprint. In his earlier (1835) work on birds, Rennie discussed this "difficulty" in the terms Wallace uses: "There are difficulties in accounting for the migration of the winter birds of passage." (279). In Rennie's 1839 volume, however, he offers the explanation that Wallace gives:

> The migration of winter birds of passage doubtless proceeds on the same general law as that which regulates the movements of those birds which spend the summer in England and leave it in winter. Birds which find the temperature and circumstances of summer in that country most congenial to their wants and habits, retire on the approach of severe weather to find something similar south; while others, which remain there in winter to avoid the extreme rigour of that season in the most northerly regions, return to their own country when that rigour has abated. (Rennie 1839, 297–298)

3. See annotations on the next page, where this passage is continued.

seeds of grasses carried by the wind or by birds
will vegetate <sup>immediately</sup> & *[illeg.]* cover it, ~~then~~ forbidding the
advance of forest. [1]Thus may be explained the
open plains of Celebes compared with the forest
plains of Borneo,—& the small flat grassy
valleys in Gilolo. Ground once taken possession
of by grasses cannot be reconquered by forest even
if surrounded by it. A clearing for a few years
only, will if left become forest, from roots &
seeds left in the earth, but if once covered
with grass all woody growth is kept down.
[2]Even the "llanos" of the Orinooko as compared with
the forest valley of the Amazon may perhaps
be accounted for in this manner. Climate or
soil will not account for it—these vary in
both districts & in parts of both are similar.
Gilolo—Jan.[uary] 20th 1858.

1. This grassy plain of Gilolo (Halmahera) is described in entry 125 of Wallace's second Malay Journal (LINSOC-MS178b): "For many miles into the interior extends a plain covered with coarse high grass thinly dotted here & there with trees, the forest country only commencing at the hills far in the interior. Such a place would produce few birds & no insects & is only good for deer which abound in the grassy plain."

The same page opens with Wallace reporting, "After a fortnight in Ternate I determined to visit the island of Gilolo for a month." This is likely when Wallace wrote the essay that he signed from Ternate. It is interesting to consider that Wallace was contemplating the curious phenomenon of the grassy tropical plains at this time, and the solution he came up with invokes a selection dynamic of sorts. Wind-borne seeds like those of grasses reach the newly exposed expanses first, resulting in a grassy ground cover inimical to the subsequent germination of the more slow-to-arrive seeds of woody species. The latter are thus outcompeted by the former by virtue of their dispersal and growth strategy.

2. The *llanos* of South America are vast grassy plains maintained by the annual flooding of the Orinoco and Arauca Rivers and their tributaries. These areas have standing water for long periods, which kills most terrestrial vegetation. In Wallace's hypothesis, the plains are colonized anew each year by species with the most mobile, far-reaching seeds—namely, grasses.

## 110

[1]From "Moniteur de l'Indes"—1847 "La Haye."
In E.[astern] Archipelago are about 600 species of Birds.
From Java nearly 300 are known
From Celebes Moluccas & N.[ew] Guinea 250.
Cockatoos not found W.[est] of Flores . . (wrong.)
Megapodius not in Timor. (query.)
Woodpeckers not found E.[ast] of Celebes. .

Mammalia of Archipelago. 175 species
    "   of Eastern part 50 sp.[ecies] 30 peculiar.
                  [2](Müller.)

[3]Land Shells in Java . .   "Zollinger's Monograph"
Helix . . . . . . . . 8          Jamaica only
                    ¹⁄₁₀th area has
   "   (Nanina) 6        more than 10 times as
Bulimus —   10   Total   many shells, 350 sp[ecies]
              32 species   being in a set sold
Succinea —   1            by Mr. Stevens.
                   Land & fresh water
                 or in proportion to area more
Clausilia —   7         than 100 to 1 !!!

Yet Java is richer in Birds Insects Plants & in fact all
other branches of Natural History.—

1. An encyclopedia of the Dutch East and West Indies published by La Haye, Belinfante Frères, each volume of which is in two parts: "Sciences, arts et industrie" and "Revue coloniale." Wallace may have mistaken the city of the publisher (La Haye) for the author; the encyclopedia was written by Philipp Franz von Siebold and Pieter Melvill van Carnbée. Its full title is *Le Moniteur des Indes-Orientales et Occidentales, recueil de mémoires et de notices scientifiques et industriels, de nouvelles et de faits importants concernant les possessions néerlandaises d'Asie et d'Amérique.*

2. The German naturalist Salomon Müller (1804–1864) published authoritative overviews of the fauna of the Malay Archipelago in papers in the *Annalen der Erdkunde* (1842) and the *Archiv fur Naturgeschichte* (1846). It is unclear if Wallace consulted one of these works, or perhaps one of the treatments of Müller's work by Heinrich Berghaus (*Atlas von Asia*, 1842–1843; *Physikalischer Atlas*, 1836–1848). Berghaus referred to plate 6, section 6, of his magnificent *Physikalischer Atlas* as a "Mammalian Monograph"; it included lists of mammalian species for major islands, based on Müller, and the accompanying text included a German translation of Müller (1842). The numbers given on this plate of the *Physikalischer Atlas* disagree slightly from those given by Wallace, however.

3. Refers to *Die Land- und Süsswasser-Mollusken von Java* (1849) by Henrich Zollinger (1818–1859) and Albert Mousson (1805–1890). Both Swiss, Zollinger was a botanist who collected in Java and environs between 1842 and 1848 (he later returned, and died there in 1859); Mousson was best known as a physicist but was also an ardent collector of freshwater mollusks.

In volume 2 of *The Geographical Distribution of Animals* (1876), Wallace commented on the "excessive and altogether unexampled productiveness of the West Indian islands in land-shells." His comparison here of the terrestrial mollusks of Java and Jamaica, and his expression of wonder at their 100-to-1 ratio as a proportion of area (if we can read the three exclamation marks as such), is very much in keeping with his on-going quest to delineate global biogeographical patterns.

## 111

*Definition of Birds –*
Vertebrated animals, with a double circulation & a double respiration, hot blood, anterior members forming wings & the skin bearing feathers. Milne Edwards
In feathers the dry matter inside the quill is the remains of the capsule from which the whole feather is developed.

Lower mandible articulated ↓ an intermediate bone (the tympanic bone). Upper mandible partly moveable. Each branch of lower mand. formed of two bones. The head articulated to the vertebral column by a single hemispheric Condyle,– below occipital hole.
12 ↓ 15 general. number of neck bones.
Bones of back, soldered together generally,–
In Ostrich partly moveable.
Ribs joined to sternum by bone not cartilage
 " have projecting pieces which lap over the one behind
clavicle shoulder blade
Coracoid analogous ↓ coracoid apophysis in Man.
furcula

---

[III]

[1]Definition of Birds.
Vertebrated animals, with a double circulation & a double respiration, hot blood, anterior members forming wings & the skin bearing feathers. Milne Edwards

---

In feathers the dry matter inside the quill is the remains of the capsule from which the whole feather is developed.

---

Lower mandible arcticulated *[sic]* to an intermediate bone (the tympanic bone). Upper mandible partly moveable. Each branch of lower mand.[ible] formed of two bones. The head articulated to the vertebral column by a single hemispheric Condyle,—below occipital hole.
12 to 15 general number of neck bones.
Bones of back, soldered together generally—
In Ostrich partly moveable.
Ribs joined to sternum by bone not cartilage
 " have projecting pieces which lap over the one behind.
*[Labeled pencil sketch]*
        clavicle or shoulder blade
        coracoid analogous to coracoid apophysis
                in man.
furcula

1. Wallace is here taking notes from the section entitled "The Class Birds" in Henri Milne-Edwards's *Manual of Zoology* (1856, 279–313). Milne-Edwards (1800–1885) was a well known French zoologist and physiologist who served for many years as editor of the *Annales des Sciences Naturelles.* His popular *Manual* first appeared under the French title *Introduction à la zoologie générale* (1851) and was republished in English in several editions, beginning with the 1856 translation by R. Knox.

Wallace's drawing is done from a figure in Milne-Edwards, labeled in the *Manual* as "Bones of the Shoulder and Sternum" (1856, 285).

## 112

From <u>Milne Edwards, Zoology.</u>
[1]Large air cellules in the thigh bones of <u>Ostrich</u>
(Proof that it is not to render the body <u>lighter</u> for
flight.)

[2]"Birds of passage feel at a certain period a
desire to change their place, as at other times
they feel a desire to prepare a nest, without
any mental effort, or any foreknowledge of
what use it will be to them." <u>I doubt this.</u>
<u>I deny this.</u>

---

[3]Notes & <u>Extracts from Bird</u> Architecture
<u>Puffin</u> burrows holes in the sand 8–10 feet deep—Powerful
bill & feet . . . <u>Sand martins</u> <u>peck</u> the sand
out clinging with their feet..
Penguins also burrow—very plentiful.
Burrowing Owls reside in burrows of marmots
& other burrowing animals.

———

Jackdaws have sometimes made nest in Rabbit burrows
<u>White</u>.

———

Lark makes nest of grass & lined with horsehair on the
ground sheltered by a Clod, or tuft of grass—

Waders, Rails &c., generally make nests on

1. Further notes from Henri Milne-Edwards's *Manual of Zoology* (1856).

Milne-Edwards states, "The extension of [air-filled cells] bears a ratio to the powers of flight of the bird: in the eagle they are found in all the bones; in the penguins the air is excluded from all, or from nearly all, the bones. The air is generally found to extend most into the bones chiefly used for locomotion, as the femur of the ostrich" (1856, 302).

Wallace interpreted Milne-Edwards's observation regarding the ostrich's leg bones as an answer to those who might cite air cells in the bones of birds as an example of good design for lightness and buoyancy in flight. This is an example of how structure can be too quickly interpreted in terms of design or adaptation for one function, when in fact it may exist for another function altogether or even be a vestige of history. If large, flightless ostriches have air cells in their bones, clearly air cells are not providentially designed to aid flight.

A similar argument is made by Darwin in *Origin of Species* (1859, 437), where he points out that skull sutures in mammals were often cited as an example of good design aiding parturition, but in fact the existence of such sutures in the skulls of birds and reptiles, which are not "born" but hatch from an egg, undermines that argument.

2. Wallace takes exception to the suggestion that birds "desire" to do these things, without either "any mental effort" or "foreknowledge of what use it will be to them." His criticism likely turns on such terms as "desire"; he accepted the concept of some instinctive or innate behavior but, far from seeing animals as automatons, he believes them to have a certain level of awareness of their behavior.

3. Here commences a long (six-page) series of notes on bird nests and nesting behavior. Most of the entries appear to have been copied out at more or less the same time. Wallace does not indicate the source, or sources, though the heading "Notes & Extracts *from* Bird Architecture" (italics added) suggests a source entitled "Bird Architecture." Indeed, James Rennie's *The Architecture of Birds* (1831) could be the source, or one of the many reprinted and revised editions of that work (e.g., *Natural History of Birds: Their Architecture, Habits, and Faculties* [1839], and *Bird-Architecture* [1844]). See also recto 108.

Note yet another parallel with Darwin, who also collected information on nest-building instincts of birds. His interest was in variation in nest building among birds of the same species. Darwin apparently did not consult Rennie, but used various other sources. A brief mention of the subject was given in the *Origin of Species* (1859, 212), but his *Natural Selection* manuscript included more than half a dozen pages of examples and references (Stauffer 1975, 498–505).

## 113

[1]the ground of grass reeds &c. in marshes—
The Eider duck and many other species
make their nests also on the ground in
remote islands or in marshes—
The anas sponsa Lin[naeus]—which perches on tree
makes its nest in hollow trees.
The redbreast & the song sparrow of America
sometimes build nests on or close to the
ground, sometimes on trees.

———

Cliff swallow & House martin, build nests
of earth or clay stuck together with
viscous saliva, & with little bits of straw
or grass intermingled—hay & feather
inside—Chimney swallow same
materials—but different shape.—

———

Flamingo heaps up mud & makes its
slight nest on the top so that it
can sit on it without fatigue
or breaking the eggs with its long legs.

1.  This sentence begins on the bottom of the previous page. The structure of Wallace's treatment of nest-building habits parallels that given by Rennie (1831; and later editions on bird architecture, e.g., 1839, 1844). Wallace categorizes the birds by type of nest built (though the categories are unnamed) in precisely the same manner as Rennie: mining birds, ground builders, platform builders, cementers, and so on.

The birds mentioned on this page are "ground builders" and "mason birds," in Rennie's treatment. Rennie (1839, 1844) even discusses the same species in much the same order as Wallace: rails, Eider ducks (*Somateria* spp.), the American wood duck (*Anas* [now *Aix*] *sponsa*, called the "summer duck" by Rennie), and the redbreast or European robin (*Erithacus rubecula*) in the chapter on ground builders. Cliff swallows (*Petrochelidon* spp.), house martins (*Delichon urbicum*), chimney swifts (*Chaetura pelagica*), and flamingoes (*Phoenicopterus* and *Phoenicoparrus* sp.) are discussed, in that order, in the chapter on mason birds.

## 114

Song thrush inside lined with fine papery
material formed of fragments of rotten
wood or cowdung..
Blackbird uses coarse clay, & inside
hay—
[1]Titmice build in holes in trees or walls
which they enlarge with their bills.
Woodpeckers all build in holes in
trees which they probably all bore
themselves—
Toucans also in holes but probably
old Woodpecker holes or those of
any rotten trees which they can enlarge.

———

[2]Psittacidae all build in holes—
cannot make them themselves. The
small white & yellow winged parroquets
of S.[outh] America build in the deserted
nests of a species of white ants which
are attached to trees & are very
friable & easily hollowed out.

1. The birds listed here—titmice, woodpeckers, and toucans—fall into Rennie's category of "carpenter birds."

2. Psittacidae is the parrot family. Parrots are not mentioned by Rennie in his discussion of carpenter birds, however, and Wallace may have been adding this information from his first-hand knowledge of South American and Southeast Asian psittacids.

## 115

[1]Pigeons build flat nests of sticks &
twigs—Eagles, Herons & Storks
also build flat nests of sticks
& coarse materials. Spoon bills
Ibisses & Boatbills also build on
lofty trees or river banks.

———————

[2]Rooks line their nests with fine roots & fibres
Crows with <u>wool</u> & <u>rabbits fur</u>.

———————

<u>Motacilla alba</u> builds on ground side of bank &c
lined with hair.
Humming bird of N.[orth] Am.[erica] nest of small pieces
of lichens outside glued with <u>saliva</u> <sup>like <u>swallows</u>!</sup>

———————

American chimney swallows' nest of small
twigs strongly glued together <sub>swifts</sub>. [3]<u>Acanthylis pelasgia</u>.

Wren builds generally entirely of moss.
Sometimes in hay with feathers &c.
lays from 6 to 18 eggs!

Goldfinch Chaffinch & Gold crested wren build
of moss lichens & wool.

———————

Magpie—large domed nest of thorns &
brambles &c. .

1. Rennie's chapters (in, e.g., 1831, 1844) on "platform builders" discuss these groups in much the same order as does Wallace.

2. Rooks and crows are included under the category of "felt-making" birds by Rennie, because these birds use soft, fine-textured building materials such as hair, spider silk, moss, and lichens. Wrens and magpies are "dome builders," as Wallace notes at the bottom of the page.

3. *Acanthylis pelasgia* is an obsolete name for the chimney swift (family Apodidae). The Latin name for this bird is now *Chaetura pelagica*. This species, abundant in eastern North America, has been aptly described as a "flying cigar" with its short oblong body and rapid wingbeats. The spindle or cigar shape of the body results from the bird's very short neck and very short legs. In fact they can seem legless on the wing—hence the family name Apodidae, the "footless" bird family. (See recto 72 for a discussion of a very different putatively "footless" bird, the greater bird of paradise.)

## 116

Caprimulgus makes no nest, eggs on ground.
Batrachostomus, a small nest on branch just large
enough to hold egg—

[1]Birds nests said to be built by instinct
because they don't improve—But they vary
according to circumstances & does man do
more. Any one race of man does not
improve in its architecture. The Indians
of each country have one mode of
building their houses which they vary
slightly according to circumstances, but
do not improve or decidedly change
any more than birds. It is only by
communication, by the mingling of different
races with their different customs, that
improvements arise & then, how slowly!
A race remaining isolated will ever remain
stationary, & this is the case with birds
Each species is generally confined to a
limited district in which the circumstances
are similar & give rise to no diversity of
habits. Man is spread over the Earth & by

*[The following sentence is written vertically in left margin
of page]*
The House martin of America has already taken to chimneys.

1. Here in the Species Notebook Wallace criticizes the claim that birds' nests are built by instinct because they do not "improve" (in unspecified ways). But they do *vary*, he points out, according to circumstances—as does human construction. His position is summed up in the opening paragraph of "The Philosophy of Birds' Nests," which he published in the July 1867 issue of the *Intellectual Observer*.

> Birds, we are told, build their nests by *instinct*, while man constructs his dwelling by the exercise of *reason*. Birds never change, but continue to build for ever on the self-same plan; man alters and improves his houses continually. Reason advances; instinct is stationary. This doctrine is so very general that it may almost be said to be universally adopted. Men who agree on nothing else, accept this as a good explanation of the facts. Philosophers and poets, metaphysicians and divines, naturalists and the general public, not only agree in believing this to be probable, but even adopt it as a sort of axiom that is so self-evident as to need no proof, and use it as the very foundation of their speculations on instinct and reason. A belief so general, one would think, must rest on indisputable facts, and be a logical deduction from them. Yet I have come to the conclusion that not only is it very doubtful, but absolutely erroneous; that it not only deviates widely from the truth, but is in almost every particular exactly opposed to it. I believe, in short, that birds do *not* build their nests by instinct; that man does *not* construct his dwelling by reason; that birds do change and improve when affected by the same causes that make men do so; and that mankind neither alter nor improve when they exist under conditions similar to those which are almost universal among birds. (Wallace 1867b, 413)

It is very instructive to explore Wallace and Darwin's respective approaches to the question of instinct as it pertains to birds. Darwin thought in terms of heritable variations in behavior subject to selection: "The nest of each bird, wherever placed & however constructed good for that species under its own conditions of life; and if the nesting-instinct varies ever so little, when a bird is placed under new conditions, & the variations can be inherited . . . then natural selection in the course of ages might modify and perfect almost to any degree the nest of a bird in comparison with that of its progenitors in long past ages" (Stauffer 1975, 500).

Wallace, in contrast, focused on learning and experience, questioning whether the cases often cited as evidence of instinct really were. Here in the Species Notebook he articulates what we could describe as cultural transmission of knowledge as a means of birds' "improving" (or at least altering) their nest building. He then suggests: "It is only by communication, by the mingling of different races with their different customs, that improvements arise & then, how slowly!" Note his curious parallel between the way humans improve their habitations with the way birds improve theirs: "A race remaining isolated will ever remain stationary, & this is the case with birds. Each species is generally confined to a limited district in which the circumstances are similar & give rise to no diversity of habits."

## 117

[1]his migrations introduces the customs of one
region among the inhabitants of another
Hence change of habits,—hence improvement
in arts.—

———

[2]The general characters of a bird's nest can ~~generally~~ in most cases
be referred to its organization. Thus web footed
birds, generally form rude nests on the ground or
on rocks,—some which habitually perch on trees
Construct their nests there. The true <u>Caprimulgi</u>
generally make no nest because both bill &
feet are quite unadapted for grasping or
arranging materials. Large & heavy birds again
make nests rude in proportion. The delicacy
& perfection of a birds nest is generally in
exact proportion to its size [activity] & to the perfect
structure of its bill & feet.

The materials of which a nest is formed will
be determined by the habits of the species & the
consequent facility for obtaining one sort rather
than another. The <u>rook</u> feeding in ploughed

1. Completing the thought from the previous page, Wallace seems to be applying the same kind of "cultural exchange" found between peoples to the way in which birds improve nest architecture—both birds and people make changes and improvements to their "dwellings" when they are exposed to changing conditions. Later we will see Wallace revisit this subject and argue against the occurrence of instinct in humans (recto 166–172).

With his focus on experience as an impetus for behavioral change, Wallace does not address the question of whether or to what extent such behaviors might change through selection, though the bird entries, beginning with "definition of birds" (recto 111) through bird architecture (recto 112–119), perhaps were written soon after his discovery of natural selection. This is not to say that he believed that nest building was not subject to modification by selection. He was thinking deeply about the nature of instinct in relation to experience and physiological and environmental need. He brought this up with Darwin in a now lost letter; Darwin replied (January 25, 1859): "I am glad to hear that you have been attending to Bird's nest; I have done so, though almost exclusively under one point of view, viz to show that instincts vary, so that selection could work on & improve them" (Darwin Correspondence Project, letter 2405). As mentioned in the notes for recto 112, Darwin discussed the subject at length in *Natural Selection* but gave only a brief mention in the *Origin of Species* (Darwin 1859, 212).

Around the same time he wrote Darwin, Wallace also touched on the subject in "The Ornithology of Northern Celebes:"

It has been generally the custom of writers on natural history to take the habits and instincts of animals as the fixed point, and to consider their structure and organization as specially adapted to be in accordance with them. But this seems quite an arbitrary assumption, and has the bad effect of stifling inquiry into those peculiarities which are generally classed as "instincts" and considered as incomprehensible, but which a little consideration of the *structure* of the species in question, and the peculiar physical *conditions* by which it is surrounded, would show to

be the inevitable and logical result of such structure and conditions. I am decidedly of opinion that in very many instances we can trace such a necessary connexion, especially among birds, and often with more complete success than in the case which I have here attempted to explain. For a perfect solution of the problem we must, however, have recourse to Mr. Darwin's principle of "natural selection," and need not then despair of arriving at a complete and true "theory of instinct." (Wallace 1860d, 145–146)

Noteworthy here are, first, his comment that through natural selection "a complete and true 'theory of instinct'" will be arrived at, and second, his reference to natural selection as "Mr. Darwin's principle," despite his independent derivation of the concept.

Eight years later, in "The Philosophy of Birds' Nests," Wallace was still troubled by the nature of instinct in relation to experience. He did not bring natural selection in at all, except perhaps indirectly by suggesting in the closing paragraph that the mental faculties shown by birds in nest construction "are, essentially, imitation, and a slow and partial adaptation to new conditions." He concluded: "that the existence of true instinct may be established in other ways is not improbable, but in the particular case of birds' nests, which is usually considered one of its strongholds, I cannot find a particle of evidence to show the existence of anything beyond those lower reasoning powers which animals are universally admitted to possess" (Wallace 1867b, 420).

2. These observations were incorporated into "The Philosophy of Birds' Nests" paper, one conclusion of which is that "the delicacy and perfection of the nest will bear a direct relation to the size of the bird, its structure and habits."

The following year Wallace published a follow-up paper entitled "A Theory of Birds' Nests: Shewing the Relation of Certain Sexual Differences of Colour in Birds to Their Mode of Nidification" (Wallace 1868). This subject was related to Wallace's ongoing argument with Darwin over the role (or not) of sexual selection in bird color and ornamentation.

## 118

[1]fields & pastures, & digging in the earth with its
bill, lines the nest with the rootlets & fibres
which it must daily meet with. The
Crow, feeding on carrion & small animals
& frequenting hills & sheep pastures & rabbit
warrens uses wool & hair & rabbits fur to
make a soft bed for its eggs & young ones.
The wren hunting insects in hedgerows & thickets
& about mossy banks constructs its nest
almost entirely of the material it must have
most constantly at hand, moss;—& its
fine bill, strong legs & great activity lead it
to form a neat & compact structure.

---

[2]At Dorey a young hen began laying probably for the
first time. I took the 2nd egg constantly till she
began to sit on the last one & lay no more when
I took that also—She still continued to sit
upon an empty nest (in a basket in the house) for
several days (10–12 till I left) with the greatest assiduity, quitting
only a few minutes each day. This was not
true instinct certainly. It only shewed some
*[Continued at the bottom of recto 119]*

1. This passage begins on the bottom of the previous page, with the idea that birds' nest-building behavior reflects their needs and their experience with the materials in their particular environment. Wallace thought a great deal about the concept of instinct—whether it existed, or could be distinguished from experience.

Wallace first explored the instinct issue in the "Philosophy of Birds' Nests" paper and its revisions, and he revisited the topic in 1883 in his "Comments on an Unpublished Paper on Instinct by Darwin." The unpublished paper by Darwin was posthumously read at the December 6, 1883, meeting of the Linnean Society by George Romanes, after which Wallace offered his perspective on the paper. The crux of his critique, based on published comments about the meeting, was his contention that there was "a considerable amount of obscurity about what was really meant by instinct. It include[s] a considerable number of phenomena, some of them mere results of muscular and nervous coordination, and in other cases simply the results of observations and experience" (Wallace 1883).

It is evident in the Species Notebook that his doubts over the nature of instinct were already well formed by the mid-1850s. Besides the undated entries on instinct in this notebook, there is an entry suggesting an experimental solution to the question dated May 1861 in his Insect Register notebook (WCP4767):

Instinct. Can a single case be shewn of an animal performing any complex act no part of which it has ever seen performed — ? or without having seen the result. For example. Will any bird [?] bred from the egg, build a nest of the same form the same material & in the same situation, as others of its species; if turned for instance into a greenhouse or a glass covered water garden, with choice of materials & situations. This is a test wh[ich] will decide the question of birds building by instinct.

Wallace paraphrased himself on this point six years later in the "Philosophy of Birds Nests" paper:

But, it is objected, birds do not learn to make their nest as man does to build, for all birds will make exactly the same nest as the rest of their species, even if they have never seen one, and it is instinct alone that can enable them to do this. No doubt this would be instinct if it were true, and I simply ask for proof of the fact. This point, although so important to the question at issue, is always assumed without proof, and even against proof, for what facts there are, are opposed to it. Birds brought up from the egg in cages do not make the characteristic nest of their species, even though the proper materials are supplied them, and the experiment has never been fairly tried of turning out a pair of birds so brought up into an enclosure covered with netting, and watching the result of their untaught attempts at nest-making. (Wallace 1867b, 417)

2. Wallace was in Dorey (Manokwari), western New Guinea, from April 11 to July 29, 1858, after which time he returned to his base at Ternate, arriving in mid-August. He departed again, for Batchian (Bacan), on October 9. In the interval he received word from Darwin and Hooker of the momentous July 1, 1858, reading of his essay, "On the Tendency of Varieties to Depart Indefinitely from the Original Type." Before his departure for Bacan he penned a letter to his mother, telling her the exciting news:

I have received letters from Mr Darwin & Dr Hooker, two of the most eminent Naturalists in England which has highly gratified me. I sent Mr Darwin an essay on a subject in which he is now writing a great work. He shewed [sic] it to Dr Hooker & Sir C[harles] Lyell, who thought so highly of it that they immediately read it before the "Linnean Society." (Wallace Correspondence Project, WCP369)

## 119

[1]True Coal is found in the <u>lower Silurian</u> formation
in Portugal—Anthracite—6 feet thick—
d[itt]o—in Virginia in the <u>lias formation</u>, 36 feet thick
bituminous—
[2]300 species of carboniferous plants known—
nearly one half are ferns—
    Johnstone, Nat.[ural] Hist.[ory] of E.[astern] borders.

---

[3]Pinnated grouse—males strut & dance
in certain spots like the Rupicola.

---

[4]Note: get information from <u>Wilson</u> & <u>Audubon</u> as to variation of
nest of same species of birds having an extended
range, where different materials will
be found & different situations available.[4]

---

[5]The Tanysiptera of Amboina builds it[s] nest
in abandoned ant's nests of papery or earthy
texture in which it can easily make
a hole.

---

[Continued from recto 118]
sensation which made <u>sitting</u> necessary to the
bird's comfort. After having once sat & hatched chickens
it is no longer <u>instinct</u> but <u>experience</u>.

1. From *The Botany of the Eastern Borders, with the Popular Names and Uses of the Plants, and of the Customs and Beliefs Which Have Been Associated with Them* (1853), by English naturalist George Johnston (1797–1855) of Berwick-upon-Tweed. Volume 1 is dedicated to botany, while volume 2 treats fossil flora. Wallace's notes here on the coal beds are drawn from the fossil volume:

> Rarely, however, has true coal been seen among the oldest rocks; indeed, the only instance we know of is that described by Mr. Sharpe, who has proved that anthracitic coal, 6 feet thick, in Portugal, belongs to the Lower Silurian formation. The most remarkable coal in the Secondary series is a rich bituminous bed, 36 feet thick, in Eastern Virginia, which Sir Charles Lyell has shown to be not more ancient than the Lias formation. (Johnston 1853, 2: 291–292)

Johnston's information regarding Lyell likely came from an 1847 paper Lyell read to the Geological Society of London: "The Structure and Possible Agency of the Coal Field of the James River, North Richmond, Virginia" (Lyell 1847, 261–282).

2. From further along in Johnston: "About 300 species of plants from the Carboniferous formation of Great Britain have been described; but, with the exception of Conifers and Ferns, few of them have a close affinity to existing families of plants" (1853, 2: 296).

3. The pinnated grouse is also known as the heath hen or greater prairie chicken, now *Tympanuchus cupido*. These birds, once widespread in eastern and central North America, are known for their spectacular breeding displays, as alluded to by Wallace here. "Rupicola" (also mentioned on recto 89) is the Guianan cock-of-the-rock, *Rupicola rupicola*, a South American passerine Wallace had collected there, whose brilliant orange males engage in similar lekking displays.

4. Wallace appears to be seeking evidence for variation in nest-building behavior in relation to the birds' finding themselves in novel environments (such as occurs with a range expansion). This is part of Wallace's inquiry into instinct in birds (see recto 116–118). Note again that Darwin sought precisely the same information about nest building, albeit for somewhat different reasons. Darwin was interested in variation in instinct because this would be consistent with his argument that instincts evolve by natural selection. Nesting behavior was often cited as an example of an instinctive and invariant behavior; Darwin took a special interest in showing that this behavior is more variable than many believed at the time. He wrote:

> Some degree of variation in instincts under a state of nature, and the inheritance of such variations, are indispensable for the action of natural selection . . . I can only assert, that instincts certainly do vary—for instance, the migratory instinct, both in extent and direction, and in its total loss. So it is with the nests of birds, which vary partly in dependence on the situations chosen, and on the nature and temperature of the country inhabited, but often from causes wholly unknown to us: Audubon has given several remarkable cases of differences in nests of the same species in the northern and southern United States. (Darwin 1859, 211)

Wallace also notes that he should consult Audubon, along with Alexander Wilson.

5. *Tanysiptera* are the paradise kingfishers (Alcedinidae), tree kingfishers of the Australasian region known for nesting in tree holes or arboreal termite nests (Wallace seems to have made an error in referring to ants' nests). In Amboyna (Ambon) Wallace collected the racquet-tailed kingfisher, *Tanysiptera nais*, which he described as "the largest and handsomest" of the paradise kingfishers.

[120]

[120]

[1]Proportionate excellence of fruits of the
  Eastern Archipelago. . . A[lfred] R[ussel] Wallace . .

1.  Mango (Mangifera indica)   Anacardiaceae.
    Mangustan (Gurcinia mangustana)   Guttiferae
    Lansat (Lansium sp.[ecies])   Meliaceae
    Durian. (Libethmus duris)   Sterculiaceae.
2.  Custard apple (Anona ~~muricata~~ squamosa)   Anonaceae
    Pine apple (Ananas sativa)
    Banana. (Musa paradisiasa)
3.  Jack (Artocarpus integrifolia)
    Sour sop (Anona muricata[)]
    Rambutan (Nephelium lappaceum)
    Shaddock (Citrus decumana)
4   Papaya. . (Carica papaya)
    Cashew (Anacardium orientale)
    Guava. (Psidium guava)
5.  Jambon (Eugenia sp.[ecies] variae)
    Blimbaig (Averhoa sp.[ecies] variae)

1. The durian is in Wallace's top tier of excellent fruits of the region. In his article "On the Bamboo and Durian of Borneo" in *Hooker's Journal of Botany* (Wallace 1856d) he calls durian the "king of fruits" in the opening sentence, and blends a studied scientific description of the fruit with a lyrical description of the sensory experience of consuming one. This is reprinted in *Malay Archipelago*:

> The Durion [*sic*] grows on a large and lofty forest tree, somewhat resembling an elm in its general character, but with a more smooth and scaly bark. The fruit is round or slightly oval, about the size of a large cocoanut, of a green colour, and covered all over with short stout spines the bases of which touch each other, and are consequently somewhat hexagonal, while the points are very strong and sharp. It is so completely armed, that if the stalk is broken off it is a difficult matter to lift one from the ground. The outer rind is so thick and tough, that from whatever height it may fall it is never broken. From the base to the apex five very faint lines may be traced, over which the spines arch a little; these are the sutures of the carpels, and show where the fruit may be divided with a heavy knife and a strong hand. The five cells are satiny white within, and are each filled with an oval mass of cream-coloured pulp, imbedded in which are two or three seeds about the size of chestnuts. This pulp is the eatable part, and its consistence and flavour are indescribable. A rich butter-like custard highly flavoured with almonds gives the best general idea of it, but intermingled with it come wafts of flavour that call to mind cream-cheese, onion-sauce, brown sherry, and other incongruities. Then there is a rich glutinous smoothness in the pulp which nothing else possesses, but which adds to its delicacy. It is neither acid, nor sweet, nor juicy, yet one feels the want of none of these qualities, for it is perfect as it is. It produces no nausea or other bad effect, and the more you eat of it the less you feel inclined to stop.

> In fact to eat Durions is a new sensation, worth a voyage to the East to experience. (*Malay Archipelago*, 85–86)

Wallace acknowledges the virtues of several other tropical fruits, but concludes: "If I had to fix on two only . . . I should certainly choose the Durion and the Orange as the king and queen of fruits." This is high praise for a fruit that produces a stench so offensive when ripe that it is banned from public buses!

As an aside, in his account of the durian in *Malay Archipelago* Wallace finds a way to make a cautionary point about anthropocentrism. Describing how durians and other heavy fruits borne on lofty trees sometimes fall and injure people, Wallace points out that the blows are rarely fatal: "Death rarely ensues, the copious effusion of blood preventing the inflammation which might otherwise take place" (*Malay Archipelago*, 86). Cold comfort, perhaps, but the main point is what comes next: "Poets and moralists, judging from our English trees and fruits, have thought that small fruits always grew on lofty trees, so that their fall should be harmless to man, while the large ones trailed the ground"—an inaccurate view of nature straight out of the natural-theology tradition. Wallace concludes, "From this we may learn two things: first, not to draw general conclusions from a very partial view of nature; and secondly, that trees and fruits, no less than the varied productions of the animal kingdom, do not appear to be organized with exclusive reference to the use and convenience of man" (87).

Papers on Natural History Geography &c..
by Alfred Russel Wallace
In "Annals & Magazine of Natural History."
1. On the Cephalopterus ornatus, Umbrella Bird
? Birds of Malacca _____ 1855.
2. On Law of Succession of Species. (Sept. 55.)
3. On a Natural Arrangement of Birds (Sept. 56.)
4. On the Orang utan or Mias of Borneo.
5. On the Habits of the Orang utan of Borneo.
6. Some Account of an infant Orang Utan.
7. On the Natural History of the Aru Islands.
8. On the Great Bird of Paradise.
9. On an error in classification of Psittacidae
10. On proposed alteration of name of Gracula pectoralis. Wall.
11. Answer to Rev S. Haughton on Bees' Cell & Darwin.

Sclater's "Ibis".
1. On Ornithology of N. Celebes & habits of Megacephalon.
2. do. of Ceram & Waigiou
3. do. of Timor.
4. Letter about Bouru.

Proc. Zool. Soc. (see)

---

[1]Papers on Natural History Geography &c..
by Alfred Russel Wallace
In "Annals & Magazine of Natural History."
1. ~~On the Cephalopterus ornatus, Umbrella Bird~~ [1852?]
   ? Birds of Malacca _____ 1855.
2. On Law of Succession of Species. (Sept.[ember] 55.)
3. On a Natural Arrangement of Birds. (Sept.[ember] 56.)
4. On the Orang utan or Mias of Borneo.
5. On the Habits of the Orang utan of Borneo.
6. Some Account of an infant Orang Utan.
7. On the Natural History of the Aru Islands.
8. On the Great Bird of Paradise.
9. On an error in classification of Psittacidae
10. On proposed alteration of name of Gracula pectoralis. Wall.[ace]
11. Answer to Rev.[erend] S.[amuel] Haughton on Bees' Cell & Darwin.

[2]Sclater's "Ibis".
1. On Ornithology of N.[orth] Celebes & habits of Megacephalon.
2. d[itt]o. of Ceram & Waigiou
3. d[itt]o. of Timor.
4. Letter about Bouru.

---

[3]Proc.[eedings] Zool.[ogical] Soc.[iety] (see)

1. Perhaps flush with the success of his Ternate essay, Wallace here takes stock of his scientific accomplishments. Close inspection of the entries reveals that an initial set of entries was made and then later added to as subsequent papers were published. The *Annals and Magazine of Natural History* entries up to number 9 appear to have been made at the same time; paper number 9 dates to February 1859. The last two entries in darker ink, number 10 concerning the proposed change in name of the bird *Gracula pectoralis* and number 11 in response to Samuel Haughton's attack on Darwin concerning honeybee cells, were both published in 1863.

2. *Ibis*, the journal of the British Ornithologists' Union, was founded in 1859 by Philip Lutley Sclater (1829–1913), who also served as editor. Sclater, a lawyer by profession, was an accomplished ornithologist and served as secretary of the Zoological Society of London from 1860 to 1902. Sclater's 1858 paper in the *Proceedings of the Linnean Society* on the zoological regions of the world prompted a critique from Wallace published in *Ibis* as "Letter from Mr. Wallace concerning the Geographical Distribution of Birds" (Wallace 1859c).

Quite a few letters from Wallace that Stevens extracted for publication in *Ibis* are not listed here. The first letter that is listed, "On the Ornithology of N. Celebes," was published in April 1860, the second (on Ceram and Waigiou) and third (on Timor) in 1861, and the fourth is Wallace's "List of Birds Collected in the Island of Bouru," published in 1863. Wallace departed the archipelago in February of 1862, arriving back in England by April 1 of that year. It is likely that the *Ibis* entries are later additions made when he was back in England.

3. Wallace had several papers in the *Proceedings of the Zoological Society of London,* including an extract of a letter to Stevens from Batchian (Bacan), Moluccas (Wallace 1859), and an extract of a letter to John Gould regarding Wallace's standardwing, the bird of paradise named for him by Gray (Wallace 1860b). Soon after returning to England Wallace read the first of many papers at the Zoological Society: his "Narrative of Search after Birds of Paradise," read at the meeting of May 27, 1862, was a topic of great interest, as he had returned with two living lesser birds of paradise (*Paradisea minor*).

[122]

In the [1]"Transactions of Ent.[omological] Soc.[iety] of
    London."
1.    On the Habits of the Butterflies of the Amazon
2.    On Insects used for food on the Amazon.
3.    Description of a new Ornithoptera from Borneo.
    (in Proceedings of Ent.[omological] Soc.[iety])
4.    [2]On the habits of an Ornithoptera in the Aru Islands.
5.    On a case of doubtful priority in Nomenclature
    Ornithoptera panthous for O.[rnithoptera] remus. .
6.    On Sexual characters in genus Lomaptera.
7.    [3]Note on habits of Scolytidae &c. .

[4]Proceedings of Linnaean Society.
1.    On varieties & Species.
2.    On Zoological Geography of Malay Archipelago.

1. The papers listed for the *Transactions of the Entomological Society of London* begin with two from his Amazonian travels (numbers 1 and 2 were published in 1853). The last two papers were published in 1860 (as "Note on the Habits of Scolytidae and Bostrichidae" and "Note on the Sexual Differences in the Genus *Lomaptera*")—the latter published in the *Proceedings* rather than *Transactions of the Entomological Society*).

2. Wallace's observations on the Aru *Ornithoptera* caterpillars and their habits that are found in this paper (Wallace 1858e) were originally recorded in the Species Notebook—see verso 34–35.

3. Scolytidae and Bostrichidae are bark beetle families. This paper includes one of the two instances Wallace observed of ill-fated attempts by bark beetles to burrow into sap-exuding trees. (The other, witnessed in New Guinea, is discussed under "instinct at fault" on verso 8.) The hapless insects may have somehow chosen the wrong tree species, or their attempts were ill-timed. In any case the error proved fatal. In the *Transactions* paper Wallace related the following:

> I had cut down a large tree in the Aru Islands, of a kind containing abundance of milky sap, which hardened on exposure to air very much like "gutta percha." A few days after I found on it dozens of a species of Scolytidæ with their abdomens protruding from the holes they had bored, but all dead. With a remarkable deficiency both of *instinct* and *reason,* the little creatures had dug their own graves, and were all glued fast by the hardening of the milky sap. In a few days more there were hundreds so killed; indeed it appeared as if not one escaped. It seems evident, therefore, that this tree could not have been the proper food of this species, or the right place to deposit its eggs. I have since observed exactly the same occurrence in another locality. (Wallace 1860a, 218–219)

Wallace reasoned that these beetles ordinarily attacked dead wood, "for if their proper and usual food was living wood, why should they all rush as to a feast directly a tree is cut and begins to dry?" His explanation of why they do this is remarkably close to the modern understanding: these beetles "attack wood in which the vital forces have ceased to act; and they are able to detect this before any external change has taken place." The conclusion he draws is insightful: "It is only at a later period that we observe the tree to be suffering, and in the parts most affected we discover the *Scolyti* to have been at work, and erroneously impute the mischief to them . . . It now becomes a question whether the supposed criminals are not really our benefactors,—teaching us, by their presence, that there is something wrong, before we could otherwise perceive it (ibid., 219–220). (See Costa 2006, chapter 15, for a discussion of bark beetle biology.)

4. The first of these papers is Wallace's watershed essay from Ternate (Wallace 1858f). It is interesting to see it listed amid the other papers as simply "On varieties and Species," belying its significance. The other paper, "On the Zoological Geography of the Malay Archipelago," was read by Darwin at the Linnean Society meeting of November 3, 1859, and published the following year (Wallace 1860c).

Note, by the way, Wallace's misspelling of the society's name. "Linnaean" refers to Linnaeus and his works and nomenclatural system, whereas "Linnean"—used in the name of several societies named in honor of Linnaeus, of which the Linnean Society of London was the first—is derived from the name Carl von Linné, which Linnaeus adopted after being ennobled in 1761. The Linnean Society of London was founded in 1788 for "the cultivation of the Science of Natural History in all its branches." It is the home of most of Wallace's surviving field notebooks, among many other precious manuscripts.

In Transactions of Royal Geog. Society.
1. On the Rio Negro..
2. On the Physical Geography of Malacca.
3. Notes of a journey up the Sadong River in
   N. W. Borneo.
4. On the Physical Geography of the Aru Islands.
   Proc. G. Soc. Vol. 2. p. 163. 1857
5. Notes of a voyage to New Guinea.
6. On Native trade with New Guinea.
7. On the physical Geography of the Malay Archipelago.

Hookers Journal of Botany
"Bamboo & Durians".
In Zoologist.
1. Note on Geographical & Permanent Varieties
2. On the Entomology of the Aru Islands—
3. Who are the Humming-birds' Relations.

[123]

In [1]Transactions of Royal Geog.[raphical] Society.
1.  On the Rio Negro. .
2.  On the Physical Geography of Malacca.
3.  Notes of a journey up the Sadong River in
    N. W. [northwest] Borneo.
4.  On the Physical Geography of the Aru Islands.
    Proc.[eedings] G.[eographical] Soc.[iety] Vol.[ume] 2. p.[age] 163. 1857
5.  Notes of a voyage to New Guinea.
6.  On Native trade with New Guinea.
7.  On the physical Geography of the Malay Archipelago.

Hookers Journal of Botany.
"Bamboo & Durians".
In Zoologist.
1.  Note on Geographical & Permanent Varieties Jan.[uary] 58/.
2.  On the Entomology of the Aru Islands—       "   "
3.  Who are the Humming-birds' Relations.

1. The Royal Geographical Society, founded in 1830, was instrumental in helping to launch Wallace's Southeast Asian expedition. The distinguished Scottish geologist Roderick Murchison (1792–1871) served as president of the society from 1851 to 1853. This was fortuitous for Wallace, as the kindly Murchison was impressed with the young man's energy and accomplishments in South America. Wallace had arrived home from his four-year sojourn in Amazonia on October 1, 1852, having survived the burning and sinking of the ship that was to have transported him and his extensive notes and collections home. In the disaster he lost what he later described as nearly everything from the two most interesting years of his travels. He made the most of what little was salvaged as he scrambled to abandon the burning ship; among the scant few fruits of his labors was his paper "On the Rio Negro," which he read at the June 13, 1853, meeting of the Royal Geographical Society.

The resulting published work (in the *Journal*, not *Transactions*, of the Royal Geographical Society; Wallace 1853b) included a detailed, beautifully executed map of the Rio Negro, one of the great tributary rivers of the Amazon. Wallace's considerable surveying skills, learned as an apprentice to his brother William, are very much in evidence in this map, the original of which is now part of the Wallace Collection at the Natural History Museum in London.

Wallace wrote Murchison on June 30, 1853—not coincidentally, days after communicating his Rio Negro paper—requesting the society's support in the form of a recommendation to the government for free passage to the Malay Archipelago, the site of his next proposed expedition. He wrote the letter in the third person: "His chief object is the investigation of the Natural History of the Eastern Archipelago in a more complete manner than has hitherto been attempted; but he will also pay much attention to Geography, and hopes to add considerably to our knowledge of such of the islands as he may visit." Wallace then reminded Murchison of both the successes and misfortunes of his South American expedition:

As some guarantee of his capabilities as a traveller, he may perhaps be excused for referring to his recent travels for nearly five years, in South America, where alone and unassisted he penetrated several hundred miles beyond any former European traveller, as shewn by the Map and description of the Rio Negro, which he has had the honour to lay before the Society at its last meeting . . .

. . . During his travels in South America he relied entirely on his duplicate collections in Natural History to pay his expenses, and he shall follow the same plan in his proposed journey. On his homeward bound voyage from Parà he suffered the loss of a very extensive and valuable collection together with all his books & instruments by the burning of the Brig "Helen" of Liverpool (in which he was a passenger) on the 6th of August 1852,— a loss which has rendered necessary the present application to the Royal Geographical Society. (Wallace Correspondence Project, WCP4308)

Murchison and the Royal Geographical Society council approved Wallace's request, and after some false starts he got under way at last on March 4, 1854, with first-class passage on a Peninsular and Oriental (P&O) steamer headed for Cairo, then overland by coach to Suez, and finally on another steamer (again with a first-class ticket) to Singapore, arriving on April 20, 1854.

## 124

[1]Plan for the arrangement of my Collection
of Coleoptera,—on return to England.—

? best a general cat.[alogue] for each locality using tickets already attached &
adding others for say first all cincindela & longicorns & so on all other
families. The whole cat.[alogue] would come in two such vol.[umes] as this giving a line
   per. sp.[ecies]

the numbers all following in regular succession would make ref.[erencing] easier.

1. Get Locality tickets printed. ? enough for Duplicates — also

2. Taking one family or group at a time, go
regularly through it, substituting the new
locality tickets, (after relaxing & setting the
specimens as to remain in the cabinet,) &
adding to the tickets the numbers referring
to my notes, with others continuing the
consecutive series for each locality.

*[The following is written vertically in left margin of page]*
perhaps better to keep old tickets
first adding no.[s] [numbers] & changing for new ones
when placed finally in Cabinet.  No. [Number]

3. [2]Arrange the families in boxes, at first all the specimens
from different localities being placed together
in a regular order, that is wh.[ere] the places
were visited, modified geographically from W.[est] to
   E.[ast]

4. [3]Form a catalogue for each family, the
numbers following consecutively (though of course
a broken series) each species standing
under the number of the locality where whence the last
first found, series of specimens or first in Alphabet order. with references
   to the others; viz.

*[The following is written vertically in left margin of page]*
Probably the best
   plan
or see above

Singepore
~~Sar.[awak]~~.           67. Eurycephalus maxillosus. Oliv.[ier]
(Sar.[awak] 26) B.[ritish] M.[useum] Cat.[alogue] p.[age] 136.
(Cerambyx)
"    41. Cincindela aurulenta. Linn.[aeus] (Sar.[awak] 535.
Lomb.[ock] 30)
Dej.[ean] Sp.[ecies] Gen.[us] p.[age]___...Smith.
Mon.[ograph] p.[age].....
varieties also should be noted.

---

1. Here we see Wallace planning ahead for the best way to organize and catalog his vast collections—Coleoptera, or beetles, in this case. We saw on recto [b-1] his draft insect cabinet designs. The notes here address practicalities like proper labeling ("locality tickets"), and having each specimen numerically keyed to his notes. While most of his specimens were sent to Stevens for sale, he retained a sizable collection for personal study: by his estimate, some 3,000 bird skins (representing about 1,000 species) and 20,000 beetles and butterflies (about 7,000 species), not to mention hundreds of ants, land snails, and mammals (George 1964, 48; Raby 2001, 165). The sheer numbers demanded some kind of systematic approach to arranging and cataloging his material.

2. Note the penciled comment to arrange specimens in a regular order, from west to east. This is reflected in the butterfly species lists on verso 72–74, which are tabulated by genus, species, and locality, with species totaled for west versus east—the biogeography of the archipelago always an integral part of Wallace's investigations. His collections were too extensive for one person to study, however, so various taxonomic groups were given to specialists to describe.

One group that he made his special study, however, was the swallowtail butterfly family Papilionidae. His paper "On the Phenomena of Variation and Geographical Distribution as Illustrated by the Papilionidae of the Malayan Region" was read at the Linnean Society on March 17, 1864, and published the following year in *Transactions of the Linnean Society of London* (Wallace 1865a). Darwin wrote to Wallace in January 1866: "I finished yesterday your paper in the [*Linnean Transactions*]. It is admirably done. I cannot conceive that the most firm believer in Species could read it without being staggered. Such papers will make many more converts among naturalists than long-winded books such as I shall write if I have strength" (Wallace Correspondence Project WCP1868).

The paper generated much interest; it was later reprinted in expanded form as "The Malayan Papilionidae or Swallow-Tailed Butterflies, as Illustrative of the Theory of Natural Selection," forming chapter 4 in *Contributions to the Theory of Natural Selection* (Wallace 1870). In this way Wallace realized his wish, stated in a letter to Bates back in late 1847, to "take some one family to study thoroughly, principally with a view to the origin of species" (Wallace Correspondence Project, WCP348). See James Mallet's lucid discussions of Wallace's paper and its role in clarifying the species concept (Mallett 2008, 2009).

3. Apart from the arrangement of the insects in drawers, a catalog had to be prepared to keep track of what species was found where. Here he suggests a primary arrangement by family, then locality, then species numbered consecutively (his discussion is continued on the next page). A further layer of complexity to his collections was duplicate species, with specimens taken in multiple localities. This is exemplified by these two beetles near the bottom of the page listed under "Singapore" but parenthetically including numbers for specimens taken in Sarawak and Lombock (Lombok).

perhaps all the no.s [numbers] had better be inserted in each locality,
in case of local varieties or closely allied species. Yes.

[1]Under each locality therefore would be found
only those species first found there. The
complete catalogue would go in a volume like
this or at most 2 such, & would be most useful to insert names
when taking portions of collection to compare
at museums at home or abroad..

The names, & authorities should be inserted as
intended to be used in naming the collection,
the generic name under which the first describer
classed the species being always given. & a reference to the best
desc.[riptio]n & figure. The

labels for the Cabinet can then be easily written
and applied according to the numbers on the locality
tickets. Small easy families as the Cincindelae may have
a systematic catalogue at once formed.

*[The following is written vertically in left margin of page]*
Place localities in Alphabetical order?. with cut index as in
Ledgers,—all numbers inserted for reference to any single specimen.

5. The arrangement of the species in the Cabinet had [2]better be also
Geographical generally, which will be also generally natural:
commencing with the Sing.[apore] & Mal.[acca] species,
then follow
those of Borneo, then of Java, then of Celebes, S.[outh]
& N.[orth] then Amboyna & the other Moluccas &
terminating
with New Guinea & the Papuan Islands, when the
genus has such an extensive range. This will be
valuable & instructive for reference & comparison.
[3]N[ota]B.[ene] for improved plan see p[age] 24 of—"Register
1858"

*[The following is written vertically in left margin of page]*
must be modified naturally.

1. The discussion of his plan for the arrangement of his beetle collections continues. The underscored reference to names and authorities touches on a topic of debate in Wallace's day—the principle of priority in taxonomic nomenclature, where the first name bestowed along with a complete description of the species has priority over any later-proposed names. This has been mentioned in annotations throughout the notebook, not infrequently in the context of names given by Wallace and his contemporaries that are now synonymized with names that were found to have been published first (see verso 24 on the antler flies Wallace collected, for example). Wallace weighed in on the debate more than once (Costa 2013). An internationally agreed-upon set of rules and conventions for the describing and naming of species and higher taxa was established only relatively recently, and naturalists are still trying to sort out the nomenclatural mess from centuries of free-wheeling species naming. (See also recto 126, 130.)

2. Historian Wilma George (1964, 49) noted that Wallace's collections "were not valuable for their extent and novelty alone," but also for the detailed locality information he provided: "His success in supplying this information led to his collections being ranked amongst the most important of all time."

There is an essential relationship between geography and species range, Wallace realized, and yet detailed locality information was not provided by collectors. He emphasized this point in "On the Monkeys of the Amazon" (1852b, 109): "In the various works on natural history and in our museums, we have generally but the vaguest statements of locality . . . Owing to this uncertainty of locality, and the additional confusion created by mistaking allied species from distant countries, there is scarcely an animal whose exact geographical limits we can mark out on the map." And yet, he declared, "on this accurate determination of an animal's range many interesting questions depend." (See also recto 133.)

3. This note refers to manuscript WCP4767, Wallace's Bird and Insect Register for 1858, now in the collection of the Natural History Museum in London. Like the recto and verso Species Notebook, the registers of birds and insects run in opposite directions in the same book. On pages 24–26 of the insect register, Wallace writes out fully what he considered his best all-around plan, arranging collections based on both taxonomic and geographical data. He suggests assigning a series of numbers to each systematically arranged set of species and genera from one locality at a time:

> The best plan therefore seems to be to take one family first, say Longicorns, & beginning with one locality, say Sarawak, relax and reset the specimens and attach new locality tickets with a consecutive series of numbers, in approximate systematic order so as to keep the species of the well marked genera together (though this is of little or no importance) . . . A second locality (say *Singapore*) is then taken & a *fresh series* of numbers begun & so on through all the localities . . .
>
> Another family, say Cicindelidae, being then taken, the numbers attached to the species are to be *in continuation* of those of the *same locality* in the former family, so that when the catalogues are completed there will be a consecutive series of numbers for each locality shewing the total number of species found there.

At the end of his discussion, he added this note: "NB. Make a list of localities, lettered A. B. C. &c. & to each species occurring in more localities than one add the letters for the other localities where found also." This way a given species could be cross-checked against those collected in other localities.

## 126

On the reference to Synonyms & the quotation of
Authorities by Naturalists..

This practice is so universal that most naturalists
look upon it as an inevitable necessity, &
some even as an important & interesting
branch of enquiry. [1]It seems hardly to
be ~~looked upon~~ considered as an evil, as something
to be got rid of, as a blot upon our
science, & as one of the causes which
decrease its popularity & deter enquiries
at the outset. If we take up any
natural history catalogue, or work describing
species, we find a considerable portion
of it occupied by names only & references
to volume & page of every work in
which the [illeg.] species have been mentioned
described or figured. A third, a half
or even three fourths of a work is often
so occupied, & the ~~labor~~ task of compiling
these references is one of the greatest
& most tedious labours of the monographer.

1. And yet one can readily surmise that "an evil" and "a blot upon our science" are precisely what Wallace considered the practice of endlessly quoting taxonomic authorities. Providing a comprehensive account of "every work in which the species have been mentioned described or figured" in each catalog is daunting. This practice, what he called "one of the greatest & most tedious labours of the monographer," was undoubtedly weighing heavily on Wallace's mind as he mulled over his notes on arranging and cataloging his collection. Although Wallace himself described a host of individual species, he soon gave up the notion of doing all the taxonomic work by himself—there were far too many new species.

## 127

Now such information is no doubt useful
& even essential to the working naturalist
but there is no occasion for repeating it
over & over again in descriptive works,
in local faunas & in works describing
species. It is a work which sh[oul]ᵈ be done once for all I
    would suggest that we should
commence by giving one reference only
to each species & let that be not to
the original describer, but to some work
in which the synonyms & references have
been already given. [1]What we want is
a series of general synonymical catalogues
which should give all the references, &
determine authoritatively & finally the true specific name to be used &
it would then be only necessary in any work
describing species, to state that the names
used in such a family or group were those
of the catalogue, & use them as names
only without reference or authority. We
do not give the etymology & derivation
of foreign or local terms every time

1.  Wallace offers a practical solution: why not eliminate all the redundant effort and use a reference catalog of sorts as the repository of such comprehensive information, enabling taxonomists and monographers to simply cite this for the litany of references pertaining to each species? A "series of general synonymical catalogues" for each taxon would become the standard reference.

Something rather like that is in practice today. New descriptions and revisions do by convention still include a review of prior taxonomic placement, synonyms, citations for earlier treatments, and so on, but they are not included in descriptive works of local fauna.

Computing has made accessing taxonomic and synonymical reference information easier than ever. For example, botanists have the wonderful online resource of Tropicos, hosted by the Missouri Botanical Garden (MBG), www.tropicos.org: "All of the nomenclatural, bibliographic, and specimen data accumulated in MBG's electronic databases during the past 25 years are publicly available here. This system has over 1.2 million scientific names and 4.0 million specimen records." Similarly, the Biodiversity Heritage Library (www.biodiversitylibrary.org), a consortium of libraries and scientific institutions, digitizes and makes available vast numbers of original works of taxonomic and natural-history literature.

Other partnerships, such as the Integrated Taxonomic Information System (ITIS; www.itis.gov), Species 2000 (www.sp2000.org), and the Global Biodiversity Information Facility (GBIF; www.gbif.org) provide access to expansive databases of taxonomic information. (See Costa 2013.)

we have occasion to use them,—the
vulgar can call a "lion" by its name
without requiring to know when it
first became an English word, by
whom & whence it was introduced,—&c.
Such information must be sought in
Etymological dictionaries if any
where not in those which describe
Lions & their habits. [1]Now it is this
absurdity that the naturalist daily
practices;—he cannot use a <u>name</u>
without ~~at~~ stopping to give its origin &
all the various errors that have been
made respecting it, & quoting every work
in which ~~it~~ the object it distinguishes has been mentioned or described.

In the present state of nomenclature, some
reference is necessary to enable persons
to recognise the species who may only
know it under one of its synonyms:
for this purpose the name of any established
author where it is figured or: described under the

1. Wallace has a knack for finding just the right examples to illustrate his points. He's right that we would consider it absurd and unnecessary to have to provide the complete etymological history of names, words, and terms each time we use them in everyday life. Yet this is the plight of the monographer when it comes to using species names. Wallace's proposed series of general reference catalogs are much like the etymological dictionaries we consult when we want or need to know such information.

same name is sufficient, or if the generic
name is now first applied then quote also
the <u>name</u> under which this author describes
it. This would at once reduce the
bulk & cost of many works very considerably.

[1]Examples:
    Papilio podalirius.
    Expands 3½–4½ inches—&c.....
    Hab.[itat] Europe .. &c ..
    Ref.[erence] <u>Godart</u> B.[ritish] M.[useum] Cat.[alogue]
        1857.
    Observations on habits, economy, &c.
Tillus elongatus.
    ... —Description ....
    Hab.[itat] Europe .. Britain ..
    Ref.[erence] <u>Fabricius</u>. <u>Curtis</u>.
    Obs.[ervations] &c. &c.
Macropteryx mystaceus.
    ... —Desc.[ription] ...
    Hab.[itat] New Guinea
    Ref.[erence] <u>Lesson</u> (<u>Cypselus mystaceus</u>) Voy.[age] de la
        Coquille.
    Obs.[ervations] &c. &c.

*[The following is written vertically in right margin of page]*
[2]N.[ota] B.[ene] Give at <u>most</u> references to 3 works.
    1st. Authority for spec.[ies] name thus "Less.[on] (Cyp.
       [selus] mystaceus). Auct.[orum]" 1824"
    2nd. where best figured—thus "Cram.[er] (Fig.[ure])"
    3rd. To <sub>some cheap & well known</sub> list where synonyms are given.
       Thus "B.[ritish]M[useum] Cat.[alogue]. Bridae 1849.
       (Syn.[onyms])"
This will give all the information necessary in a very small
    space.

1. For his examples Wallace chose two European insects he knew well and a Southeast Asian swift: *Papilio podalirius* Linnaeus (Papilionidae), the scarce swallowtail; *Tillus elongates* Linnaeus (Cleridae), a checkered beetle; and *Macropteryx (Cypselus) mystaceus*, now *Hemiprocne mystacea*, the mustached tree swift (Hemiprocnidae). The authorities he mentions here are:

> Johann Christian Fabricius (1745–1808), Danish zoologist; author of many works on entomology.
>
> Jean Baptiste Godart (1775–1825), French lepidopterist; co-author, with P. A. J. Duponchel, of *Histoire naturelle des Lépidoptères de France* (1821–1842).
>
> John Curtis (1791–1862), British entomologist; coauthor, with A. H. Haliday and Francis Walker, of *British Entomology* (1824).
>
> René Primevère Lesson (1794–1849), introduced on recto 91 and 105. Wallace refers to Lesson as the authority for the mustached tree swift, citing his *Voyage autour du monde entrepris par ordre du gouvernement sur la corvette "La Coquille."*

2. This is a practical recommendation. In listing references, consider not only the authority for the species but also where naturalists can consult the best illustration. The third reference would be a catalog dedicated to synonyms—a "synonymical catalog" of the kind he proposed on recto 68–69 and 127 (see also recto 158).

## 130

[1]The beauty & advantage of the binomial
nomenclature is in fact completely neutralised
if we are obliged to quote a host of
synonyms in addition. The old specific
phrase would be better than this;—it
would occupy less room & would in
the majority of cases ensure the determination of the species.

In the meantime Naturalists should combine
to check the further increase of synonyms by
adopting the plan proposed at p.[age] 67.
[2]The commencement of that plan would
be an Era in Natural Science & it would
then be only necessary to form the catalogues
before mentioned complete up to that
date & Naturalists might boast of a
universal language,—brief definite
& unchangeable;—which they cannot
do with justice at the present time.

May. 12th 1858

[3]Thompson changes Aphies Dej.[ean] (a coleopteron) into
Amillarus Thomp.[son] on account of Aphis a genus
of Hemiptera. This sh[oul]d not be allowed

1. Wallace is not finished railing against the burdensome convention of citing synonyms. His point is well taken: the Linnaean binomial system was a vast improvement over previous means of naming species with multisentence descriptions. Proxy for all the verbiage was a simple Latin binomial (occasionally trinominal), but the more elaborate formal description was available in reference works. Here, too, Wallace suggests relegating the minutiae of synonyms, authorities, and the like to reference works, enabling the working botanist or zoologist to give a name and citation where all the detail associated with the name can be found.

2. Referring to his "Plan to stop the further increase of synonyms" on recto 67, Wallace declares that adopting such a plan will usher in a new "Era in Natural Science." He speaks as one with all too much firsthand experience with nomenclatural minutiae.

3. Here Wallace rejects a name change made by beetle specialist James Thomson [not Thompson; see recto 133], on the basis of the standing generic name (*Aphies*, given by French entomologist Auguste Dejean in 1837) being too similar to the aphid genus *Aphis*. Thomson accordingly substituted a genus he coined, *Amillarus*, in his treatise *Archives Entomologiques ou recueil contenant des illustrations d'insectes nouveaux ou rares* (1857–1858). The beetle in question is a species of round-headed wood-borer, family Cerambycidae. Thomson wrote in volume 1 of his treatise that Dejean's name *Aphies* was a *nom déjà employé*, or "name already employed" (1: 312).

Aphis and *Aphies* are, of course, different names. Making nomenclatural changes on the basis of mere similarity in names would indeed introduce another level of confusion to an already confusing taxonomic system. This would be of concern to anyone who, like Wallace, had to keep straight the names of thousands of taxa. Taxonomic synonyms were already legion—naturalists collecting the same species in different times and places often gave them their own respective names, unaware that a description and name existed already. Redundancy was further exacerbated by geographical variation or polymorphism, and the subjective nature of deciding if one had a new species or merely a variety of an already-named species. This led to a proliferation of synonyms, and taxonomists then had to work out which names were valid.

In any case, there may be more to Thomson's action than meets the eye. Debates over nomenclatural rules raged in the nineteenth century, and one point of disagreement was whether names should follow a principle of priority, whereby the first name to be bestowed by an author had priority over any later names given, or a principle of convenience, whereby the most widely used or best established name is adopted regardless of when the name was given. (Wallace was a champion of priority—see recto 67; Wallace 1858a, 1863d, and 1874.) The renowned French coleopterist Auguste Dejean (1780–1845) was an outspoken critic of the priority principle. Advocating the use of what he considered to be the most commonly cited names, he proceeded to change hundreds of names in his published catalogs. Was Thomson registering his displeasure with Dejean by deciding to eliminate a genus that the Frenchman had created?

Wallace may have disagreed with Dejean's principle of common usage, but he clearly felt that Thomson's stated reason for making this name change—based on similarity to another name—was unacceptable.

A few years later another French entomologist, Louis Alexandre Auguste Chevrolat (1799–1884), published a critique of Thomson and attempted to resurrect the genus *Aphies* (Chevrolat 1861). This did not stand either. As it happens, like many of the names Dejean conferred, the name *Aphies* is not recognized as valid today. It is a *nomen nudum* ("naked name"), in the jargon of the International Code of Zoological Nomenclature, meaning it is a designation unacceptable as a scientific name because it was not published with an adequate description.

[131]

*[Pencil sketch]* [1]end of lower mand.[ible] of Cockatoo

*[Pencil sketch]* same with sheath removed
shewing nervous covering of bone
with comb teeth points entering
the sheath & connected with
nerves traversing the bone wh.[ere] they
enter at a. a.—*[Refers to diagram at left]*

---

In the upper mand.[ible] exists the same arrangement.

---

C.[harles] Allen observed
♂. ♀ simple [2]form of windpipe in
a ♂ spec[ime]n of Phon.[ygama] keraudreni
all others had it simple
*[Pencil sketch]* (in a 2nd. nearly similar)
& a 3rd.

---

a ♂. Phon.[ygama] viridis
had it descending
equally low but with
a single bend.

---

Ph.[onygama] ~~viridis~~ atra in Waigiou
had a short bend only of
about 1/2 an inch.

*[Feather]* from throat of P.[honygama] viridis. Linn.[aeus]

1. Wallace may have been interested in the ways that cockatoo mandibles differ from those of other birds. On the lower mandible are two points that, when combined with the pointed top mandible, give the cockatoo the advantage of being able to hold, manipulate, and tear in three separate places. His notes relate to the enervation of the mandibles.

2. Here Wallace is commenting on the elongated looped tracheae of certain species of manucodes (*Manucodia*, formerly *Phonygama*): *M. keraudrenii*, *M. viridis*, and *M. atra*. The name "manucode" is from a latinized form of the Malay *Manukdewata*, "bird of the gods," the name applied to birds of paradise generally for more than two centuries before the modern family name Paradisaeidae was established in 1825. Manucodes are known for their loud and powerful vocalizations, as reflected in the common name of species like *M. keraudrenii*, the trumpeter manucode.

Five manucode species are found in New Guinea and northern Australia, all with a remarkably long, looped tracheal passage (like that of their relatives, the birds of paradise). This anatomical oddity was (and is) of considerable interest to ornithologists. It was commented on by René Lesson, who named several manucode species, and it was the subject of papers read by zoologist William Alexander Forbes at the Zoological Society of London in 1882 (Forbes 1882a, 1882b).

Clench (1978) discussed tracheal elongation in birds of paradise, and Frith (1994) reviewed the literature on tracheal modifications in nonpasserine birds. Frith concluded that those with elongated, looped tracheae "produce calls that are louder, lower in pitch, or carry farther than the calls of close relatives that lack this anatomical modification" (Frith 1994, 553).

Wallace's *Manucodia (Phonygama) keraudrenii* and *M. atra* specimens were collected in Waigiou, western New Guinea. On this page Wallace mentions Charles Allen, his erstwhile assistant from the first year of his travels in the archipelago. Allen later collected independently for Wallace (between 1860 and 1862), including on a trip to New Guinea from February to June 1861, when he would have had an opportunity to collect these birds. Wallace's entry was likely made following that trip. He had collected *M. viridis* previously, in the Aru Islands.

*Note for determining species population of Globe* 132.

[132]

[1]Note for determining species population of Globe.
In making a systematic calculation of the number of
species of Coleop.[tera] in countries only partially known,
it will be necessary to determine an approximate
law for the total number of species common to two
countries when only a portion of each is known.
For instance if there are 2000 British species found
also in France or about ⅗ of the whole,—it will
probably be found that another proportion will exist
between partial collections in England & France.
The first 100 species collected by a person in the
two countries, the first 500 or the first 1000
will probably present different ratios. A series
of results of collecting in different countries
compared in every possible way will shew
whether a greater proportion of the common species
of each country (which will be those first obtained)
or of the rarer ones (which will enter only in a more
complete collection) have a wider distribution.
If any rule however approximate can be established
we shall have some foundation to go upon in
determining the species population of a region from

1. This idea, which may have been inspired by Alexander von Humboldt's "botanical arithmetic" (Browne 1980) is novel, but chimerical. There are too many taxonomic and geographic variables; arriving at even an "approximate" law for determining the total number of species common to two areas when only a fraction of each is known is impossible without more information.

Wallace is onto something, however, in that partial collections amount to samples, and samples and sampling effort can be used to generate species-accumulation curves. The rate at which new species are added to a list is a function of the number of samples taken. Initially, new species are found at a rapid rate with each successive sample, but the rate declines as a greater and greater proportion of the actual number of species has been sampled. When the curve plateaus this indicates that something near the full complement of species in the area has been sampled.

This is not something that Wallace appears to have pursued, though he was keenly interested in population estimates. Those he reported in various writings were, of course, little better than educated guesses. In *The Geographical Distribution of Animals*, for example, he wrote of Coleoptera (a group of special interest for him): "As the total number of species at present known to exist in collections is estimated . . . at 70,000 species, we may be sure that were the whole earth as thoroughly investigated as Europe, the number would be at least doubled, since we cannot suppose that Europe . . . can contain more than one-fifth of the whole Coleoptera of the globe" (Wallace 1876, 189).

partial collections made at two or more localities
in it, or of a group of islands only partially explored.

---

**1** ? French plan better—first par.[agraph]—<u>colour</u> only. second par.[agraph] form &
see Thomp.[son] Arch.[ives] Ent.[omologiques]     sculpture

**2**<u>Note for descriptions in "Coleoptera</u> Malayana
In first line give the <u>prominent characters</u> ? in larger type so
that by glancing the eye over these, the species
may be approximately determined,—then follow
the specific characters in detail: thus from
Whites Cat.[alogue] of Long.[icornia] p.[age] 341.

---

**3**<u>Pascoea Idae</u>                              Long.[icorns] lin.[es] 11
<u>Viridescenti</u>—niger; elytus metallcis subsulcatis fasciis 4
    albidis:
capite thorace pedibus que nigris, elytris sulcis punctatis,
    medio
obsoletis, fasciis macularibus inconspicuis transversis pilosis;
abdominis lateribus pilis brevibus adpressis irridescentibus
irroratis, tarsis tibiis que ad apices subtus pilis subferringeniis.
Hab.[itat] Ceram, Amboyna.
Ref.[erence] Auct.[orum] White, ([Auct]) B.[ritish] M.[useum]
    Cat.[alogue] Longicornes, cum fig.[ure]
Obs[erved]   Found on fallen trees & in new plantations in the
virgin forest, *[illeg.]* not common.

*[The following is written vertically in left margin of page]*
Easier than French plan with essential spec.[i]fic words
    <u>italicised.</u>

---

Here the dimensions opp.[osite] the name & the general
    characters
in the first line, with subsections of the genera will
assist greatly in quickly determining species . .

1. Considering different approaches to the catalog he envisions, Wallace approves of the layout in *Archives Entomologiques ou recueil contenant des illustrations d'insectes nouveaux ou rares* (1857–1858). Published by the Société Entomologique de France, this catalog was authored by James Thomson (1828–1897), a wealthy American naturalist who resided in France.

2. Wallace planned on publishing taxonomic and biological information based on his collections once he returned to England, as indicated in his March 2, 1858, letter to Frederick Bates, brother of his friend Henry Walter Bates:

> Now with regard to your request for notes of habitats &c. I shall be most willing to comply with it to some extent, first informing you that I look forward to undertaking on my return to England a "Coleoptera Malayana" to contain descriptions of the known species of the whole archipelago, with [an] essay on their geog. distribution, and an account of the habits of the genera & species from my own observations. (Wallace Correspondence Project, WCP367)

Birds, beetles, and butterflies were groups of special interest to Wallace, both because they sold well and, more important, because, being conspicuous and abundant groups, detailed knowledge of their diversity, habits, and variations was within reach. Thus they could be used to test his ideas about the nature of species and varieties.

Despite his good intentions, once back home Wallace was overwhelmed by the biological riches of his travels, and while he did publish a number of species descriptions, the bulk of his collections were given to others to describe and publish. Specialists worked on particular groups. For example, while a comprehensive *"Coleoptera Malayana"* was not produced, Francis Polkinghorne Pascoe (1813–1893) published *"Longicornia Malayana; Or a Descriptive Catalogue of the Species of the Three Longicorn Families,"* treating Wallace's cerambycid beetles (Pascoe 1864–1869).

3. Wallace adapts the format of Adam White's 1855 *Catalogue of Coleopterous Insects*, modifying White's entry for the long-horned beetle *Pascoe Idae* according to his own formatting preferences (see A. White 1855, 8: 314).

## 134

[1]Mites, generated in spirit..
Some Oryctes had been in arrack 6 months.
on being taken out in a few hours they
were covered with mites!!! crawling out of
the joints!!

---

Same in Borneo to after a week's soaking

[2]Natives of Gilolo seen at Sahoe;—
Called Alfures;—colour, of the pure Malay
brown quite as light as the hill Dyaks of
Borneo & lighter than most pure Malays.
Stature,—tall but rather slight (Galela men stout & strong) hair various
rough waved or rather frizzly,—never either
the smooth straightness of the Malay or the
perfect frizzle of the Papuan race. .
Features much more Papuan than Malay.
The nose larger, apex produced;—brows
prominent,—lips thick, more or less bearded
& with much curly hair on the body. .
A mixture of Dyak or Dyak Chineese Celebes Malay, with a
true Papuan race would produce such variety
& if the men were the former & the women
the latter the light colour would be probably the
result. .

Sahoe. Sept.[ember] 1858.

1. *Oryctes* is a genus of rhinoceros beetle (Scarabaeidae: Dynastinae), which are among the largest beetles worldwide in terms of body mass. *Oryctes rhinoceros* is the well-known rhinoceros beetle of Malaysia. Wallace marvels at the emergence of mites through the joints after the beetle specimens had been in arrack, an alcoholic drink typically produced in Southeast Asia from fermented sugarcane. Arrack was a cheap and easy way to pickle specimens; mites are hardy indeed, in this case they likely survived in sinuses within the beetle's body that had not become filled with the preservative.

By "generated" Wallace may have meant "reproduced"—but he could have meant "spontaneously generated." The possibility of spontaneous generation was hotly debated in the Victorian period (extensively reviewed in Parley 1977, Strick 2000), and in keeping with his iconoclastic tendencies Wallace was sympathetic to the idea. Proponents argued for two forms of spontaneous generation: "archebiosis," or "abiogenesis," was the production of living organisms from inorganic materials, and "heterogenesis" was the production of living organisms from dead organisms or their remains. Could Wallace have been suggesting that the mites were "generated" from the dead tissues of the preserved beetles? Was he invoking heterogenesis?

It is unclear what Wallace thought of spontaneous generation at the time he was in Southeast Asia, but his review of Henry Charlton Bastian's book *The Beginnings of Life* in 1872 shows that by then he entertained it as a real possibility: "[Bastian's book] brings together a large body of facts, either new or hitherto almost ignored, which, unless they can be otherwise explained, prove much more than the mere production of low living organisms from dead matter," Wallace wrote in part one of his two-part review. He continued in part two:

> But if the facts of Archebiosis and Heterogenesis are true, and all the lower forms of life are continually being produced de novo, under the influence of unknown laws of development, then we may fairly conclude that, when once the earth had arrived at conditions favourable to the production of living organic matter, the process of devel-opment would be rapid, and an immense variety of low forms of animals and vegetables would soon people it. It is a fair inference, too, that if such complex organisms as Ciliated Infusoria, Rotifers, Nematoids, and even simple Acari, can be developed independently of the slowly modifying influence of natural selection, the same laws of development will continue to act a subordinate part much higher in the scale. (Wallace 1872, 303)

Note Wallace's inclusion of Acari—mites—in his list of organisms that might be developed "independently of the slowly modifying influence of natural selection." Much of his review discusses the implications of spontaneous generation for the Darwinian view of life. Coming just a few years after revealing his apostasy over the evolution of human cognition, one might imagine Darwin and his circle rolled their collective eyes at the appearance of this review. Still, Wallace was no crank mystic; he maintained that naturalists should apply the highest standards of scientific scrutiny to such phenomena rather than dismissing them outright.

2. The village of Sahoe, in the interior of Halmahera (Wallace's Gilolo), was the second village that Wallace visited on that island (coastal Dodinga was the first). This is the very area where Wallace "discovered the exact boundary-line between the Malay and Papuan races" (*Malay Archipelago*, 323). Note the language he uses here in the notebook. In modern terms he seems to believe he has found the contact zone between these two human races, where they intermix. The last sentence is curious: it suggests that Wallace believes that the males have a greater influence on offspring than the females, as lighter-skinned children are expected to arise from a lighter-skinned Malay father and darker-skinned Papuan mother.

135

[135]

[1]Paradisea n.[ew] s.[pecies] Semioptera Wallacei G.[eorge] R.[obert] Gray.

This bird inhabits the virgin forests of the island of Batchian one of the western Moluccas. It frequents the lower trees of the forest, and like ~~most~~ many of its allies is in almost constant motion. It flies from branch to branch, clings to the twigs, & even to the vertical smooth trunk almost as easily as a woodpecker. It continually utters a harsh creaking cry, something between that of the Paradisea apoda & the more musical voice of the P.[aradisea] regia. The males at short intervals open & flutter their wings erect the long shoulder feathers & expand the elegant shields on each side of the breast. Like the other Birds of Paradise the dull coloured females & young males far outnumber fully plumaged birds which makes it probable that it is only in the 2nd. or 3rd. year that the extraordinary accessory feathers are fully developed. Feeds on fruit,—probably insects occasionally.

1. Wallace's standardwing, *Semioptera wallacei*, is endemic to the Indonesian islands of Halmahera, Bacan, and Kasiruta. Also known as the standardwing bird of paradise, the name *Semioptera* is derived from Greek terms for "military standard" and "wing," referring to the remarkable erectile shoulder plumes on the wings of courting males.

Wallace obtained the first specimens of this species that was to bear his name through the efforts of his assistant Ali, who shot one in October 1858. He wrote to Samuel Stevens on October 29: "I believe I have already the finest and most wonderful bird in the island. I had a good mind to keep it a secret, but I cannot resist telling you, I have got here a new <u>Bird of Paradise</u>!! of a new genus!!! quite unlike anything yet known, very curious and very handsome!!! When I can get a couple of pairs, I will send them overland, to see what a new Bird of Paradise will really fetch" (Wallace Correspondence Project, WCP1705).

He soon collected others, and on January 30, 1859, he sent a consignment to Stevens that included "four males, one female, and one young bird of the new Batchian Paradisea, besides one red-ticketed private specimen," as stated in his accompanying letter (WCP4749). George Robert Gray provisionally named the species at the March 22, 1859, meeting of the Zoological Society, using Wallace's description of the bird from a letter.

> This Paradise-Bird proves, as Mr. Wallace remarks in his lettre, to be a new form, differing from all its congeners . . . I have endeavoured to transform the rough sketch into the probable appearance of the living bird; and I further add the provisional specific name of *Paradisea wallacii*, which appellation I think is justly due to Mr. Wallace for the indefatigable energy he has hitherto shown in the advancement of ornithological and entomological knowledge, by visiting localities rarely if ever travelled by naturalists. I wait for the arrival of the specimens before venturing to give more detailed accounts of its subgeneric characters, or a full description of its coloration, &c., which I hope to have the pleasure of laying before the members at some future meeting of the Society. (Gray 1859b, 130)

Curiously, the bird's distinguishing features were described more fully in a *report* of this meeting that appeared in the *Athenaeum* (Gray 1859c), and this is considered today to be the official or original description.

Wallace subsequently sent John Gould a letter (September 30, 1859) excerpted as his "Notes on *Semioptera Wallacii*, Gray," published in the *Proceedings of the Zoological Society of London* for 1860 (Wallace 1860a). Several elements of the description here in the notebook are found in this letter, extracts of which was read at the Zoological Society meeting of January 24, 1860.

Wallace's descriptive notes on the behavior of *S. wallacei* were reprinted several times, included in Gould's *Handbook to the Birds of Australia*, volume 2 (1865), and eventually incorporated into chapter 38 of *Malay Archipelago* (565–566). As the notebook entry gives the species name and authority, we know these notes were entered sometime between Gray's paper in March 1859 and Wallace's September 30, 1859, letter to Gould.

Wallace's standardwing bird of paradise (*Semioptera wallacei*). From Alfred Russel Wallace, *The Malay Archipelago*, 4th ed. (London: Macmillan & Co., 1872), 329.

## 136

[1]Note on Longicorns.

Almost the only genus of Lamiidae always found on
foliage is Glenea (Lebaron). & Colobothea. Others when found
on leaves have only visited them from their regular
haunts, dead trees, fallen branches &c. Isosules is
very often seen on the wing & on foliage but
I think always in the neighbourhood of fallen
timber.    Of the Cerambycidae, the
beautiful Callichromes are true leaf
haunters wherein they differ from
Xystrocera which are always found
on the under side of timber.

Rh Xoanodera trigona I always found
resting on the midrib of a leaf its legs
& antennae rigid as if dead. Hammaticherus always on timber. The
    beautiful
Distenia also lays flat on leaves, with
antennae laid flat & beautifully curved.
Selethris is also a leaf genus

1. Lamiidae and Cerambycidae were formerly treated as separate families. Currently the lamiids are treated as a subfamily: they are the flat-faced longhorn beetles (Cerambycidae: Lamiinae).

Wallace found the greatest diversity of beetles associated with felled trees, which is to be expected, as the larvae of many groups feed within the wood of dead or dying trees. Other species consume sap, leaves, flowers, fruit, or living bark, and still others feed on fungi. Here he notes a few long-horned beetles whose "haunts" are leaves. He does not describe them as leaf-feeding, but notes their quiescent resting posture along the leaf midrib. This is quite common, especially among nocturnal insects, their coloration often cryptically matching their background substrate—whether green leaves or mottled gray bark—as they hide by day.

Wallace later became interested in protective coloration (both camouflage and mimicry) in the context of adaptation and natural selection, and discussed these at length in his books *Contributions to the Theory of Natural Selection* (1870) and *Darwinism* (1889). (See also Wallace 1867a.)

In the parlance of the day, the term "protective resemblance" was often used to describe cases of matching background color, while "protective imitation" described mimicry of objects like leaves or twigs. Phasmidae (leaf and stick insects) were among Wallace's favorite examples of mimicry—several of his specimens are preserved in Drawer 28 of the Natural History Museum's Wallace Collection, in London. Another favorite was leaf butterflies of the genus *Kallima*, which is illustrated in *Malay Archipelago*.

Wallace's admiring friend, the naturalist Edward Poulton (1856–1943), published a landmark survey of mimicry and related phenomena in 1890 entitled *The Colours of Animals*.

Incidentally, in 1862 Henry Walter Bates (1825–1892) published his ideas on the form of "protective resemblance" that we now call Batesian mimicry (where a harmless species imitates the coloration of a harmful species), based on his study of Amazonian heliconiid butterflies. Wallace and Darwin were enthusiastic; Wallace put together an exhibit of butterfly specimens illustrating the concept, a display that is now in the Natural History Museum (London).

[137]

Ants & Aphides

Observing a swarm of small black ants. (Bac.[an] 17.)
on the leaves of a tall grass which were covered
with minute aphids I cut off a portion of a
leaf & though the ants mostly scuttled away, one
remained intent on a little aphis cluster &
allowed me to bring him into the house &
examine his proceedings under a lens. There
were about a dozen minute white aphids, closely
clustered together each with his trunk buried
in the leaf~ves~ quietly sucking up the juices. They
had each two minute bristles under the end
of the abdomen & finely annulated antennae
rather thick & about ½ the length of the body. These
were laid over the back but occasionally raised
with a wavy motion & the whole group looked
like a lot of long eared white rabbits nibbling
at some very short grass. Occasionally one of
them raised up his abdomen with a stretch
of the body & legs as if to get his rostrum
more deeply into the food stratum.

1. Wallace's careful observations of the interaction of an ant and honeydew-producing aphids continue to the top of recto 140. The observations, which were made in December 1858 on the Moluccan island of Batchian (Bacan), are not mentioned in his letters, journals, or books, such as *Malay Archipelago*. He does not identify the ant, but in his introductory remarks to Frederick Smith's "A Catalogue of the Aculeate Hymenoptera and Ichneumonidae of India and the Eastern Archipelago" (F. Smith 1873) Wallace noted that "*Formica lactaria* and *F. circumspecta* were observed in company with Aphides, and feeding on their sweet secretion." *Formica* is the type genus of the entire ant family, Formicidae, and includes more than 200 species worldwide. Wallace can be forgiven for not having identified the ant in his field notes—these "most abundant and omnipresent of tropical insects," as he described them in his introductory remarks, often require detailed study under a microscope to identify them beyond the genus level.

Wallace's description of the aphids as white suggests a woolly aphid, family Eriosomatidae. Many woolly aphids feed in discrete circular clusters, heads on the inside and abdomens on the outside, creating an "aphid buffet" and facilitating the ants' harvest of honeydew. Wallace notes that the aphids form clusters but does not indicate if the clusters are circular.

[138]

[1]The ant was in a state of great excitement,
moving round & round the little group wh.[ich]
he could cover with his outstretched legs,
gently touching them with a rapid motion of his
antennae, running over them from side to
side, then standing still & opening his
powerful jaws to their greatest extent
so as almost *[illeg.]* to touch the back of one
of the little motionless animals, as if he
were going to devoured it. Then commenced
another rusling *[sic]* & tapping & ^expectant^ gaping quite
ludicrous to look at till at last an
aphis gently raised up his tail & out
came a drop of transparent viscous liquid
& at the same instant down came the
ants open mouth & seized the luscious drop
which hung for some seconds between his
extended mandibles while he slowly sucked
it in. He then brushed his face & cleaned
his jaws with his fore feet & again commenced
watching his flock. For several minutes

1. Wallace does not interpret the behaviors he observes in this case, unlike Darwin's use of similar observations in the *Origin of Species* (1859, 210–211). Seen through the lens of natural theology, the honeydew-producing aphids could be viewed as performing a service to the ants. Darwin argued against the idea that species in any way behave for the *exclusive* good of other species, and so in this case argues that the ant-aphid relationship is a mutualism (to use a modern term): the aphids need to have the liquid removed, he suggests, and the ants can use this as a resource. It is now understood that the ants afford the aphids protection from predators and parasites in exchange for the honeydew resource.

Incidentally, Wallace noted but did not discuss another curious example of ant tending that he came across. In his introductory comments to Frederick Smith's catalog of his hymenopteran collections (F. Smith 1873) Wallace noted that the ant *Formica badia* "makes a small fragile nest under palm leaves; and in it I observed the larvae of a small Homopterous insect (a species of Cercopiae?), the perfect insects being found on the same plant." Chances are he did not realize what was going on at the time, but cercopids (frog hoppers, family Cercopidae) and their relatives the membracids (tree hoppers, family Membracidae) often engage in the same relationship with ants as aphids, and for the same reason. What Wallace noted was an ant "corral"—these ants take homopteran tending one step further and provide a shelter, rather like a barn for their honeydew-producing "cows." There are even some sugar- and amino-acid secreting caterpillars that are sheltered in this way by ants. See Costa (2006, 264–265 and 600–601) for further examples and discussion of this phenomenon.

no more [1]milk was given till he became
quite impatient rushing madly about &
continually passing his antennae over the
abdomens of the aphides to ascertain
which were most likely next to produce
the desired luxury. At last another drop
was given out & at short intervals afterwards
three more, all of which were seized &
sucked up in the same eager manner.
The velocity with which the ant w[oul]d. turn
& catch the drop before it fell was
remarkable, & also the manner in which
he kept his jaws asunder at their widest
gape holding the drop between them by
capillary attraction, while it was sucked
up & gradually disappeared. The ants
swarmed on the leaves but had no nest
there. They were travelling across the
ground in several directions to other shrubs
& herbage where I could not trace them.
The saccharine drop was not emitted from the

1.  The aphid's "milk" is honeydew, a viscous mixture of water and sugars from the plant sap. Aphids and some other sap-sucking insects have a gut structure called a filter chamber that aids in concentrating the nutritional component of sap by allowing water to pass through the gut more quickly. The process is inefficient, however, so some of the sugars are excreted along with the water.

## 140

[1]subterminal spines or tubucles of the aphis but from the extreme apex of the abdomen.

[2]December. 1858.

---

[3]Agassiz on Lake Superior.      p.[age] 377.
"There are in animals peculiar adaptations which are characteristic of their species, and which cannot be supposed to have arisen from subordinate influences. Those which live in shoals cannot be supposed to have been created in single pairs. Those which are made to be the food of others cannot have been created in the same proportions as those which live upon them. Those which are everywhere found in innumerable specimens must have been introduced in numbers capable of maintaining their normal proportions [a] to those which live isolated, & are comparatively & constantly fewer. For we know that this [b] harmony in the numerical proportions ~~of~~ between animals is one of the great laws of nature. The circumstance that species occur within definite limits where no obstacles prevent

1. Wallace is correct that the honeydew is not excreted from the "subterminal spines or tubercles." Termed cornicles, these are paired tubular structures found at the posterior end of the abdomen, functioning in the secretion of waxes, defensive compounds, or alarm pheromones. Honeydew is excretory, produced at the anus.

2. Wallace was on the island of Bacan (Batchian in his time) in December 1858, the Moluccan island where he found the new bird of paradise *Semioptera wallacei* (see recto 135). He collected there until April 1859.

3. The passage Wallace quotes here, which continues on recto 141, is from the 1850 book *Lake Superior: Its Physical Character, Vegetation, and Animals, Compared with Those of Other and Similar Regions*, coauthored by the renowned Swiss-born Harvard paleontologist and comparative anatomist Louis Agassiz (1807–1873) and American James Elliot Cabot (1821–1903). Agassiz had moved from the University of Neuchâtel in 1846 to head the Lowell Institute at Harvard. In 1859 he founded the Museum of Comparative Zoology at Harvard. A staunch anti-transmutationist, Agassiz never accepted Wallace and Darwin's theory.

Note that Agassiz's language here is reflective of the natural theology tradition: he speaks of "harmony in the numerical proportions" of species, but he is no biblical literalist. Agassiz points out that group-living fish cannot have been created as single pairs, their social group being of central importance to their mode of life. Similarly, Agassiz is ecologist enough to recognize that predator populations cannot exceed that of their prey. Each species exists in the right proportion to fill its role in the balance of nature. It is the notion of harmony and balance, however, that raises Wallace's ire, and he criticizes this view in language almost identical to that used in his critique of Lyell (see recto 49) over the same issue.

Note that the terms Wallace labels "a" and "b" are taken up on the bottom of recto 141. Agassiz and *Lake Superior* are also discussed on recto 142–147 and verso [d-1] of the Species Notebook.

their wider distribution leads to the further
inference that these limits were assigned to
them from the beginning, & so we should come
to the final conclusion that the order which prevails
throughout nature is intentional that it is regulated
by the limits marked out on the first day of
creation & that it has been maintained unchanged
through ages with no other modification than
those which the higher intellectual powers of man
enable him to impose on some few animals
more closely connected with him."

---

[1]! What are the <u>normal proportions</u> & <u>harmony</u> spoken of.
The proportions have continually varied & are varying.
Are the horses in S.[outh] America <u>harmonious</u> or not?
In the tertiary period there were horses,—then none
now they are again. Whatever <u>exists</u> must be
in <u>harmony</u> or it could exist no longer. The
<u>proportions</u> of all animals are self regulating,
& constantly varying,—it has not been maintained
<u>unchanged</u> for any great period as Agassiz well
knows . .

1. Reminiscent of his criticism of Lyell in the *Principles* about the "balance of species" (recto 49), Wallace is dismissive of Agassiz's assertion that proportions of species represent a harmonious balance. Of Lyell Wallace asks: "Where is the balance?" Of Agassiz he essentially asks: "Where is the harmony?" To Wallace, species proportions "have continually varied & are varying," and have "not been maintained unchanged for any great period."

The case of the horse was of intense interest and importance to naturalists. Fossilized teeth and bones of horses had been found in North and South America (Darwin found some in South America while on the *Beagle* voyage), indicating that horse species once flourished in the Americas but had gone extinct thousands of years ago. Horses did not again roam the American plains until Spanish colonists introduced them in the sixteenth century. This was a profound fact; as Darwin expressed in his *Journal of Researches:* the fossil horse tooth "greatly interested me," he wrote, continuing, "Mr. Lyell has lately brought from the United States a tooth of a horse . . . Certainly it is a marvelous fact in the history of the Mammalia, that in South America a native horse should have lived and disappeared, to be succeeded in after-ages by the countless herds descended from the few introduced with the Spanish colonists!" (Darwin 1845, 130).

Wallace may have had Darwin's passage in mind when he rhetorically asks, "Are the horses in S. America harmonious or not? In the tertiary period there were horses,—then none now they are again."

## 142

[1]Erratic drift theories.
Lyell imputes them to ~~Gla~~ icebergs floating
Southwards. Agassiz to glaciers travelling
Southwards. The former supposes the whole area
of the drift to have been either at once or
successively under the ocean,—the latter
supposes the continent to have been nearly
as now at the time of the dispersion of
the drift & to have been partially sunk
since in the places where marine deposits
have been formed.

Facts mentioned by Agassiz observed near
Lake Superior seem to prove the glacier theory
such as shallow E.[ast] & W.[est] vallies being gro[o]ved
across continuously which could never have
been produced by floating ~~glaciers~~ icebergs stranded
on shallows. Many other facts show that
the drift & the furrows were produced on
dry land. There is however an argument
not referred to either by Lyell or
Agassiz which seems quite fatal

1. This entry also pertains to *Lake Superior* (1850). "Erratic drift" refers to boulders and unconsolidated rocky debris, or "drift" in geological parlance. Wallace compares Lyell's view of the origin of drift (icebergs) with Agassiz's argument that ice sheets are responsible. Agassiz has the better of the argument, in Wallace's view: observations reported in *Lake Superior* (395–416) supported the action of continental glaciation in moving and depositing drift.

Agassiz, following his countryman Jean de Charpentier (1786–1855), was a strong advocate of the "glacial theory." He interpreted such phenomena as erratic boulders, moraines, striations, and other landscape features in terms of ice sheets, or glaciers. Lyell, in contrast, invoked icebergs as the causative agent in early editions of *Principles of Geology*. During cold epochs when the land subsided and the ocean extended over what is now dry land, icebergs were posited to float southward a considerable distance. Debris dropped by them would be found on the modern landscape, the seas having receded, as out-of-place boulders seemingly plunked down many miles from the current coast.

Darwin soon became a convert to the glacial theory, holding that some continental phenomena stemmed from glacial action while certain coastal phenomena stemmed from icebergs. (His paper "On the Distribution of the Erratic Boulders and on the Contemporaneous Unstratified Deposits of South America" was read in 1841, and published in 1842 in *Transactions of the Geological Society of London* 6: 415–431.) When he visited Cwm Idwal in Snowdonia, Wales, Darwin viewed the landscape through the lens of glaciation; as he wrote later in his *Autobiography*: "Yet these phenomena are so conspicuous that . . . a house burnt down by fire did not tell its story more plainly than did this valley" (Barlow 1958, 70).

We see in this Species Notebook entry that Wallace, too, was convinced that Agassiz and others were correct about a former ice age. He discussed the topic often over the years in books, articles, and book reviews, and in contexts ranging from geological to astronomical to climatological.

to the iceberg theory. [1]It is that if the drift around Lake superior were brought by icebergs the depression forming that lake must have been under the ocean, & when eliminated would have been a salt lake. As the period of the drift is admitted to have been recent how did the lakes get rid of ~~the~~ their salt? The only possible explanation is that the lake cavities have been altogether formed since the surrounding country has been elevated above the ocean, a most improbable supposition—as the grooving extend to its banks.

The fact of all the lakes being perfectly fresh proves however that they have been principally formed since the last elevation of the country which must therefore probably be at a very ancient epoch.

---

[2]On Instinct & Reason, a poet says
"In this 'tis God that acts, in that tis Man."
But errors of instinct show this can not be the case.

1. The argument Wallace calls "quite fatal" to the iceberg hypothesis is a characteristically insightful one: under Lyell's hypothesis, the depression that is the modern Lake Superior would have existed prior to and during the period of submersion when icebergs were posited to have moved over the area and deposited their load. However, being a depression the lake bed would have been left filled with saltwater when the sea receded—creating a salt lake like that in Utah. This is clearly not the case, and Wallace finds it more likely that the depression that constitutes the lake bed formed after elevation of the land.

2. The poet in this case is Alexander Pope, and the quote is from *An Essay on Man* (Pope 1734, Epistle III, part 2):

> Whether with reason, or with instinct blest,
> Know, all enjoy that power which suits them best;
> . . .
> And reason raise o'er instinct as you can,
> In this 'tis God directs, in that 'tis man.

Here Wallace puts his finger on the idea of anomaly and variation in nature and what this teaches. *Errors* of instinct give the lie to the idea that "in this God acts," the deity presumably being infallible and not capable of errors. (See also recto 177.)

## 144

Embryology to assist classification.
All birds have web feet in early stages.
Palmipedes are therefore not naturally or
necessarily a distinct group but may cont.
the lower parts of several groups.
Raptores only the highest development of
raptorial aquatics. Embryo birds have all
hooked bills. Raptores are therefore lower
than Passeres.

Aquatic mammals are low, not because
they are aqquatics but because they have
webbed feet—embryos have all webbed feet.
Bats are therefore low because they have
webbed fore feet though extremely developed.
Manatus is not cetacean but a low
Pachyderm, as seals are low Ferae.

In Insects the lower stages are mandibulate
the higher haustillate;—the mandibulate
orders are therefore lowest Coleoptera below
Lepidoptera & Homoptera. &c. &c.
A catterpillar just before becoming pupa

---

[144]

[1]Embryology to assist classification.
All birds have <u>web feet</u> in early stages.
<u>Palmipedes</u> are therefore not naturally or
<u>necessarily</u> a distinct group but may cont.[ai][n]
the lower parts of several groups.
<u>Raptores</u> only the highest development of
raptorial aquatics. Embryo birds have all
<u>hooked</u> bills. <u>Raptores</u> are therefore lower
than <u>Passeres</u>.

———

Aquatic mammals are low, not because
they are aqquatics but because they have
webbed feet—embryos have all webbed feet.
Bats are therefore low because they have
webbed fore feet though extremely developed.
<u>Manatus</u> is not cetacean but a low
Pachyderm, as seals are low Ferae.

———

In Insects the lower <u>stages</u> are mandibulate
the higher haustillate;—the mandibulate
orders are therefore lowest <u>Coleoptera</u> below
<u>Lepidoptera</u> & <u>Homoptera</u>. &c. &c.
A catterpillar *[sic]* just before becoming pupa

1. Wallace continues his notes on Agassiz, who pioneered the use of embryology to inform classification. Most of these observations are found in *Lake Superior* (see Agassiz and Cabot 1850, 193–199), and are based on a series of twelve lectures entitled Comparative Embryology that Agassiz delivered at the Lowell Institute in Boston in 1848–1849.

In the passages to which Wallace refers, Agassiz presents his argument for how traits in evidence at different developmental stages can be used to "order" taxa. If a trait like webbed feet is found in all birds in embryo, for example, then those groups that have retained webbed feet as adults would be "lower" or "less developed" than those that have lost this trait as adults. By using enough such traits, Agassiz hoped for a complete ordering of species or species groups that would in turn revolutionize classification of those groups. Not incidentally, he also felt that this ordering somehow reflected the sequence of taxonomic groups in the fossil record—the "law of parallelism" merges with the "law of succession." This became known as the "threefold parallelism" between embryology, classification, and paleontology (see, e.g., Bowler 2003, 121–123; Bryant 1995). In *Lake Superior* we can see Agassiz's early expression of this idea:

> If there is any internal evidence that the whole animal king-
> dom is constructed upon a definite plan, we may find it
> in the remarkable agreement of our conclusions, whether
> derived from anatomical evidence, from embryology, or
> from paleontology. Nothing, indeed, can be more gratify-
> ing than to trace the close agreement of the general results
> derived from the investigation of their embryonic changes,
> or from their succession in geological times. (Agassiz and
> Cabot 1850, 196–197)

When it comes to interpreting embryology, it is interesting to note the contrasting conclusions drawn by different naturalists. For Agassiz, embryology could be used to gain insight into relationships, but relationships in the context of unchanging archetypes. This same pattern was viewed in a radically different way by Wallace and Darwin, who saw the affinities revealed by embryology in an evolutionary context. (This concept was also at the heart of the transmutational argument of *Vestiges of the Natural History of Creation*, the book that convinced Wallace of species change.)

As Wallace put it in his collection of essays titled *Contributions to the Theory of Natural Selection* (1870, 301):

> Agassiz . . . insists strongly that the more ancient animals
> resemble the embryonic forms of existing species; but as
> the embryos of distinct groups are known to resemble
> each other more than the adult animals . . . this is the
> same as saying that the ancient animals are exactly what,
> on Darwin's theory [and his own!], the ancestors of exist-
> ing animals ought to be; and this, it must be remembered,
> is the evidence of one of the strongest opponents of the
> theory of natural selection.

has a coleopterous form.

In Molluscs the naked gasteropoda &
Cephalopoda are higher than the testaceous.
[1]This agrees with Geolog.[ical] succession
<u>Ammonites</u> before <u>Belemnites</u>.
In crustacea <u>crabs</u> in early stages have a
lobster form—∴ [therefore] Brachyarea higher than
Macroura and are geologically later. .
    From <u>Agassiz</u> in <u>Lake Superior</u>. .

[2]There is more resemblance or analogy in many
points between the flora of Eastern N.[orth] America
& E.[ast] Asia than between the former & Europe.
The pacific coasts of America have more
correspondence to Europe. The U.[nited] S.[tates]
Rhododendrons &c. are more fully represented in
N.[orth] China & Japan than in Europe.
correspondence of Climate accounts in part
for this—     "Agassiz"—
The Tertiary fossil flora of O[e]ningen resembles
that of E.[astern] N.[orth] America more than any other.

1. Here is an example of how Agassiz related embryological development to classification and the order of appearance of particular forms in the fossil record. He suggested that with groups such as cephalopods, some forms (the "naked" or shell-less, say, like slugs) are "higher" than those with shells (testaceous, like snails), based on the observation that, developmentally, the naked form goes through a stage where a shell begins to form, but that shell is lost at later developmental stages. Turning to the fossil record, Agassiz noted that belemnites—extinct squidlike cephalopods outwardly naked but with a hard internal structure (the guard, or rostrum) for rigidity—are found in more recent strata than ammonites, extinct cephalopods known for their well-developed spiraling shell. Thus development in modern gastropods, which have shells, is seen as paralleling the sequence seen in the fossil record.

The molluscan and crustacean examples that Wallace notes are found in *Lake Superior* (Agassiz and Cabot 1850, 197, 199). Agassiz's interpretation of parallelism is there too: "Now if this arrangement be the real order of succession of the Cephalopods according to their structure and development, is it not remarkable, does it not indicate the maintenance of the same plan throughout the creation[?] . . . [W]e might almost conclude, that in this class the order of succession of their fossil types is a safer guide for our classification, than anatomical investigation" (199).

2. Agassiz touches on the remarkable correspondence between the eastern North American and East Asian flora in *Lake Superior* (150). He states, too, that the flora of western North America corresponds to that of Europe.

His interpretation is that eastern North American and eastern Asia are of greater antiquity than western North America and Europe, and he supports this with the "extraordinary and unexpected" fact that fossil European plants resemble modern trees and shrubs of eastern North America. The plants are "old-fashioned" and "bear the mark of former ages" compared with European and western North American plants. The Tertiary fossil flora Wallace mentions is found at Oeningen, a Swiss Miocene site studied by Oswald de Heer (1809–1883), a professor of en-

tomology and botany at the University of Zurich. Oeningen is renowned in paleontological circles for its rich beds of insect and plant fossils. Hundreds of species were described by de Heer, many in the three-volume *Flora tertiaria Helvetiae* (*Tertiary Flora of Switzerland*; 1855–1859).

Paleoecologists today would not fully agree with Agassiz, but he is not so far off. The current understanding is that the eastern North American and Asian flora are remnants of approximately 50-million-year-old circumboreal flora that extended right around the Northern Hemisphere. The formation of the Central American land bridge about 7 million years ago resulted in the drying of western North America and provided a means for South American plants to move northward, breaking up the formerly continuous circumboreal flora. As a result the eastern North American and eastern Asian flora are more similar to each other than either is to the western North American flora.

Agassiz's Harvard colleague, the botanist Asa Gray, analyzed this botanical distribution pattern a few years after *Lake Superior* appeared. He was prompted to undertake the study by Darwin, as it happens. Darwin was interested in species migration routes at the time, and asked Gray to compare a list of species of European plants with those found in Tierra del Fuego. "Without multiple creation, I think we must admit that all now in T. del Fuego, must have traveled through N. America," Darwin wrote to Gray in May 1856 (Darwin Correspondence Project, letter 1863). Gray's subsequent *Statistics of the Flora of the Northern United States*, published in 1856, detailed the striking similarity between Asia and eastern North America. In an October 1856 letter to Gray, Darwin wrote: "Nothing has surprised me more than the greater generic & specific affinity [of E. North America] with E. Asia . . . Can you tell me . . . whether climate explains this greater affinity? or it is one of the many utterly inexplicable problems in Bot. Geography?" (Darwin Correspondence Project, letter 1973).

Wallace's interest in the subject pertains more to the "law of succession" than species migrations, though he notes that Agassiz seems to think that "correspondence of Climate accounts in part for this."

## 146

Agassiz gives 31 + 74 [105] species of flowering plants
found on N.[orth] shores of Lake Superior as
identical with those of Sub alpine &
Northern Europe. All the mosses but
one were identical with European species,
& the ferns all but 3.

———

In all 350 sp.[ecies] of flowering plants were collected.

———

Of the above number he thinks however
that many may prove distinct & some few
have been introduced.

———————————————

[1]The road side weeds of the Atlantic
States are all European. The native
weeds have disappeared westwards.
Agassiz gives a list of 130 species
now common in cultivated districts
of Atlantic U.[nited] States all of which are
European. What becomes of the "Harmony
of distribution",—the "balance of species";—
the "proofs of intelligence in the nat.[ural] distrib.[utio]n
of species" &c. &c. Did the "wonderful order"
Agassiz speaks of exist before the country was

1. This passage bears on the supposed balance or harmony of nature (see recto 49, 140–141). A significant proportion of eastern North American plants are introduced from Europe. If there ever was a "harmony of distribution" or "balance of species," there certainly is no longer, with the "native weeds" driven westward. How can the natural distribution of species be inferred to reflect a "wonderful order" or "proof of intelligence" to begin with, if such human introductions can be accommodated, displaced, or adapted to so readily?

These data from Agassiz were later presented by Wallace in *Darwinism* (1889, 15), in an essay on the struggle for existence.

overrun by these strange plants,—or does it
exist now? [1]If there was a "thoughtful
adaptation" of in the assemblage of plants
before Europeans came to N.[orth] America there
can be no such adaptation now.

[2]The true freshwater fishes of N.[orth] America are
different from those of Europe without a single except..[io]n
Most of the species of Lake Superior are peculiar.

Agassiz argues that each species must have been
created where it now exists, but he does not
take into account former Geolog.[ical] changes, separations
& junctions of river basins by elevation &
depression & possible or probable former
connection of N.[orth] America & Europe towards their
northern extremities.

[3]M[oun]t. Washington—Lat[itude][/?]
Above 4350 feet no tree. Flora commences resembling
Greenland. Many species same as on shores of Lake
Superior. Summit (6280 f[ee]t.) produces plants many
of which are found only in Labrador or Lapland.
                    "Agassiz."

1. Wallace continues his critique of Agassiz: If ever there was a "thoughtful adaptation" in the assemblage of plants before the European colonists introduced a veritable army of their plants, "there can be no such adaptation now" as the country is overrun with the introduced species.

2. This comment reflects Wallace's interest in geographical distribution, and, more important, his sense that species distributions change dynamically over time, hand in hand with the dynamically changing earth. Wallace developed this conception most fully in his landmark treatise *The Geographical Distribution of Animals* (1876), in which he pioneered the use of fossil fauna to infer historical species distributions and distributional changes over time. This vision is in stark contrast to that of Agassiz, who held that species were created by and large where they now exist. In his words (Agassiz and Cabot 1850, 375–376): "If we face the fundamental question which is at the bottom of [the observed distribution of freshwater fish], and ask ourselves, where have all these fishes been created, there can be but one answer given . . . The fishes and all other freshwater animals of the region of the great lakes, must have been created where they live." Agassiz continued:

> It cannot be rational to suppose that they were created in some other part of the world, and were transferred to this continent, to die away in the region where they are supposed to have originated, and to multiply in the region where they are found. There is no reason why we should not take the present evidence in their distribution as the natural fact respecting their origin, and that they are, and were from the beginning, best suited for the country where they are now found. (376)

Wallace could not disagree more.

3. The highest peak in the northern Appalachians at 6,288 feet (1,917 meters), located in Coos County, New Hampshire (44° N latitude). Although higher peaks are found to the south (the highest in eastern North America being North Carolina's Mount Mitchell, at 6,684 feet, or 2,037 meters), Mount Washington's higher latitude results in a tree-line, alpine tundra, and flora with a boreal affinity, as Wallace mentions here.

Wallace does not elaborate, but at the time he kept the Species Notebook he likely believed that dispersal explained the plant species found in Greenland and on Mount Washington. He later expressed this view in *Island Life* (1880a), in connection with the high-elevation flora of isolated African mountains. Darwin corresponded with Wallace over this question, preferring a model that had the movement of plant populations in latitude and elevation as a result of ever-changing climate and advancing and receding ice sheets. As he put it in *Origin of Species:* "As the tide leaves its drift in horizontal lines, though rising higher on the shores where the tide rises highest, so have the living waters left their living drift on our mountain-summits, in a line gently rising from the arctic lowlands to a great height under the equator" (1859, 382).

Wallace came to accept Darwin's view, as seen in his essay "English and American Flowers" published in the *Fortnightly Review* (1891a, 802):

> As the cold period gradually passed away, these hardy plants kept close to the gradually retreating ice, and in this way mounted to the higher peaks of many mountains from which the ice and even perpetual snow wholly passed away. Thus it is that so many species are now common to the Rocky Mountains and the European Alps; and, what seems more extraordinary, that identical plants occur on the summits of the isolated Scotch and Welsh mountains, and also on the White Mountains of New Hampshire and some of the mountains to the south of them.

## 148

[1]Beetles,—forms habits distribution & statistics
It is a melancholy fact that many of our fellow
Creatures do not know what is a beetle! They
think cockroaches are beetles!—
Tell them that beetles are more numerous more
varied & even more beautiful than the birds or
beasts or fishes that inhabit the earth & they
will hardly believe you;—tell them that he
who does not know something about beetles misses
a never failing source of pleasure & occupation
& is ignorant of one of the [most] important groups
of animals inhabiting the earth & they will think
you are joking;—tell them further that he
who has never observed & studied beetles passes
over more wonders on every field & every copse
than the ordinary traveller sees who goes round
the world & they will perhaps consider you
crazy,—yet you will have told them only the
truth & less than the truth as they themselves
will admit if ever a change of task or
occupation should lead them to enquire seriously

1. Wallace's lament may have been intended as notes for a popular article or letter on the subject. It is likely that this was aimed not at his Malay companions but at folks back home. Lyell commented in the second volume of his *A Second Visit to the United States of North America* (1849, 2: 206) that the Americans he met seemed to call all insects "bugs," and that "Londoners, by way of compensation, miscall the cockroach a black beetle." It is possible that somewhere in the multitude of articles that Wallace read he came across a reference to cockroaches as "black beetles," which would have roused his ire.

His lament is shared by many an entomologist. Blinded by the large and obvious, most people are profoundly ignorant of insects and their arthropod relatives, groups that are vastly more abundant and diverse than all vertebrates (indeed all other animals) combined. The two groups that get most of what little attention insects get at all—butterflies and beetles—are themselves on the more impressively large and lovely end of the insect spectrum. These are the groups that Wallace was chiefly interested in collecting, not so much reflecting his own bias as the market realities concerning what insects were in demand and could be sold by his agent.

But there is, perhaps, a deeper significance to an appreciation of beetle diversity in the case of Wallace, and indeed of Darwin. As pointed out by Berry and Browne (2008), at the Linnean Society's 1908 celebration of the fiftieth anniversary of the Wallace and Darwin papers, Wallace gave an address that asked "why us?"—"Why did so many of the greatest intellects fail, while Darwin and myself hit upon the solution of this problem?" He came to an unexpected conclusion: "In early life both Darwin and myself became ardent beetle-hunters." In retrospect it is easy to see how beetles, which are wonderfully diverse even in species-poor England, could inspire admiration for diversity and serve as the ideal test case for probing the distinction between species and varieties and their geographical distribution.

into the matter.

---

[1]"The sudden passage from an irrational to a
rational being is a phenomenon of a distinct
kind from the passage from the more simple
to the more perfect forms of animal
organization & instinct. To pretend that
such a step or rather leap can be part of
a regular series of changes in the animal
world is to strain analogy beyond all reasonable bounds." <u>Lyell's Principles</u>"
Here the absolute distinctness of <u>reason</u> &
<u>instinct</u> is assumed;—the argument depends
on the terms <u>rational</u> & <u>irrational</u> which
imply no gradation.

---

[2]In Lyell's recapitulation of facts as to reality of
species—he says "4thly. The entire variation from
the original type which any given kind of
<u>change</u> can produce may usually be
effected in a brief period of time, after

1. This Lyell quote comes from *Principles* (1835, 1: 239) and is part of Lyell's argument for the recent appearance of humans on the planet. He is suggesting that the human mind differs not only in degree but in kind from the minds of nonrational beings and therefore could not have arisen by gradual steps.

Wallace seems to be having none of this, arguing that Lyell has started from an assumption of "absolute distinctness" in reason and intellect between humans and other animals; the very terms of the argument, he suggests, by definition imply distinct and separate categories. During the period of the Species Notebook Wallace was pondering the relationships among instinct, intellect, learning, and experience, and entries on the subject are scattered throughout the notebook (see recto 112, 116–119, 143–144, 149, 155, 166–177). The question of differences in degree versus kind among animals (including humans) was central. He was likely not a strict materialist at this time when it came to the human mind, and as his certainty about this grew in later years he became an outspoken critic of the idea that the human mind could have evolved by purely natural means, first revealing his views publicly in 1869 (see Kottler 1974, C. H. Smith 1992).

2. The quoted passage that begins here and continues to the next page is found in *Principles* (1835, 3: 21). In it Lyell reiterates his point that species vary only within limits, and so this variation can be effected in a short amount of time. Wallace's response follows on the next page.

which no further deviation can be obtained
by continuing to alter the circumstances
though ever so gradually; indefinite divergence
either in the way of improvement or
deterioration being prevented & the least
possible excess beyond the defined limits
being fatal to the existence of the individual."

---

[1]Here rests the whole question & Lyell assumes
it in his favour. He assumes than [that] only
change of circumstances produce[s] variety—how
then do varieties constantly occur in the same
place & under the same circumstances as
the original species? How can he prove
that variation may not go on at a rate
commensurate with Geological changes?
How can man's hasty experiments settle
this—his "though ever so gradually" is a
gratuitous assumption. What are "the
defined limits",—he assumes that they
exist.

1. Adept at putting his finger on the crux of an argument—and its weak points—Wallace perceptively notes that Lyell assumes that it is ever-changing circumstances (in the environment) that engender variation but that this variation has its limits, despite the fact that environmental conditions continue to change. But if changing environment ("change of circumstances") is the only way variety is produced, Wallace rhetorically asks, why do we see variety continually generated in species in the same place under unchanging conditions? And on what basis does Lyell argue that variation has its limits? These passages echo points made on recto 43.

Note that Wallace makes reference to the generation of new varieties from the same "original species" and "under the same circumstances." Insofar as he understands at this point that "varieties are incipient species," as Darwin later put it, this sounds like speciation in situ—sympatric speciation, in modern terms.

Wallace's comment about "man's hasty experiments" may refer to experiments mentioned by Lyell immediately before the "recapitulation" that includes the quoted passage. Lyell cited German physiologist Friedrich Tiedemann (1781–1861), who suggested that the fetal brain appears to pass through stages, piscine, reptilian, avian, and so on, during development (Tiedemann 1826). Anticipating the studies of German-Estonian zoologist Karl-Ernst von Baer (1792–1876; see Baer 1828), this ontogenetic pattern was later interpreted in an evolutionary context by Chambers, Darwin, and others. Of course Lyell did not see it this way, and in arguing as much he also mentioned experiments on hybrid sterility discussed earlier in the *Principles*: "On the contrary, were it not for the sterility imposed on monsters, as well as on hybrids in general, the argument to be derived from Tiedemann's discovery, like that deducible from experiments respecting hybridity, would be in favour of the successive degeneracy, rather than the perfectibility, in the course of ages, of certain classes of organic beings" (1835, 3: 20).

Note, finally, that it was Tiedemann's concept of embryological development that inspired Robert Chambers's transmutational interpretation of this process. Chambers included a diagram in *Vestiges* (Chambers 1844, 212) based on a very similar one in Tiedemann, illustrating different vertebrate groups in a treelike pattern of relationship based on embryological development. Chambers explained how he thought changes in embryological development could lead to transmutation: "It is apparent that the only thing required for an advance from one type to another in the generative process is that, for example, the fish embryo should not diverge at A, but go on to C before it diverges, in which case the progeny will be, not a fish, but a reptile" (213).

## 151

## Von Buch on Flora of Canaries.

[1]Von Buch on Flora of Canaries.

| | | |
|---|---|---|
| Species peculiar— | 164 } | 48 sp[ecies] found also at |
| Found elsewhere— | 213 } | Madeira. |
| Probably imported | 158 | Primitive Flora |
| Total flowering plants. | 535 | *[Written vertically]* |

| | | | |
|---|---|---|---|
| African region to. | 1200 f[ee]t. | 182 sp.[ecies] | 82 peculiar. |
| Mediterranean " | 2600 " | 264 " (106) | 38 " |
| Forests _____ " | ?4500 " | 68 " | 28 " |
| Pines _____ " | 5900 " | 10 " | 8 " |
| Alpine _____ " | 10380 | 11 " | 8 " |

[2]At S[ain]t. Helena the introduced are more numerous than
the primitive plants. At S[ain]t. Michel almost
all the plants are either Portuguese or Brazilian
and at Bermuda most of the present flora has
been introduced.

[3]In the Canaries are African, Spanish, Majorcan,
Grecian, Brazilian & Peruvian plants now
wild. Of most of the genera there are only
single species. Pyrethrum & Cineraria many
closely allied sp.[ecies] Sempervivum 20 sp.[ecies]

Compositae ⅟₇ of the indigenous plants.

V.[on] Buch supposes Canaries to have risen separately
from the sea, & never to have been joined together
nor to the continent of Africa.

1. Another reference to Leopold von Buch's *Flora of the Canary Islands* (*Physikalische Beschreibung der Canarische Inseln*, 1825). On recto 90 Wallace noted von Buch's observation that sheltered valleys often have their own groups of closely allied plant species. Here he is concerned with a related observation: levels of endemism on islands. At this time Wallace certainly interpreted such endemism in terms of transmutation stemming from long isolation of chance ancestral migrants.

Note that he is also interested in the proportion of endemics found at different elevations. The "African region" extending to 1,200 feet has 45 percent endemism, the "Mediterranean region," extending to 2,600 feet, has 14.3 percent, and so on. The highest elevation, the alpine zone, has few species but high endemism: eight out of eleven species, or 72 percent. (The correct maximum elevation for the Canaries is 12,198 feet, or 3,718 meters, at the peak of Tenerife's Mount Teide.)

2. Wallace notes that several distant islands in the Atlantic Ocean, including Saint Helena and Saint Michel, have more introduced than endemic ("primitive") species. It is hard to know to what extent this is a result of human disturbance, as these islands were discovered and colonized by fifteenth- and sixteenth-century mariners. Note Wallace's comment that most of the flora is Portuguese or Brazilian; midway between Portugal and Brazil, these islands were heavily used by the Portuguese ships traveling back and forth between the Old and New Worlds.

3. The introduced genera are species poor, while native genera, like *Pyrethrum*, *Cineraria*, and *Sempervivum*, are speciose. Some of these groups were mentioned previously, on recto 90. Wallace's comment about the Compositae (now Asteraceae) comprising fully one-seventh of the flora of the Canary Islands resonates with the next entry, von Buch's conclusion that the archipelago has never been physically connected with the African mainland. Composites as a group tend to have small, wind-dispersed seeds, making this among the most likely of plant groups to colonize remote islands (as indeed is seen with even more remote archipela-

goes—endemic tree-sized composites, the *Scalesia*s, are found in the Galapagos Islands, for example).

Wallace has aligned a great many significant points here, and we can be sure they are connected dots in his mind. He later wrote at length about the patterns of endemism found on continental versus oceanic islands—in his 1880 book *Island Life*, for example, and in papers such as the address "Oceanic Islands" for the American Geographical Society (1887). (See also recto [a-1].)

Wallace's attention to these and other intriguing passages from *Flora of the Canary Islands* (recto 90) would suggest that von Buch must have been key for Wallace, yet von Buch and the Canary Islands are not mentioned in many papers or other publications by Wallace. He did, however, include von Buch in his list of those who had articulated a belief in species change, writing in his book *Darwinism*:

These opinions of some of the most eminent and influential writers of the pre-Darwinian age seem to us, now, either altogether obsolete or positively absurd; but they nevertheless exhibit the mental condition of even the most advanced section of scientific men on the problem of the nature and origin of species. They render it clear that, notwithstanding the vast knowledge and ingenious reasoning of Lamarck, and the more general exposition of the subject by the author of the *Vestiges of Creation*, the first step had not been taken towards a satisfactory explanation of the derivation of any one species from any other. Such eminent naturalists as Geoffroy Saint Hilaire, Dean Herbert, Professor Grant, Von Buch, and some others, had expressed their belief that species arose as simple varieties, and that the species of each genus were all descended from a common ancestor; but none of them gave a clue as to the law or the method by which the change had been effected. (Wallace 1889, 5–6)

[152]

[152]

[1]Volcanoes. On W.[estern] extremity of N.[ew] Guinea or near Salwatty—Dampier—doubtful.

---

1. Wawani—in Amboyna. W.[est] side of Hitoe 2 miles from coast. eruption in 1674 after great earthquake destroyed a village—1694 another eruption.
   1797. Great heat & vapour. (Tuckey).
   1816 & 1820 eruptions. In 1824 new crater opened.
2. Banda—1586 eruption & 9 more up to 1820.
3. Seroa. (S.[outh] of Banda) In 1693—inhabitants all came to Amboyna (Phil.[osophical] Trans.[actions])
   Nila—has a crater also?
4. Makian—1646. great eruption enlarged crater.
5. Motir—Volcanoes, threw out stones in 1778. "Forrest"
   Ternate eruption in 1608 & frequently since.
   Gammaconora. In 1673 a mountain was elevated & threw out pumice stones. (Valentyre)
   Tolo in Moutay—a volcano.
   At Sanguir many villages destroyed & men killed in 1711—in 1856, 2000 men killed.

---

? Tyfore & Meyo—no—

1. Wallace lists recorded volcanic eruptions in the archipelago, based on information compiled from multiple sources:

William Dampier (1651–1715) was a legendary explorer-adventurer (read: "pirate") in his time, and also had a taste for natural philosophy. Dampier published several accounts of his travels, including the three-volume *A New Voyage Round the World* (1697–1703). Naturalists consulted Dampier for information on the exotic organisms, peoples, and landscapes he encountered, but his accounts did not always withstand scrutiny, as Wallace's comment "doubtful" signals here. Indeed, there are no active volcanoes in western New Guinea where the island of Salwatty is found.

James Hingston Tuckey (1776–1816) sailed to the Malay Archipelago aboard the HMS *Suffolk* in the late 1790s and witnessed the volcanic eruption at Amboyna in 1797.

Lyell, in *Principles of Geology*, recounted the Sorea (Wallace's Seroa) eruption of 1693 and reported that most of the inhabitants fled to Banda, citing the *Philosophical Transactions of the Royal Society* for 1693–1694.

Thomas Forrest (?1729–?1802), a British East India Company captain, reported the Motir eruption in his *Voyage to New Guinea* (1780).

Some of this information is given in *Malay Archipelago*: Forrest's report from 1778 of stones thrown from the volcano of Motir, for example, and the crater-enlarging 1646 eruption of the volcano of Makian (*Malay Archipelago*, 327). In a footnote Wallace reported that a devastating eruption of the latter volcano occurred in 1862, shortly after his departure from Southeast Asia.

Gammaconora (now Gamkonora) is the highest peak on the island of Halmahera. This large volcano erupted on May 20, 1673. Tolo may refer to the Gulf of Tolo, located between the eastern and southeastern peninsulas of volcanically active Sulawesi. Sanguir (now Sangir) Island in northern Sulawesi saw highly destructive eruptions of the volcano Gunung Awu in 1711 and March 1856.

## 153

[1]In interior of Batchian, 400–500 elevation large
<u>Chama</u> & <u>Tredacria</u> are found. A bed of coal
with dicotyledonous leaves & shells of recent forms.

———

In S.[outh] pen.[insula] of Batchian is a Geyser.
Behind village of Batchian is raised coral about
100 feet high.
Near Kaserota is a conglomerate with volcanic
fragments imbedded in the beach, inland crystalline
basaltic rocks cont.[ainin]g copper.

———

[2]<u>N.[orth] Celebes</u>. . very volcanic = active volcano W.[est]
Ratahan—hot springs & boiling mud—volcanic rocks
every where—with a few intermediate bits of stratification.

———

Gagie Is.[land] is volcanic—very rugged & bare to S.[outh]

———

1.  Wallace notes signs of uplift and volcanic activity in northern Celebes (Sulawesi) and Batchian (Bacan), a small island to the southwest of Gilolo (Halmahera). Here is his description in *Malay Archipelago* of the geology in the vicinity of Batchian village:

> The village or town of Batchian is situated at the head of a wide and deep bay, where a low isthmus connects the northern and southern mountainous parts of the island. To the south is a fine range of mountains, and I had noticed at several of our landing-places that the geological formation of the island was very different from those around it. Whenever rock was visible it was either sandstone in thin layers, dipping south, or a pebbly conglomerate. Sometimes there was a little coralline limestone, but no volcanic rocks. The forest had a dense luxuriance and loftiness seldom found on the dry and porous lavas and raised coral reefs of Ternate and Gilolo; and hoping for a corresponding richness in the birds and insects, it was with much satisfaction and with considerable expectation that I began my explorations in the hitherto unknown island of Batchian. (*Malay Archipelago*, 332)

The coal seams of Bacan are indicative of a former depositional environment, and the coal and marine deposits seen at 400–500 feet elevation on this island are evidence of uplift. *Chama* and *Tredacria* (perhaps *Trinacria*?) are genera of marine bivalves, both with extant and extinct species. Woodward (1851), for example, lists fifty extant and thirty fossil species of *Chama*.

2.  The hot springs and boiling mud of Ratahan, a district near the distal end of the northern peninsula of Sulawesi, are indicative of the active volcanism of the area. There are at least eight or nine active volcanoes on the northern arm of Sulawesi, some reaching nearly 2,000 meters in height. Eruptions of some (Lokon and Empung) have occurred as recently as 2011. Ambang volcano, the westernmost volcano of the northern arm of Sulawesi, experienced a major eruption shortly before Wallace's arrival in the Malay Archipelago, in the late 1840s.

## 154

[1]The Present Age.

"I believe that no age has seen so far as
ours into the true theory of the universe & of
humanity; I think that there are some
thousands of our contemporaries who have
more mental penetration, acuteness, true
philosophy & even moral delicacy than
all past ages together could produce;
but this rich cultivation to which in my
opinion no epoch has any thing comparable
is beyond the age & has but little influence
upon it. [2]A gross materialism, valuing
things only for their immediate utility tends
more & more to seize upon humanity, &
to throw into the shade all that serves to
satisfy a love of the beautiful or a
simple curiosity desire of knowledge. All
religions & all philosophies alike teach us
that man should have a higher aim than
his physical enjoyments & interests. Has
our great material progress brought us nearer
to these [illeg.] higher objects of existence? Have
we on the whole become more intelligent

1. The following is originally from an essay entitled "M. de Sacy et l'École Liberale," in *Essais de morale et de critique,* by French philologist and philosopher Ernest Renan (1823–1892). Wallace likely encountered it in translation, such as that which appeared in *Blackwood's Edinburgh Magazine* for November 1861 (9: 626–639).

Wallace's interest in this passage is telling—it can be seen as an early expression of his interest in spiritualism, a "theory of the universe & of humanity" that holds that there is a transcendent quality evident in human moral, aesthetic, and creative attributes. In these passages Renan critiques the materialism of the time, asking if it truly ennobles or "brings us nearer to these higher objects of existence."

2. Concerns over excessive materialism and its pitfalls were expressed elsewhere by Wallace, such as in his address to the Royal Geographical Society regarding the physical geography of the Malay Archipelago (1863c), which warned that "if [the establishment of nature preserves] is not done, future ages will certainly look back upon us as a people so immersed in the pursuit of wealth as to be blind to higher considerations." He also wrote critically in *Malay Archipelago* of "money-getting" peoples. (See also annotations for recto 155.)

more honest more desirous of liberty
more sensible to the beautiful? In this
lies the whole question."
<u>Ernest Renan</u>" in Revue des deux Mondes.

---

[1]"A fine thought, a noble sentiment, an
act of virtue, make man the king of
creation far more than the power to bring
what he wants almost instantaneously from
the ends of the earth."    Id.[entify]

---

[2]<u>Instinct</u>. "The young swallow, when it
migrates on the approach of the first winter of
its life is impelled by an instinct implanted
by the deity, & it neither knows the causes
that prompt it to fly nor the end to be
attained by its flight." Combe, Const.[itutio][n] of Man
    p.[age] 1

---

[3]<u>Fullom's Marvels of Science</u>—
Absurd book full of error. He makes the <u>Insects</u>
the <u>lowest</u> & the <u>Annulosa</u> the highest of the
<u>Articulata</u>. He says Birds seem to be linked
to the terrestrial animals by the <u>Vespertilio</u>.
"The sight of Bats is very imperfect & of the same
character as that of the Owl!"
"Orang utan native of [regions N.[orth] of] <u>Coromandel</u>"

1. This statement reflects Wallace's sensitivity to "higher considerations." The quote is a contemporary reflection on the "power to bring what [one] wants almost instantaneously from the ends of the earth"—a reference to the global shipping and rail network that could quickly bring such rarities as Wallace's collections from the remotest islands of the East to the salons of London. The sentiment reflects the unease many people felt at the time, as today, with the ever-accelerating pace of modernity. As Spencer Tracy's character in the 1960 film *Inherit the Wind* expressed it: "You may conquer the air, but the birds will lose their wonder and the clouds will smell of gasoline." Wallace may have relied on and marveled at the technological innovations of his day, but he was aware that they came with a price. This passage suggests that it is the "higher considerations" of virtue and aesthetic appreciation that truly distinguish humans from the rest of the animals, not merely our technology—a sentiment that very much resonated with him.

2. From George Combe's *Constitution of Man* (1828, 1: 3). Combe (1788–1858), a Scottish phrenologist, natural philosopher, and reformer, promoted phrenology in his popular *Constitution of Man*. Its materialistic view of the brain's faculties, seeing the mind as the product of physiological processes, anticipated modern views of the brain, consciousness, and behavior. Wallace was an advocate of phrenology, and said in his autobiography (1905, 1: 234) that is was Combe's *Constitution of Man* that piqued his interest in "phreno-mesmerism."

Combe advocated understanding the world through natural processes, but argued that these exist for the attainment of ultimate ends. Thus here he attributes the migratory instinct of swallows to knowledge "implanted by the Deity," since the swallow knows neither "the causes that prompt it to fly nor the end to be attained." Wallace likely disagreed with Combe's interpretation of the swallows' behavior, but it is easy to see how Combe's discussion is consonant with his efforts to understand the relationship of instinct and experience in behavior.

3. Wallace had a low opinion of Stephen Watson Fullom's book *Marvels of Science and Other Testimonies to Holy Writ* (1852). It is indeed a marvel that as late as 1852 an author could venture that birds "seem to be linked to the terrestrial animals by the Vespertilio, better known [as] bats," or that the eyesight of bats is on a par with that of owls (Fullom 1852, 308). And Wallace had first-hand knowledge that orangutans were found far, far to the east of Coromandel, which is in the Tamil Nadu region of the Indian subcontinent.

Fullom's book is more apologia for natural theology than treatise on natural history. Indeed the book's very first sentence declares its intent: "Undoubtedly the most precious of man's gifts, invaluable and indispensible as they all are, is revealed religion" (1852, 1). Clearly Fullom did not exercise his mental gifts in learning much about the subjects he ventured to write about. As for revealed religion, Wallace firmly rejected its claims, as he expressed in a letter to his brother-in-law Thomas Sims in 1861:

> To the mass of mankind religion of some kind is a necessity. But whether there be a God and whatever be His nature. Whether we have an immortal soul or not, & whatever may be our state after death, I can have no fear of having to suffer for the better off in a future state, who have lived in the belief of doctrines inculcated from childhood & which are to them rather a matter of blind faith than intelligent conviction . . . In my solitude I have pondered much on the incomprehensible subjects of space[,] eternity[,] life & death! I think I have fairly heard & and fairly weighed the evidence on both sides, & I remain an utter disbeliever in almost all that you consider the most sacred truths. (Wallace Correspondence Project, WCP 3351)

## 156

[1]Batocera—carry antennae, curved right & left
    of head & rather low.

Gnoma—opened at < of 45° point bent downwards

Monoham[*illeg.*]—d[itt]o  d[itt]o  " upwards.

Tmesisternus—set on leaves & dead twigs with antennae
    l[*illeg.*]d close
    to body—fall if disturbed—

Hammaticherus. antennae elevated & opened moderately
    straight with ends curved downwards.

See Cat.[alogue] of Paris Mus.[eum] p[age] 39 (opp[osite][*?*])
    for another plan

---

5. [2]Trichius abdominalis

Descrip.[tion] &c. .

| T.[richius] abdominalis. | Schmidt (a) | France, Germany |
| Cet.[onia] fasciatus | Oliv.[ier] (a) fig.[ure] | |
| | | Italy—not uncommon |
| T.[richius] abdominalis | Blanch (d) syn.[onym] | |

Habits & oeconomy _____

1. Wallace describes the resting postures of several long-horned beetles, family Cerambycidae. They hold their long, slender antennae in different positions. Most of these are cryptically colored and blend in with their substrate to avoid detection by vertebrate predators.

*Batocera* species are the flat-faced long-horns of the cerambycid subfamily Lamiinae. *Batocera wallacei*, Wallace's long-horned beetle from the Aru Islands, was named in Wallace's honor by American entomologist James Thomson in 1859.

Long-horned beetles, called "longicorns" at the time, for their long antennae, were perhaps the single most diverse group of beetles that Wallace collected in the Malay Archipelago. His finds were published as *Longicornia Malayana* between 1864 and 1869 by Francis Polkinghorne Pascoe (1813–1893). This treatise included "no less than a thousand species, upwards of up to 800 which were regarded as new to science." In a published commentary on the localities given in *Longicornia Malayana*, Wallace illustrated the richness of the cerambycid collecting in the archipelago with a table summarizing his catch for each region. In Sarawak alone, from November 1854 to November 1855, he listed 1,700 beetles collected, 270 of which were longicorns (Wallace 1869b, 693). (See also recto 125 and the illustration on the annotation page for verso 6; the uppermost two beetles shown in the figure are longicorns.)

2. This is a continuation of ideas sketched in "Plan for references in Synopses" on the next recto page.

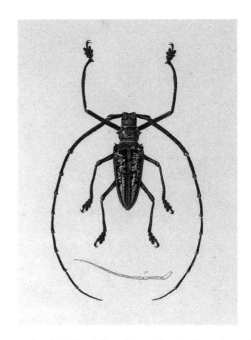

*Batocera wallacei*, Wallace's long-horned beetle. From James Thomson's *Arcana naturae* (1859), 1: plate VI.

## 157

[1]Plan for references in Synopsis
Give an alphabetic list of Authors quoted
with a letter referring to each of their works.
Thus
Gray, G.[eorge] R.[obert]
a. Genera of Birds
b. Catalogues of Birds in B[ritish] Museum
c. Birds of voyage of _____

Then Give species thus . .

Accipiter ruficeps. ~~G~~
.......... Desc[riptio]$^n$ ..........
[~~Auct.~~] Gray. (c) ~~Syn.[onym]~~ Strickland. (b) for Syn.[onym]
~~Fig.[ure]~~ Gray (a) for Fig.[ure]
Habitat . . . &c. &c.

[2]Gymnetis Lanius. ~~Linn[aeus]~~.
Description &c. ........
Scarabaeus lanius. Linn.[aeus] (a) Syn.[onym] Brown (d),
Fig.[ure] Drury (b)
Jamaica.

[3]Trichius abdominalis
Description locality &c.
Schmidt. (a)    Oliv.[ier] (Cet.[onia] fasciatus) (a.) for
fig.[ure]
Blanch.[ard] ~~Syn.[onym]~~ (d) for Synonyms.

[4]Forms opp.[osite] best keeping sp.[ecies] & authors names in
columns
[Above statement refers to bottom of recto 158]

1. Here and through recto 159 Wallace outlines candidate formats for systematic or synoptic catalogs, a subject of practical as well as theoretical interest to him, given the size of his collections (see also recto 68–69, 126–130). This first page shows Wallace's plan for presenting authors alphabetically by name and listing their published works. Following up on his idea on recto 129 (see the marginal note printed at the bottom of that transcription page) to give three references for each species—the authority, the work where it is best figured, and synonyms—he experiments here with formats. As a space-saving device, the author and the appropriate letter designating the published work can be given (rather than the full title of the work). Thus, under *"Accipiter ruficeps,"* Wallace has a description followed by Gray reference "(a)" for authority, Strickland reference "(b)" for synonyms, and Gray reference "(a)" for the best figure.

Incidentally, *Accipiter ruficeps* may have been a name made up as an exemplar for Wallace's musings. Gray did not bestow this name on any *Accipiter* species, but in 1858 he did describe the Fijian hawk, *Accipiter poliocephalus*. It is curious that *ruficeps* means "rufous-crowned" while *poliocephalus* means "gray-headed."

English zoologist George Robert Gray (1808–1872) was the younger brother of John Edward Gray (1800–1875), also a noted zoologist. G. R. Gray was keeper of the ornithological section of the British Museum for more than forty years and authored the monumental *Genera of Birds* (1844–1849) and other bird (and insect) monographs. Gray described Wallace's standardwing, *Semioptera wallacei* (see recto 135), the bird of paradise Wallace found on the Moluccan island of Batchian (Bacan).

2. Following are a few insect examples. *Gymnetis lanius* is a scarab beetle found on Caribbean islands. Karl Hermann Burmeister (1807–1892), author of many works on entomology as well as herpetology, is recognized as the author of this species.

3. *Trichius abdominalis* belongs to a group of scarab beetles (Scarabaeidae) known as the bee beetles, subfamily Trichiinae. A genus of about seven species, the bee beetles have fuzzy bodies with yellow elytra bearing prominent black spots or bands—on the whole, reminiscent of bumblebees. Another *Trichius* species, *T. fasciatus*, captured by Wallace near Neath and duly reported in the April 1847 issue of the *Zoologist*, occasioned the first appearance of his name in print (Wallace 1847). (See also verso [d]).

4. Refers to the following page, recto 158, where species names and authors are in tidy columns.

[158]

[1]Form for a Synonymical Catalogue
Cetonia aruginosa.

~~Cetonia aruginosa a Drury—Illus.[trations] &c. [illeg.] .. (1770)~~
~~Scarabaeus speciosissimus~~

| | 2nd. column. | 1st. column. | | |
|---|---|---|---|---|
| Drury . . . . | | Cet.[onia] aruginosa | Illus. &c. Fig | 1770 |
| | | *[Illustrations of Natural History]* | | |
| Scopoli. . . . . . | | Scar.[abaeus] speciosissimus | Del. Fl. &c. | 1776 |
| | | *[Deliciae Flora et Fauna Insubricae Ticini]* | | |
| Herbst . . . . . | | Cet.[onia] speciosissimus | Col[?] &c. | 1790 |
| Olivier . . | | Cet.[onia] aurata. var.[iety] | Entom. &c. | 1789 |
| | | *[Entomologie]* | | |
| Fabricius— | | Cet.[onia] fastuosa | Ent. Syst. &c. | 1801 |
| | | *[Systema entomologae]* | | |
| " | | " " | Syst. El. | 1801? |
| | | *[Systema eleuthatorum]* | | |
| Gory et Percheron | | Cet.[onia] aruginosa | Mon. &c. | 1833 |
| | | *[Monographie des cétoines et genres voisins]* | | |
| Burmeister ___ | | Cet.[onia] aruginosa | Hand. der Ent. | 1842 |
| | | *[Handbuch der Entomologie]* | | |

[2]The above would be an immense labour &
of no necessity. A catalogue determining
authorityatively the name to be used &
giving references to the 2 or 3 best figures
& original descriptions would be ample.
This would be possible, the other
impossible!

1. Here Wallace sketches a possible "Form for a Synonymical Catalogue," with an orderly listing of synonyms, authors, and references. The model on this page is the scarabaeid beetle *Cetonia aruginosa* (now *aeruginosa*), the emerald fruit beetle.

2. Creating a comprehensive record of the nomenclatural history of each and every species would indeed be immensely laborious. Wallace was no doubt dismayed contemplating the prospect of cataloging his thousands of specimens. On recto 68–69 he suggested that "a series of synonymical catalogues" giving all the citations for the nomenclatural history of each species would be a useful reference tool, obviating the need to repeat the information in every species description. In this later entry Wallace seems to conclude that such a synonymical catalog would be a lot of unnecessary work. We can imagine him throwing up his hands at the prospect of documenting all the synonyms and their authorial sources just for his own collection. Cutting to the chase, he suggests again that simply including the accepted name and two or three references for the best figures and the original descriptions should suffice—a more tractable approach, in any case. (See recto 129 and discussion in Costa 2013.)

Dorcopsis asiaticus.

| Pallas. | Didelphys asiaticus | n.a. Petrop. Fig. | 1777 |
| Gmelin | Didelphus Brunii | S.N. 1.109 | 1798 |
| Cuvier | " " | Tabl. Elem | 1798 |
| Gray. | Halmaturus asiaticus | List. Mam. B. M. | 18 — |
| Illiger | Halmaturus Brunii | Prodrom — | 18 — |
| Muller | Hypsiprymnus Bru | Verhand — | 18 — |
| Gray | Dorcopsis Brunii | Verhand. — | 18 — |
| Gray — | Dorcopsis asiaticus | Voy. Samarang. | 18 — |

---

[1]Dorcopsis asiaticus.

| Pallas. | Didelphys asiaticus | n.a. Petrop. Fig. | 1777 |
|---|---|---|---|
| | [Acta Academiae Scientiarum Imperialis Petropolitanae] | | |
| Gmelin... | Didelphis Brunii | s.n. 1.109 | 1798 |
| | [Systema Naturae] | | |
| Cuvier... | " " | Tabl. Elemen. | 1798 |
| | [Tableau élémentaire de l'histoire naturelle des animaux] | | |
| Gray. | Halmaturus asiaticus | List. Mam. B.M. | 18__ |
| | [List of the specimens of Mammalia in the collection of the British Museum] | | |
| Illiger | Halmaturus Brunii | Prodromus— | 18__ |
| | [Prodromus systematis] | | |
| Muller | Hypsiprymerus Brunii | Verhand— | 18__ |
| | [Verhandlungen] | | |
| ~~Gray~~ " | Dorcopsis Brunii | Verhand— | 18__ |
| | [Verhandlungen] | | |
| Gray __ | Dorcopsis asiaticus | Voy. Samarang | 18__ |
| | [Zoology of the Voyage of H.M.S. Samarang] | | |

1. This catalog example uses the Aru kangaroo *Dorcopsis asiatus*. The format is similar to the descriptive synonymy given by John E. Gray in his paper "A List of Species of Mammalia Sent from the Aru Islands by Mr. A. R. Wallace" (J. E. Gray 1858). Ironically, Gray criticized Wallace for neglecting to send an actual specimen of this kangaroo, which Wallace felt was distinct from the Papuan forms. Gray's comment that "it is to be hoped that he did not neglect it, thinking it a common Kangaroo" may have stung Wallace, whose stock-in-trade was the ability to recognize even subtle differences among populations in different geographical locales. As early as his paper on the monkeys of the Amazon (Wallace 1852b) Wallace chided collectors for not paying close attention to the precise geographical locale in which they collected their specimens.

## 160

[¹]Psittacus lory. L.[innaeus] is a great nectar eater;—When
nectariferous trees are in flower they are frequented
by flocks of Trichoglossus & Eos & small finches
2–3 at a time of Lorius. When these are
shot a stream of nectar often pours out of
their mouths & their heads are often quite
dusty with pollen. They must be in
these countries an important agent in crossing
the individuals of hermaphrodite flowers &
in fertilising those with separate sexes . .

1. *Psittacus lory,* now *Lorius lory* (parrot family, Psittacidae) was described by Linnaeus in the tenth (1758) edition of *Systema Naturae.* Lories (modern genera *Chalcopsitta, Eos, Lorius,* and *Vini*) and lorikeets (genera *Charmosyna, Glossopsitta,* and *Neopsittacus*) are broadly distributed throughout the South Pacific and Australasian regions. (See recto 75 for Wallace's reference to another brush-tongued parrot, *Eos garrula.*)

Wallace notes here how these nectar-feeding birds must be important pollinators. Indeed, they are the main pollinators of a host of ornithophilous (bird-loving) plant species, the flowers of which exhibit a syndrome of bird-attractive characteristics such as red or orange color, long tubular shape, abundant nectar, and absence of scent.

*Lorius lory* is one of sixty-plus species of flower-visiting brush-tongued parrots, so-called for the specialized hairlike structures (papillae) at the distal end of their tongue. The papillae stand on end as the tongue is extended to feed, increasing the exposed surface area to facilitate the consumption of nectar and pollen. They also use their tongue to feed on the juices of soft fruits. (See Willmer 2011, chapter 15, "Pollination by Birds.")

## 161

[1]Paradisea rubra.

In the island of Waigiou takes the place
of P.[aradisea] papuana in New Guinea & P.[aradisea] apoda
in Aru. Its cry is very similar to these
species but hardly so loud & shrill,—or
quite so well defined. It seems very abundant
by its voice being clearly heard in the forest
but the birds are so continually in motion &
generally singly so that they are but seldom
seen & the adult males very rarely indeed.
They feed on the Waringin or Indian fig & as these
are very lofty trees they are difficult to shoot. They
however often descend into the low trees & shrubs
& even very near the ground creeping among the twigs
& running along the trunks apparently in search of
insects when fruit is scarce. At such times
they utter a low cluck-cluck, very different
from their usual loud note which they give
utterance to when perched on some lofty
tree.

1.  The red bird of paradise, *Paradisea rubra*, is found in the Moluccas, just west of New Guinea, while the lesser bird of paradise (*P. papuana*) is found in New Guinea and the greater bird of paradise (*P. apoda*) in the Aru Islands. Wallace encountered *P. rubra* on the western Papuan island of Waigiou (now Waigeo), off the Doberai Peninsula. He arrived there on July 4, 1860, after an arduous journey from Ceram during which he was nearly shipwrecked and, having reached the village of Muka, had to send a boat back to rescue two hired assistants marooned on an island. The rescue took the better part of a month. In *Malay Archipelago*, Wallace concluded the account of his Ceram-to-Waigiou ordeal by declaring himself "much relieved that my voyage, though sufficiently unfortunate, had not involved loss of life" (*Malay Archipelago*, 527).

On this and the next few pages Wallace describes the plumage and habits of the red bird of paradise. Some of this description was incorporated into *Malay Archipelago* (see chapter 38, "The Birds of Paradise").

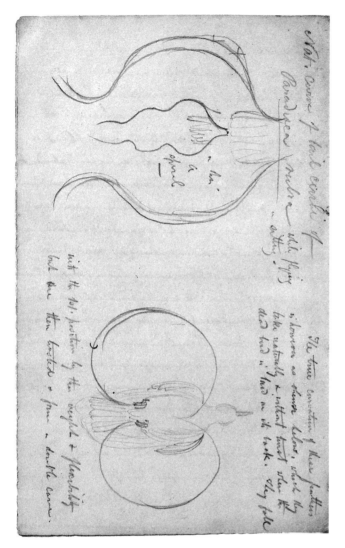

## 162

*[The following is written vertically on the page]*
[1]Nat.[ural] curve of tail cirrhi of
<u>Paradisea rubra</u> while flying
       or sitting.
*[Sketch of Paradisea rubra tail cirrhi or specialized feathers]*
*[The following is written vertically on the sketch]*
*[illeg.]* this
a
<u>spiral</u>

The true curvature of these feathers
is however as shewn below, which they
take naturally and without twist when the
dead bird is laid on its back. They fall
*[Sketch of ventral view of Paradisea rubra showing natural
curvature of tail cirrhi]*
into the top position by the weight & flexibility
but are then twisted & form a double curve.

1. The "tail cirrhi" Wallace sketched are the specialized feathers of this fabulous bird of paradise, the *Paradisea rubra*. Wallace notes that the feathers form a graceful circle by their natural curvature, but that this is not ordinarily evident in living birds, as the feathers hang down (or stream behind, when flying) in a corkscrew fashion (see illustration on the next annotation page and description on recto 165).

The breathtaking plumage and courtship displays of birds of paradise and other birds are understood today in the context of sexual selection, which Darwin saw as a special form of selection. Wallace was slow to accept sexual selection, particularly as it applied to what we now call female choice (see discussion in annotations for recto 89). Even in his book *Darwinism* (1889) we find Wallace attributing the ornamentation of males to their putatively greater "vital force," or sexuality. This may be seen as an example of interpreting observations through the lens of prevailing cultural mores.

Ironically, though now deemed essentially incorrect there are elements of Wallace's view on the subject that conform nicely with sexual-selection theory. He pointed out in *Darwinism*, for example, that the dramatic ornamental "accessory plumes" of many birds are of no real use, and indeed "must be rather injurious than beneficial in the bird's ordinary life. The fact that they have been developed to so great an extent in a few species is an indication of such perfect adaptation to the conditions of existence, such complete success in the battle for life, that there is, in the adult male at all events, a surplus of strength, vitality and growth-power which is able to expend itself in this way without injury" (Wallace 1889, 293).

This is precisely one of the modes of operation of sexual selection in the modern view: in intersexual selection, plumage that is energetically expensive and cumbersome, and that may therefore place the male at greater risk of predation, can be selected for because it is an indicator or signal of fitness (Cronin 1991).

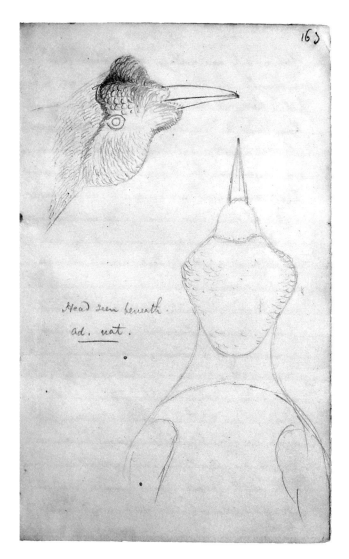

[163]

*[Sketch of Paradisea rubra head, lateral and ventral views]*

[1]Head seen beneath.
<u>ad. nat.</u>[uram]

1. A pencil sketch cannot, of course, convey the vivid coloration of the bird. Its yellow and iridescent-green head and red tail feathers were later gorgeously painted by the well-known ornithologist and bird artist John Gould (1804–1881) and included in *The Birds of New Guinea, and the Adjacent Papuan Islands* (Gould 1875–1888, 1: plate 30). While Gould was working in Australia he obtained a specimen of *P. rubra* from Wallace; this specimen is now one of the treasures of the Museum Victoria in Victoria, Australia.

Red bird of paradise *(Paradisea rubra)*. From Alfred Russel Wallace, *The Malay Archipelago*, 4th ed. (London: Macmillan & Co., 1872), 527.

## 164

There is only one village where these birds
are caught by the natives, in the others in
fact their existence is scarcely known, &
there are in all only six or eight men
who are acquainted with the art. [1]I resided
six weeks at this place in a little native
hut, surrounded by rugged & sharp coralline
rocks to walk over which is almost impossible.

While here I had many of the birds
brought me alive & kept several in a large
bamboo cage for some days, but owing to
not being able to procure their proper
food they all died very quickly. I was
however able to observe their habits &
admire the beauty of their plumage. The
head is the most beautiful part of the
bird owing to the rich swelling emerald
throat & the two little crests on the head
contrasting with the delicate pale yellow
of the neck breast & shoulders. The
tuft of blood red [side] plumes rise behind the

1. Wallace's residence in Bessir, Waigiou, is described in chapter 36 of *Malay Archipelago*:

It was quite a dwarf's house, just eight feet square, raised on posts so that the floor was four and a half feet above the ground, and the highest part of the ridge only five feet above the floor. As I am six feet and an inch in my stockings, I looked at this with some dismay; but finding that the other houses were much further from water, were dreadfully dirty, and were crowded with people, I at once accepted the little one, and determined to make the best of it. At first I thought of taking out the floor, which would leave it high enough to walk in and out without stooping; but then there would not be room enough, so I left it just as it was, had it thoroughly cleaned out, and brought up my baggage. The upper story I used for sleeping in, and for a store-room. In the lower part (which was quite open all round) I fixed up a small table, arranged my boxes, put up hanging-shelves, laid a mat on the ground with my wicker-chair upon it, hung up another mat on the windward side, and then found that, by bending double and carefully creeping in, I could sit on my chair with my head just clear of the ceiling. Here I lived pretty comfortably for six weeks, taking all my meals and doing all my work at my little table, to and from which I had to creep in a semi-horizontal position a dozen times a day; and, after a few severe knocks on the head by suddenly rising from my chair, learnt to accommodate myself to circumstances. We put up a little sloping cooking-hut outside, and a bench on which my lads could skin their birds. At night I went up to my little loft, they spread their mats on the floor below, and we none of us grumbled at our lodgings. (1869a, 534–535)

Wallace's stilt house at Bessir, Waigiou, New Guinea. From Alfred Russel Wallace, *The Malay Archipelago*, 4th ed. (London: Macmillan & Co., 1872), 535.

wings & the curved white ends droop over the
tail, but [1]I had no opportunity of seeing
them properly displayed when they no doubt
rise up at a considerable angle as in
the allied yellow plumed species & must
have a very beautiful effect though being
so much shorter they cannot equal the
grace & ~~beauty~~ elegance of P.[aradisea] apoda & P.[aradisea]
    papuana.

The singular tail cirrhi have however
a very unique appearance hanging down
in a graceful spiral curve owing to their
section being semi cylindrical & ~~the~~ having
naturally a large outward curve forming
in fact in the first killed bird laid on its back a complete
circle. [2]These singular, whalebone like feathers
undergo many ~~singular~~ curious changes of form
in successive moults before arriving at
their perfect condition.

1. Wallace had a difficult time collecting the secretive red bird of paradise, but he eventually took several that had come to feed on the fruits of an Indian fig tree near his house (see recto [b]). As these birds had feeding and not breeding on their minds, he did not observe them in full regalia, with their breathtaking ornamental plumes deployed as they would be in courtship.

Although he states here that these birds cannot equal *P. apoda* and *P. papuana*, in *Malay Archipelago* he wrote that while *P. rubra* "wants the grace imparted by their [*P. apoda* and *P. papuana*'s] long golden trains, [it] is in many respects more remarkable and more beautiful" (530).

2. Wallace calls these specialized, hard, shiny feathers "whalebone like" for good reason: they are made of the same substance. "Whalebone" is baleen, the stiff, densely packed bristles inside the mouths of baleen whales that are used to strain krill from the water. Baleen, like birds' feathers and mammalian hair (and claws and nails), is largely composed of the protein keratin.

## 166

[1]Instinct—Birds' Nests.
In investigating instinct we proceed by degrees,
from the easy & near to the diff.[icult] & remote. It is
a mental characteristic & we must study
first animals not too far removed from
ourselves.

We are not even yet decided what
are the senses possessed by insects.
What relation their powers of seeing
hearing & feeling have to ours we
are quite ignorant. Their sight may
be microscopic or telescopic, perfect or
imperfect—we cannot decide. They
may appreciate by sound & by feeling
what to our senses are quite incomprehensible.
Their antennae may be instruments of a
sense of wh.[ich] we know nothing. [2]How
then in the midst of this ignorance
can we pretend to fathom their minds
to say what & how much they perceive
or remember or reason or reflect?

1. These notes and those on the next two pages are draft arguments for Wallace's essay "On Instinct in Man and Animals," which was included in his *Contributions to the Theory of Natural Selection* (1870, 91–97). Here Wallace is not so much addressing birds' nests as he is the concept of instinct generally, and these ideas may have grown out of further reflection on the instinct concept when he was working on birds' nests (his "Philosophy of Birds' Nests" was published in 1867; see recto 112–117).

It is noteworthy that the topics he addresses on recto 166–172 exactly parallel the order in which they appear in "On Instinct in Man and Animals." Consider the following section headings as taken in order from the essay and their corresponding pages in the Species Notebook:

"How Instinct May Be Best Studied"—recto 166–167

"Definition of Instinct"—recto 168

"Does Man Possess Instincts"—recto 168–169

"How Indians Travel through Unknown and Trackless Forests"—recto 170–172

Could these entries have been made after—even well after—Wallace's return home in 1862? He noted in the preface to *Contributions* that the essay had not been previously published, but he gives no indication of when he first outlined the ideas for it. The notebook entries are not dated, but there is, after all, no reason to think that they were *not* penned in the eastern archipelago, though it is possible, if improbable, that he used this notebook a half dozen years after his return as a place to sketch out his ideas. (The later entries, after these, by and large correspond to the years 1860–1862, with at least one made soon after his return to England.)

In the 1870 essay he expanded the entries in the Species Notebook in several ways, for example by discussing Benjamin Thompson Lowne's studies of the sensory organs of different insect groups. We shall see that, partly in connection with his Owenite philosophy of self-improvement and the importance of learning and environment, Wallace downplayed the importance of inherited behavior, whether in birds, dogs, horses, or humans

(Jones 2002). (See Wallace 1870, 1873; Species Notebook recto 168–172.)

Here he starts with the suggestion that we must investigate instinct systematically, working from the "easy & near" to the "diff.[icult] & remote." He is right that other organisms experience the world in quite different ways from humans. In day-active insects, for example, the visual spectrum is shifted relative to that of humans and other vertebrates, such that they cannot perceive wavelengths at the red end of the spectrum but can see well into the purple to ultraviolet wavelengths. This is why red coloration in insect-pollinated flowers is rare, while it is quite common in bird-pollinated flowers. Many insect-pollinated flowers have markings called nectar guides that are visible only in ultraviolet wavelengths. The insect world is also very much more of a chemosensory world than that of vertebrates; the antennae that Wallace mentions are indeed "instruments" that would be understood in this respect only with the benefit of twentieth-century electronic technology.

2. Wallace finds it presumptuous for humans to make pronouncements about the "consciousness" or "mind" or "reason" of insects when we are so profoundly ignorant of their sense perceptions.

## 167

To leap at once from our own consciousness
of mind & reason to that of insects
is as absurd as if we were to attempt
to comprehend the higher brains of mathematics
with only the preliminary knowledge of the
multiplication table. [1]In other sciences
we do not proceed so. How have
Cuvier & Owen & Huxley acquired their
knowledge of animal forms & all the
beautiful homologues of the skeleton.
Not by passing at once from the
most simple to the most complicated
but by tracing modifications step by
step. Suppose that Prof.[essor] Owen had
confined his attention solely to the
skeleton of <u>man</u> & of a <u>mackerel</u>
a life time of study would never
have shewn him their [true] relations & he
would inevitably have come to the
conclusion that there was not continuity[?]
of Plan between them.—
So with instinct—we use the term to shew our ignorance
of the [illeg.]. plan of all intellect

1.  The argument on this page sounds like an endorsement of the idea of continuity in the organic world—that humans and other animals (insects) differ in *degree*, not *kind*, if they can be linked by intermediate steps (albeit a great many of them). Richard Owen would not have seen the essential continuity between humans and mackerels if he had confined his analysis to just these extremely different organisms, Wallace argues. The "continuity of Plan" between them is made apparent only by tracing the modifications step by step.

In other words, Wallace argues that an appreciation of homology stems from a careful, step-by-step comparative study of different forms, not from simply comparing the simplest with the most complex form, and that, by analogy, "instinct" and "intellect" can be understood only through careful study of all manner of organisms (such as the insects mentioned on recto 166–167). That we need more data is the main point.

This argument and that on the previous page (166) can be found in the opening section of the essay "On Instinct in Man and Animals," titled "How Instinct May Be Best Studied" (Wallace 1870).

## 168

[1]Define instinct. The performance of a
complicated act absolutely without previous
instruction or knowledge of it. Thus it is said
& repeated, that birds & insects build nests
gather & store food & provide for the[ir] future
wants without any instructions fr[om] the[ir] fellows
& without even knowing that such acts have
been performed by others. This however is
assumed. It has never been tried. Bees
have never been carefully secluded from the
larva state, & then loosed in an enclosure
where they could not communicate with
other bees. Birds reared from the egg
in confinement have not been shewn to
make the same nest as their fellows.

[2]Has man <u>instincts</u>? No. He
may perform some simple operations without
teaching but never compound ones. It
is sometimes absurdly stated that the
infant seeks the breast & sucks by
instinct. But sucking is as mechanical [& human/?] an
act as breathing or swallowing, & we think
few infants would survive were their natural
food left to their own instinct to discovered *[sic]*
in the first 24 hours of their existence,

1.  Again, these arguments represent in draft form those in Wallace's essay "On Instinct in Man and Animals" (see recto 166), the second section of which is entitled "Definition of Instinct." Wallace critiques prior definitions, using the examples of birds' nests and honeybee cells. His position as expressed in the essay is that "instinct should be defined as—'the performance by an animal of complex acts, absolutely without instruction or previously-acquired knowledge.'" The key issue is that birds were commonly viewed as being able to build their nests and honeybees their waxen cells "without ever having seen such acts performed by others, and without any knowledge of why they perform them themselves" (Wallace 1870, 204). Wallace disputes this, pointing out that no one had yet done the experiment of raising birds or bees in isolation from conspecifics to determine if they instinctively know how to build. Learning from nest mates may be essential, he concludes.

Today it is understood that "instinct" is a problematic concept. There are indeed some seemingly "hard-wired" fixed-action-pattern behaviors, but most behavior is shaped by the interplay of genetics and the environment. The degree of influence of experience and learning varies widely among behaviors and among taxa. (See also discussion on recto 172.)

2.  This corresponds to the third section of the 1870 essay: "Does Man Possess Instincts." Wallace dedicated several pages of the essay to considering the question of human instinct, arguing against the idea that humans have any instinctive behaviors. In the sucking infant example, he draws a distinction between "instinct" and what we might term a mechanical behavior driven by necessity: "This is one of those simple acts dependent upon organization, which cannot properly be termed instinct, any more than breathing or muscular motion" (Wallace 1870, 206). Eating, he seems to suggest here and in the essay, is simply a mechanical act done out of necessity and not instinct per se, any more than breathing or walking are instinctive.

## 169

[1]for it cannot be seriously imagined that the
newly born infant if totally unassisted
could make the slightest attempt to find its
natural food or is not wholly & absolutely
dependant on its *[illeg.]* giving him others. As to
the mere act of sucking which has been
said to be instinctive,—it is on the contrary
involuntary! at least if breathing is so,
for breathing is but sucking in air & if
we *[illeg.]* stop the mouth with any porous[?]
or perforated mass containing an agreeable
liquid it is drawn into the mouth instead of the air
& the sensation produced by its absorbtion
being pleasant the infant continues the
act of sucking liquid broken only by the
necessary involuntary suction of air.

What are very commonly called instincts
in man are only habits. It is said, when
in danger of falling we instinctively grasp
at some object to sustain us—, & we
put out our hands to save our body,—but

1. The argument continues from the previous page. Note the statement near the bottom of the page: "What are commonly called *instincts* in man are only habits." What motivates this critique is perhaps Wallace's determination to separate humans in a cognitive sense from the rest of organic nature. That is the essence of the bombshell he dropped in 1869, the year he publicly declared that the human brain could not be explained by natural selection, and that therefore human consciousness and cognitive abilities must be the product of a divine plan of some kind.

In this light, it would be of interest to know the answer to the question posed on recto 166, namely just when these passages were penned. If indeed they were made at the end of his journey in the Malay Archipelago, or very soon after his return, this would have bearing on contemporary thinking about when and how Wallace developed his contrary ideas about human evolution. Other passages in the Species Notebook, such as those copied out from Ernest Renan and George Combe (recto 154–155) suggest an awareness and interest in "higher" considerations when he was in the East.

Wallace's earliest public announcement of his position that natural selection could not explain human cognition was, as noted above, made in 1869. It was concerning this astonishing statement about human evolution that a disappointed Darwin wrote to Wallace in 1869, hoping, he said, that Wallace had not "murdered too completely your own and my child" (Darwin Correspondence Project, letter 6684). This famous break with Darwin (which strained but did not end their friendship) was of a piece with Wallace's growing commitment to spiritualism through the mid- to late 1860s (Gross 2010, Jones 2002, Kottler 1974, C. H. Smith 1992).

In 1870 Wallace was defending his apostasy, and in his usual spirited style he continued to attack arguments for physical continuity between humans and other animals. True instinct in humans would indeed represent an example of such continuity—supporting the idea that humans differ in degree and not kind from other animals. Thus, an important object of Wallace's 1870 essay can be seen as a sustained attack on this idea, such as where he sets up the argument by noting that "many of the upholders of the instinctive theory maintain, that man has instincts exactly of the same nature as those of animals, but more or less liable to be obscured by his reasoning powers." Was the argument largely sketched out here in the Species Notebook as much as a decade earlier, or were these passages added sometime well into the 1860s?

## 170

[1]these are the results of <u>habit</u>,—a child
does not do them & suffers in consequence
till <sup>repeated</sup> experience has rendered the habit confirmed.
So when any thing is thrown towards <sup>what *[illeg.]*</sup> us which
we wish to catch, a man closes his
knees, a woman opens hers,—both it may
be said <u>instinctively</u>, but probably the result of
the habits caused by the form of the dress of
each.
? What do <u>chicks</u> hatched by stream do
when first hatched—must see—
<u>ducks do</u>.—Do they for instance <u>scratch</u>
& then <u>step</u> <u>backwards</u> which they may soon
learn to do from the <u>hen</u>. The action
of the muscles may lead to <u>pecking</u> as it does
to walking—but with it lead to destroying
wholesome food from other sorts—&c. &c.

———————

[2]Indians said to find their way in the
forests by instinct. Nothing of the kind. They
can find their way in <u>their own</u> forests
because from infancy they have been used
*[Continued bottom of recto 171]*

1. The examples on this page are not included in Wallace's 1870 essay.

2. This interesting bit of Victorian-era lore is discussed in the fourth and final section of the 1870 essay: "How Indians Travel through Unknown and Trackless Forests." The idealized "noble savage" image of Native Americans included their seemingly uncanny ability to traverse the vast and trackless wilderness without aid of any kind—by instinct, as it were. Wallace is dismissive of this idea here, and he addresses the matter in greater detail in his essay. "This is much misunderstood," Wallace writes, "for I believe it is only performed under such special conditions, as at once to show that instinct has nothing to do with it" (1870, 207). The point he makes here—indigenous people may be able to find their way in *their own* forests—reflects his belief that this is more familiarity from experience than instinct.

## 171

[1]Corydon Sumatramus. .
Skin form, flesh with the sweet smell of fruit eating
thrushes or parrots, none of the rank fetid odour
of all fissirostres, legs strong, thigh bone 1¾ in.[ch]
tibia 1¼ in[ch]: thighs fleshy & powerful as in the typical
perchers,—flesh of thigh skinned 1¾ in[ch]: in circumference.
Scull firm hard, not cellular, orbits round entire
above; brain rather large. .
Wings moderate—fore arm 1¾ in.[ch]
Stomach muscular, with fruit & soft insects (larvae &c) locustae[?]
larynx simple at fork with lateral muscles attached
high up.
Sternum truly passerine.
Nest, suspended from a branch.

---

*[Continuation of recto 170]*
[2]to wander in them. They know the highlands &
lowlands, the course of the main streams,
the direction of the dividing ridges, & the
general slope & lay of the country, even
though to a stranger it is apparently
level. Thus, when required to cross this
forest country even along a line they have
never been before, they always know where
they are. The direction of the streams, the
nature of the soil, the slope of the ground,

1. The dusky broadbill of Sumatra is a member of the Eurylaimi-dae (broadbill family)—a small family of mostly insectivorous passerines (perching or song birds—note that Wallace recognizes it as a passerine from its sternum). These birds have a broad, flattened beak that is as wide as it is long; Wallace refers to it as the "great gaper" in *Malay Archipelago*. Wallace reported collecting his first specimen of the dusky broadbill on Sunday December 8, 1861, in the Maura Dua region of Sumatra (manuscript WCP4767, 77).

"Fissirostres" is an obsolete term for insect-eating birds with such wide beaks, including swallows and goatsuckers. Their remarkably wide gape helps these birds eat insects on the wing. Wallace notes that fissirostres typically have a fetid odor but the dusky broadbill is rather sweet-smelling, maybe because it also eats fruits (see mention of stomach contents). John Gould included a lovely illustration of this species in his seven-volume *The Birds of Asia* (London: Hullmandel & Walter, 1850–1883).

The Latin name of the dusky broadbill is now spelled *Corydon sumatranus*. The species was named in 1822 by Sir Thomas Stamford Raffles (1781–1826), the "father of Singapore." A naturalist as well as statesman, Raffles was an avid natural-history collector. He set up the Botanic Garden of Singapore, and his collections were used to establish a natural history museum (now the Raffles Museum of Biodiversity Research). *Rafflesia*, a genus of parasitic plant renowned for having the largest flowers in the world, is named in his honor, as are a host of other plant and animal species.

2. The following is a continuation, from the bottom of recto 170, of the discussion of Indians' purported ability to find their way in "trackless" areas.

## 172

[1]the character of the vegetation tell them at
every step where they are. They are never
confused, they have no fear of being lost,
for they are at home. As they approach a
tract they have been in before many minute
indications guide them. They cross some
stream perhaps which they know must be
one they have already reached in another place.
Here on they take a new course, & come out
at the exact spot desired. The European
to whom they have acted as guide,—is astounded
asks them if they have ever made the same
journey before. They answer "no", & he then
imputes their general local knowledge to
"instinct" & thinks they have come into
a true straight line, with no material or
visible objects to direct them. But take
these same men into another country with
other streams & hills & soil & vegetation, &
after bringing them by a circuitous course
from a given point tell them to return
through a forest of some extent, & they will
be certainly decline the attempt. Their instinct
will not act out of their own country.

101 *[Written upside down]*

1. Conclusion of the discussion of Indians' ability to find their way in "trackless" areas. For Wallace the bottom line, literally and figuratively, is that experience teaches native peoples how to navigate through their world; they do not have some special instinct that allows them to do so.

Modern behaviorists would agree with Wallace generally, that humans exhibit no instinctive behaviors (not to be confused with reflexes), but they would also point out that this is no longer considered an informative way to frame the question. Through much of the twentieth century, "instinct" behavior was defined in terms of fixed or innate patterns of behavior reflecting a "hard-wired" genetic basis, in contrast to malleable or flexible behaviors contingent on experience and learning. Domjan (2012) discussed "instinct" as a focus of the European ethological research tradition, largely stemming from the work of ethologists Nikko Tinbergen (1907–1988) in the Netherlands and Konrad Lorenz (1903–1989) in Austria, while the "behaviorist" research tradition in the United States focused more on behavior as flexible and shaped by environment or experience.

Today the distinction between learned and instinctive behavior is no longer seen as a valid framework for understanding behavior and its evolution, owing to, among other things, a growing awareness of a subtle and complex interplay between genetics *and* environment in the expression of behavior (Bouton 2007, Champagne and Mashoodh 2009).

## 173

[1]Honey comb, is said to be constructed of
such form as to combine with mathematical
precision the greatest strength with the
least materials—quote authors.

This is erroneous on two points:
First a honey comb, being suspended vertically
from its upper edges it is evident that
when full of honey the strain upon the
structure will increase regularly from the
lowest to the highest row of cells:—
But these cells are all alike, of equal
size form & thickness; it is evident
therefore that the upper part being strong
enough the lower parts are unnecessarily
strong:—There is therefore a waste
of material, owing to a simple mechanical
law not being applied.
2ndly. The double series of horizontal cells
of which each comb is composed are so
placed that every cell sta is
opposed to parts of three opposite cells
and the bases of these cells

[Sketch of cells]
100 [Written upside down]

1. Starting here and running to the top of recto 177 is an extended discussion of the construction by honeybees of their waxen cells, in lore said to combine mathematical precision with an optimal balance of strength and economy of materials. These entries were likely inspired by Wallace's formulation of arguments rebutting a critique by Samuel Haughton titled "On the Form of the Cells Made by Various Wasps and by the Honey Bee" (Haughton 1863). Wallace replied to Haughton with a paper entitled "Remarks on the Rev. S. Haughton's Paper on the Bee's Cell, and on the Origin of Species," published in the *Annals and Magazine of Natural History* (Wallace 1863e).

Samuel Haughton (1821–1897) was a professor of geology at Trinity College, Dublin, and the author of several geological works. He was deeply religious (ordained in 1847, though he never became a vocational priest), and he became a severe critic of Wallace and Darwin's theory of evolution by natural selection. Bowler (2009) recounted how Haughton had the dubious distinction of being the first to attack the theory in print, when his criticisms, voiced at the Geological Society of Dublin meeting of February 9, 1859, were published soon afterward in the society's journal. It was Haughton who dismissed the theory, saying, "If it means what it says, it is a truism; if it means anything more, it is contrary to fact." (In his autobiography Darwin later dismissed Haughton in turn, writing that his criticism amounted to saying, "All that [was] new in there was false, and what was true was old.")

In his 1863 rebuttal of Haughton, Wallace defended Darwin's argument regarding bees' cells and the more economical use of wax by hive bees as compared with bumblebees. In chapter 7 of the *Origin of Species* Darwin argued that selection favored the economy of wax in the evolution of the instincts of hive bees, and that related bees, such as *Melipona* and bumblebees, could be seen as exhibiting transitional forms of cell construction. The main point discussed here in the Species Notebook relates to the claim about bees' construction of their honeycombs—and this is also the way Wallace opened his 1863 paper. The bee is said, he wrote, to be "enabled to construct [its] cells on true mechanical and mathematical principles, so as to combine the requisite accommodation for rearing its brood and storing its honey, with the greatest amount of strength and the utmost economy of material" (Wallace 1863e, 304).

Honeycomb was commonly cited in the natural-theology tradition for its mathematical properties: its regular hexagons that allow for packing in a maximum number of cells per unit area (this is the "recondite problem" solved by bees that Darwin mentions in chapter 7 of the *Origin*), and its structural or "engineering" properties, as honeycomb was thought to exhibit the strongest structure possible for the smallest mass of building materials. Wallace's memorandum to "quote authors" may refer to Lord Brougham's writings on the subject. Henry Lord Brougham (1778–1868), 1st Baron Brougham and Vaux, Scottish statesman and education reformer, discussed animal instinct and the ability of bees to construct mathematically perfect cells in *Dissertations on Subjects of Science Connected with Natural Theology* (1839) and *Dialogues on Instinct* (1844), books that Wallace and Darwin were familiar with.

Unlike the Species Notebook entries bearing on man and instinct (recto 166–172), which are draft notes for his paper on the subject, the discussion of honeycomb and bees' cells appears to be an interesting side issue that occurred to Wallace while he worked on his rebuttal to Haughton. Contrary to received wisdom, he is going to argue here that the bees are in fact wasteful.

## 174

[1]are not horizontal but each composed
of <u>three planes</u> sloping down to a
point at a certain angle wh.[ich]
is said to be the angle of greatest
strength with least material.

[2]But it is evident that in this part
of the structure strength is not required
since the pressure can never be more
than that ~~of~~ produced by one cell full of honey, & when
the opposite cells are both full, there
is <u>no pressure</u>. But if strength
were wanted, a plane at right angles
to the cells, of equal thickness would
be stronger, having <u>less surface</u> with the
same pressure. For example if a
square vertical tube be required to bear
a great pressure of a fluid on its
bottom, that bottom would be stronger *[illeg.]*
horizontal ~~than if~~ oblique (if of the
same thickness) because presenting less
surface to the fluid pressure.

99 *[Written upside down]*

1. The double series of back-to-back cells in a honeycomb, where each cell is opposite to parts of three adjacent cells on the other side, is mentioned in Wallace's 1863 paper (1863e, 305). The bases of the three cells do not meet in a flat plane but form a little pyramid, "sloping down to a point at a certain angle [which] is said to be the angle of greatest strength with least material." This is the point he will attack.

2. Wallace argues that, while strength is not an issue in this part of the structure, even if it were, a *flat* base, not a pyramidal one perpendicular to the axis of the cells, would be stronger per area of pressure. This makes sense: if the cell's base extends down to a point, whatever pressure is exerted in the cell is concentrated at the point, whereas the pressure would be spread evenly over the larger surface of a flat plane.

Wallace builds this argument into his 1863 paper, where he points out that the primary function of the cell is not honey storage but brood rearing:

> The primary use of the cell, however, is not the storing of honey—but the accommodation of the larva and pupa; for this it must have a certain *diameter*, and the triangular cell must therefore circumscribe the circular one, and will then be found to require more materials even than the circular cell with solid intervals, without taking into account the fact that the sides of the triangular cells, being without support in their whole length, would have to be thicker than those of any other form, if of equal strength. The same argument will apply in a less degree to the walls of a square cell. (Wallace 1863e, 306)

## 175

[1]The bee therefore uses more material
to make a complicated base of inclined
planes to its cells, & by that very
circumstance makes them <u>weaker</u>
instead of stronger!

If this is correct,—then it follows
that the <sup>horizontal layers of</sup> hexagonal cells, with their pyramidal
bases & definite angles, have no mechanical
advantages for the particular uses to
which they are applied, but result probably
from some necessity of the animal's physical
& mental organization.

[2]What has always been considered the
most wonderful part of the structure, the
pyramidal base *[illeg.]*, seems to have
been ~~considered~~ <sup>quite misunderstood</sup> by mathematicians. Their
error has been in not considering that
the ~~edges~~ <sup>margins</sup> of each of the 3 planes forming<sup>*[illeg.]*</sup>
the pyramid are fixed & supported
by the edges of the opposite cells.
The true way ~~of making them~~ is therefore
to make them at r[igh]t. angles, with the
98 *[Written upside down]*

1.  Here is Wallace's key point. Far from representing the behavioral epitome of mathematical precision, the bees use more material to make a base that is more complicated than needed to store honey, and in the process they make the cells weaker, not stronger.

2.  I have suggested elsewhere (Costa 2009, 235) that in Wallace and Darwin's day the exquisite intricacy of honeycomb and individual cells put them on a par with the eye as an irresistible example of a complex structure unattainable by purely natural processes. It is thus no surprise that both Wallace and Darwin discussed honeycombs at some length and took pains to refute the Paleyan arguments for their perfection.

least surface. [1]Just as the Engineer
makes his double lock gates (which
are each supported on one side only)
meet at an oblique angle to support
each other, but a single lock gate
which is supported on both sides he
~~makes~~ places square with the walls
& thus with the least possible surface.

[2]If it were required to make the
thinnest & strongest base possible to
a <u>vertical</u> hexagonal tube to be
filled with a fluid, it is probable
the 3 planed pyramid rising <u>inwards</u>
would be very near the thing. But
in the bees cell you have 3 rigid
walls meeting in the centre to support
the base, & moreover the ~~base~~ tube
is <u>horizontal</u> not <u>vertical</u>, which
completely alters the condition of the
problem, as regards [fluid] pressure.

These facts were it seems to me

*97 [Written upside down]*

1. With his practical experience with design and engineering, Wallace surely appreciated the apt parallel between locks and bees' cells, which he may have read in C. M. Willich's 1860 paper "On the Angles of Dock-Gates and the Cells of Bees." The locks Wallace is referring to are canal locks—we can think of the section of canal between locks as analogous to the bees' cell, with liquid pressure brought to bear on one side. The doors of a double-gate lock meet in the middle, closing at a 109-degree inner angle to provide the greatest structural strength against the water pressure on the "upstream" side of the gates (which is maximal when the water level on the "downstream" side is lower). Normal single-lock gates, in contrast, swing open and shut horizontally and are perpendicular to the water pressure.

2. The bees' cells are horizontal, not vertical, and so Wallace in essence points out that the various arguments about the pyramidal-base design being optimal are moot, since it is only a good design, first, if the cells are vertical, and second, if the pyramid rises upward.

## 177

[1]quite unknown to or at least unnoticed
by the great mathematicians who have
made all the world believe that the
bees, by ~~pra~~ practically solving an
abstruse mathematical problem, have
given the most convincing proof ~~that~~ of
the supremacy of <u>instinct</u>, leading the
Poet when comparing it with <u>reason</u>
to exclaim

  "In this 'tis God that works in that 'tis man"

---

[2]<u>N.[orth] Africa & S.[outh] Europe</u>
Africa has most species common to Italian
Islands Sicily & Sardinia, fewer with Italian
main land, as is the case with Spain &
Morocco.—Spanish-Moroccan & Italian-
Algerian sp.[ecies] often analogous. <u>Erichson</u>
Submarine ridge from Sicily to Africa.

---

[3]Faunas of Europe—1st. African long extinct
2nd. the arctic, still existing on mountains
3rd. the Lusitanian & Asiatic—

96 *[Written upside down]*

1. Wallace believes that he has hit on a point concerning honeybees that even "great mathematicians" have missed. He concludes that the bees have not been endowed with an instinct to solve an "abstruse mathematical problem," but have a simple cell-building capacity stemming "from some necessity of the animals' physical and mental organization." The bees are not an example of the "supremacy of instinct."

He concluded his masterful rebuttal with a quote of Haughton's own words turned back on him: "No progress in natural science is possible as long as men will take their rude guesses at truth for facts, and substitute the fancies of their imagination for the sober rules of reasoning" (Wallace 1863e, 309).

It is on a related and ironic note that he concludes his discussion here. He seems to suggest that to maintain that there is a chasm between "supremacy of instinct" (in the animal world) and human reason is to infer a false distinction. Wallace paraphrases Alexander Pope (who is also cited on recto 143 in this context) as saying, "In this 'tis God that works, in that 'tis Man." What Pope wrote, rather, in *An Essay on Man* (1734; Epistle III, part 2) is: "In this 'tis God directs, in that 'tis man." The irony is that Wallace subsequently did argue for a chasm between humans and other animals, though when it came to instinct versus experience he was prepared largely to deny there were significant differences, believing that both humans and nonhumans rely on experience, not instinct.

2. These brief notes, or at least their theme, were incorporated into Wallace's anniversary address to the Entomological Society of London, given on January 27, 1871, subsequently published in the March 30, 1871, issue of *Nature* (3: 425–438) and reprinted as chapter twelve in *Studies Scientific & Social* (Wallace 1900). "Erichson" is German entomologist Wilhelm Ferdinand Erichson (1809–1848), who published extensively on beetle taxonomy and systematics.

These themes are familiar ones for Wallace: underlying causes of current biogeographical relationships. Note the reference to the submarine ridge linking Sicily to North Africa. In *Principles of Geology* Lyell discussed a similar ridge (perhaps the same one?) running from Gibraltar to North Africa.

3. Note, too, the theme of faunal succession in Europe, from a long-extinct African fauna to an arctic fauna that prevailed during glacial periods, eventually giving way to the modern fauna (with the arctic species persisting at high elevations).

## 178

[1]Spencer's First Principles

1. Religion & Science

All general beliefs have some basis of truth.

Religious ideas either a special creation, or

have arisen by a process of evolution. If

the first the matter wants[?] no further enquiry—If

the 2nd we must inquire into its growth & uses.

The religious feeling is *[illeg.]* produced by natural

causes, & in some way conducive to human

welfare.—

95 *[Written upside down]*

1. Social philosopher Herbert Spencer's book *First Principles of a New System of Philosophy* (1862), which articulated a vision of evolution whereby all structures in the universe trend from a simple, undifferentiated, homogeneous state to a complex, differentiated, heterogeneous state. This book was an elaboration of Spencer's essay "Progress: Its Law and Cause," published in 1857. Wallace's notes are from the very beginning of the opening chapter, "Religion and Science." The passage that caught Wallace's attention reads:

> § 1. We too often forget that not only is there "a soul of goodness in things evil," but very generally also, a soul of truth in things erroneous. While many admit the abstract probability that a falsity has usually a nucleus of reality, few bear this abstract probability in mind, when passing judgment on the opinions of others. A belief that is finally proved to be grossly at variance with fact, is cast aside with indignation or contempt; and in the heat of antagonism scarcely any one inquires what there was in this belief which commended it to men's minds. Yet there must have been something. And there is reason to suspect that this something was its correspondence with certain of their experiences: an extremely limited or vague correspondence perhaps; but still, a correspondence. (Spencer 1862, 3)

Wallace, who was back in England by early spring 1862, wrote to Darwin in September of that year that he was "now reading Herbert Spencer's "First Principles," which seems to me a truly great work, which goes to the root of everything" (Wallace Correspondence Project, WCP1852). This notebook entry can thus be dated to about this time. Inspired by his own and Darwin's evolutionary insights and .his recent reading of Spencer, Wallace and his old friend Henry Walter Bates paid Spencer a visit that year:

> Soon after my return home, in 1862 or 1863, Bates and I, having both read *First Principles* and been immensely impressed by it, went together to call on Herbert Spencer . . . Our thoughts were full of the great unsolved problem of the origin of life–a problem which Darwin's *Origin of*

*Species* left in as much obscurity as ever–and we looked to Spencer as the one man living who could give us some clue to it. His wonderful exposition of the fundamental laws and conditions, actions and interactions of the material universe seemed to penetrate so deeply into [the] nature of things . . . that we both hoped he could throw some light on that great problem of problems. (Wallace 1905, 2: 23–24)

Spencer had a profound influence on Wallace's thinking at this critical juncture. His enthusiastic embrace of Spencer's arguments in *First Principles* and *Social Statics* (1851) surely resonated with his nascent sense that transmutation was, for humans at least, the result of a "persistently directed force." Wallace was so impressed with Spencer that he named his first child for him. (Herbert Spencer Wallace was born in 1867 and died at the age of seven, in 1874). Wallace also readily adopted Spencer's phrase "survival of the fittest" and preferred this to Darwin's "natural selection." He systematically crossed out the words "natural selection" in his copy of the *Origin*, inserting "survival of the fittest" instead (Paul 1988). Darwin finally gave in to Wallace's exhortations to adopt the phrase, though he was not much of a Spencer fan himself. Beginning with the fifth edition of the *Origin of Species* in 1869, he noted, after defining natural selection, that "the expression often used by Mr. Herbert Spencer of the Survival of the Fittest is more accurate, and is sometimes equally convenient."

[179]

[1]94 *[Written upside down at bottom of page]*

1. The upside-down number "94" on this otherwise blank page is from the reverse (verso) notebook pagination. Note that upside-down page numbers run from 92 (on recto 181) to 101 (recto 172). Wallace numbered the recto and verso pages in advance, and in the case of the Species Notebook his entries on the recto side ran far longer than those on the verso side, encroaching on several of the pages originally numbered with verso pagination. As has been noted previously, Wallace used most of his field notebooks in this tête-bêche fashion, including his bird and insect collection registers (WCP4766, WCP4767).

[1]Zoologist Critique on Darwin
Classification—paralleled by arrangement
of [2]Planets & Satellites in Solar system.
Yes—but these have probably also had
a common descent from the sun.

———————

[3]Plants & animals have separated before
sexuality appears. Therefore sexes
in plants and animals have been
independently produced.
Qy. [Query?] are there no rudiments of these
in early organisms.

———————

How could [4]mammiferous nutrition
have been perfected by short stages?
I think it might. It seems much
more difficult to conceive how a
foetal animal should be born like the
kangaroo, by selection.

———————

[5]The most perfect animals of each
species are the strongest & therefore would
propagate the race true.
So they do till some modifi-

93 [Written upside down]

1. This concerns the publication in the *Zoologist* of a lengthy review of the third edition of the *Origin of Species*, along with consideration of several related works supportive of and opposed to Darwin's theory. The review, by geologist and botanist George Maw, appeared in *Zoologist* 19 (1861): 7577–7611. Darwin commented on this review in a letter to Wallace dated May 24, 1862:

> I write one line to thank you for your note & to say that the B. of Oxford wrote the Quarterly R (paid £60), aided by Owen. In the Edinburgh Owen no doubt praised himself. Mr. Maw's Review in Zoologist is one of the best; & staggered me in part, for I did not see the sophistry of parts—I could lend you any which you might wish to see; but you would soon be tired. Hopkins in Fraser & Pictet are two of the best. (Darwin Correspondence Project, letter 3570)

2. In the *Origin* (1859, 57) Darwin pointed out the symmetry in the relationships of varieties to parent species, species to parent subgenera, subgenera to parent genera, and so on, with groups of species clustered like satellites about other species. The *Zoologist* review stated that "there is not a single principle in organic classification that Mr. Darwin enlists as favorable to his theory that is not strikingly paralleled in some unrelated part of creation," going on to cite the analogy between the orbiting heavenly bodies and the "groups subordinate to groups" arrangement of classification. Wallace's point is interesting: he notes that the planets and satellites have "also had a common descent from the sun"—a fact that underscores and strengthens Darwin's parallel between the relationships of satellites, planets, and stars and that of varieties, species, and genera, since in both cases there is common descent, of sorts. This view of planetary origins was articulated in Robert Chambers's *Vestiges of the Natural History of Creation;* it was thus quite natural for Wallace to see "common descent" in both astronomical and organic contexts, given his great enthusiasm for *Vestiges.*

3. Maw, in the *Zoologist* review, maintained that sexuality in plants and animals is genealogically independent, an unrelated analogy. Darwin argued that males and females essentially have a common ancestor, in the sense that bisexuality evolved from a unisexual state. This was the basis of many of Darwin's botanical interests. He was fascinated by cases of dioecious plants, and cases where rudimentary stamens or pistils are in evidence in otherwise unisexual flowers underscored, for him, the essential relationship.

4. Maw seems to view mammary glands and suckling in mammals as an example of a complex system in which all the parts need to work for functionality; a partial or incomplete system would be nonfunctional: young mammals "could not live during the imperfected stages of the organs of supply," as in a transition from an egg-laying to live-bearing and suckling condition. "However rapid the strides of improvement might be," Maw declared, "the extinction of the improved progeny would at once prevent the results of the improvement being accumulated and realized." In modern "intelligent design" terms, this would be an example of "irreducible complexity," the idea being that it would be impossible to evolve such a complex system in a stepwise manner, because all parts need to be present for any functionality.

Wallace dismisses this, pointing out that it is much more difficult to see how selection could have resulted in the curious marsupial system, where essentially fetal animals must crawl to the safety of the marsupium to suckle. Wallace does not appear to be suggesting that the marsupial system itself represents an early (linking) stage in mammalian evolution, nor does he mention monotremes in this context.

5. Wallace agrees with Maw that the most "perfect" animals of each species are the strongest (fittest) and that therefore they propagate the species, but he goes on to point out (on the next page) that they are in turn superseded by new varieties with features that are superior under whatever new conditions arise, which leads them to become the new "true and dominant type."

cation of circumstances makes a
slight variation advantageous & then
that variety becomes the true &
dominant type.

———————

[1]A type which has given rise to two
ofr more distinct branches, could not
continue to exist, because the
successive steps each having a dominant
advantage over the preeceding one
must have led to their extinction.
          A existing with E & e.

———————

*[Diagram]*

          But does A ever exist
with E & e—is it not
always some third[?] modification
of A say  2)  in another
direction—A being extinct.

*92 [Written upside down]*

1. George Maw suggested that the coexistence of type A of a species with its descendant forms (rather than its having been long since exterminated) was a problem for Darwin (and Wallace), since their theory suggests that each new form supplants its parent form.

Wallace here seems to be arguing for a way out of this problem, suggesting that it is not A itself but yet another descendant form of A, an A modified in a different direction that he labels "2," that coexists with the descendant forms E and e. In modern terms, the persistence of such "living fossils" is accepted. Darwin argued in the *Origin of Species* that species representing ancestral forms persist in isolation, as on islands, but this is not a line of argument that Wallace takes up here.

Verso Notebook

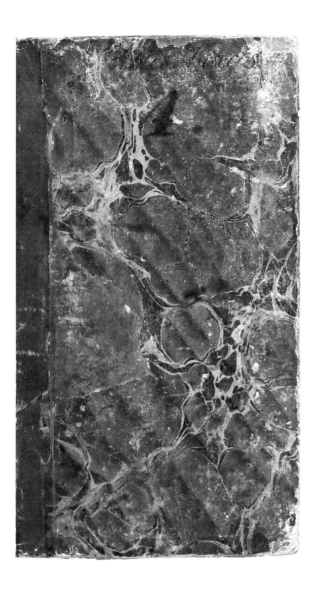

[1]Notes. Insects.
4.

1. The verso side of the Species Notebook is titled "Notes. Insects." Most of the entries, though not all, relate to insects and insect collecting.

The numeral "4" indicates that this was the fourth of Wallace's specimen notebooks. The "Insect, Bird and Mammal Register, 1855–1860" (WCP4766) is labeled notebook no. 3, but which notebooks Wallace designated nos. 1 and 2 are unclear. Linnean Society ms. 179, a field notebook with torn paper on the front cover boards, was likely one of these. The other may be the early notebook that Wallace reported having lost:

> The meagreness and brevity of the sketch I have here given of my visit to Singapore and the Malay Peninsula is due to my having trusted chiefly to some private letters and a note-book, which were lost. (*Malay Archipelago*, 45)

B *[Written upside down]*

*[The next two notebook pages were left blank by ARW and are omitted]*

(Note that the following two blank pages are omitted)

N.[ota]B.[ene] Mr. Felder of Vienna sent Coleop.[tera] of
   White Nile to
   [1]Dr. Mohnike—I will exchange for N.[ew] Guinea
   sp.[ecies]

Birds of Paradise—All but P.[aradisaea] apoda (wh.[ich] too
   large) to
be mounted separately under bell glasses with plumes displayed
will be superb & much better than a large case in group.

*[The next notebook page was left blank by ARW and is omitted]*

1. Otto Gottlieb Johan Mohnike (1814–1887) was a German physician and naturalist in the employment of the Dutch East Indian Army from 1844 until his retirement in 1869. Though he collected widely he specialized in beetles, an interest he shared with Wallace. The two met on Wallace's arrival in Amboyna (Ambon) at the end of November 1857, as Wallace recorded in his second Malay Journal (LINSOC-MS178b, entry 115): "Twenty hours from Banda brought us to Amboyna on the 30th of Nov. where I presented a letter of introduction from my friend Dr. Bouier of Macassar, to the chief medical officer of the Moluccas Dr. Mohnike, a german & a naturalist who I understood could speak English. I found however he could only write & read it, so we had to exchange compliments in French a matter of some difficulty to us both." The two became friends, and Wallace recorded his appreciation for the assistance and advice provided by Mohnike. He also admired the doctor's entomological interests, and noted that "Dr. Mohnike collects & studies the *Coleoptera* & has formed a magnificent collection during many years residence in Java Sumatra, Borneo, Japan, & Amboyna" (ibid.).

(Note that the following blank page is omitted)

[1]Dr. Mohnike possesses about 30 species
of Japan Longicorns mostly _fine_.
*[Frivald written and then wiped away]* Fridvalsky—Pesth—
    collects & will
sell or exchange Hungarian & E.[astern] European Insects.

---

Erichson   Faun[i]e Ent.[omologie] d'Allemagne

                                    complete

---

Peters, W.[ilhelm] a German naturalist, collected in
Mozambique—1842–1848—? coll.[ecte][d] insects

---

[2]Comte, A.[uguste] Philosophie Chimique et Biologique. 12/6

---

Note. Soap is said to destroy the brilliancy of
mother of pearl., whiting & water good
to wash it with. Will not the
same probably apply to the elytra of beetles

---

[3]THERATES LABIATA . . . . . . . . . [4]Long 8.10 Lin[es]
Nigro cyanea, nitidae, labro mandibulis femoribus eoxis, ab-
domine que pallida rufis, mandibuli ad apican nigris
&c. &c. . . . . . .
Ref. Fab.[ricius] Syst.[ema] Eleuth.[atorum]  Smith
    mon.[ograph] *[illeg.]*. (Fig.[ure])
Hab.[itat] N.[ew] Guinea, Aru Islands Waigiou
On foliage in shady woods, often near the sea. This actively
emits a fine odour like that of the genus Callichroma.

---

? the above form—giving a scale of the length as well
as the dimensions. would it cost much extra
to print? I fear it would.
a single line thus _____ would do
& for the large species underneath the name.

1. Mohnike was stationed in Decima, Japan, in 1849, during which time he was instructed by the Dutch East India Company to make "investigations and collections in the interest of the natural sciences."

Other naturalists mentioned on this page: Fridvalsky von Fridval, an entomologist in Pest, Hungary; Wilhelm Ferdinand Erichson (1809–1848), a German entomologist, curator of Coleoptera at the Museum für Naturkunde in Berlin from 1834 until his death in 1848; Wilhelm Karl Hartwig Peters (1815–1883), a multitalented German naturalist and explorer also based in Berlin, where he served as director of the Museum für Naturkunde.

Wallace refers to a French translation of Erichson's work on German entomology. This is likely the *Naturgeschichte der insekten Deutschlands*, a series of entomological volumes initiated and edited by Erichson.

Peters collected extensively in Africa between 1842 and 1848, and documented his collections and explorations in his monumental *Naturwissenschaftliche Reise nach Mossambique auf Befehl Seiner Majestät des Königs Friedrich Wilhelm IV in den Jahren 1842 bis 1848 ausgeführt* (Scientific Travel Performed in Mozambique on Command of His Majesty the King Friedrich Wilhelm IV in the Years 1842 to 1848), published in several volumes between 1852 and 1882.

2. The influential French philosopher Auguste Comte (1798–1857) was a founder of the positivist movement in philosophy and the study of sociology as a discipline. Comte, like Wallace, held utopian socialist views, which is likely what attracted Wallace to his writing. *La Philosophie chimique et la philosophie biologique* (1838) was the third volume in Comte's six-volume *Cours de philosophie positive*, published in Paris between 1830 and 1842.

3. *Therates labiata* is a large and beautiful tiger beetle (family Cicincelidae) that Wallace collected on the Ké (Kai) Islands in January 1857. His entry here nicely captures the alert, active nature of tiger beetles. Quick and very visual, these fierce predatory beetles can often be seen raising themselves up as if on "tiptoes"

as they scan their vicinity for prey. He also notes the curious floral scent emitted by this species. Such secretions often play a defensive role in insects, but Wallace suggests that in this species the odor could play the role of fatal attractant for its insect prey. (This hypothesis does not appear to have been pursued and so remains an intriguing but untested possibility.) The observations given in this entry were eventually included in *Malay Archipelago* (429):

> In the forest itself the only common and conspicuous coleoptera were two tiger beetles. One, *Therates labiata*, was much larger than our green tiger beetle, of a purple black colour, with green metallic glosses, and the broad upper lip of a bright yellow. It was always found upon foliage, generally of broad-leafed herbaceous plants, and in damp and gloomy situations, taking frequent short flights from leaf to leaf, and preserving an alert attitude, as if always looking out for its prey. Its vicinity could be immediately ascertained, often before it was seen, by a very pleasant odour, like otto of roses, which it seems to emit continually, and which may probably be attractive to the small insects on which it feeds.

The term "otto of roses" refers to rose oil, the essential oil extracted from the petals of certain rose species. The term comes from "attar of rose" (*attar*, originally an Arabic word, in Turkish means someone who sells flower oil or essence. It is related to the Turkish *ıtır*, or geranium).

4. This is Wallace's shorthand for giving body length of the species, in this case 8.10 lines. "Lines" (or "lignes") are units of measurement widely used before metric measure came into standard use in the sciences. One "line" was equal to one-fortieth of an inch. The term is still used in button manufacturing.

The only families which are absent in
Britain are
Paussidae.        Rutelidae, Anoplognathidae
Glaphyridae, Horiidae, Pimeliidae, Brenthidae[?]

---

[1]The yellow spotted trimerous beetles found about
fungi &c. are <u>Eumorpha</u>—(<u>Endomychidae</u>)
The narrow elongate smooth shining beetles
often blue with red head & thorax are of the
genus <u>Languria</u> (<u>Erotylidae</u>)
The small cylindric woodborers, are <u>Platypus</u>
Tomicus or Scolytus (Scolytidae) near <u>Curculionidae</u>.
The Bostrichidae are near these but are <u>pentamerous</u>
have the thorax produced over the head & antennae
ending in a 3 jointed club.
The Scydmaenidae come near these ? <u>no,</u> I think

---

The Pselaphidae, near Staphylinidae doubtful—
? both closely allied.

---

[2]3500 Longicorns in Thompsons coll.[ection] 1857.
He supposes collections of Europe to contain 7000 species!

---

[3]Ja[c]quemont was 3½ years in India (1829–32)
had 6000 francs a year at first, afterwards more
? Is there any account of his collections.
He had great assistance from the English Government &
the native princes.

1. This and the following are miscellaneous notes on insects of several families.

2. A reference to coleopterist James Thomson (not Thompson) (1828–1897), an expatriate American living in France. The year given, 1857, is the publication date of volume one of Thomson's *Archives entomologiques, ou Recueil contenant des illustrations d'insectes nouveaux ou rares* (see recto 130). That year also saw the publication of Thomson's *Monographie des Cicindélides*.

3. Victor Jacquemont (1801–1832) was a French naturalist and explorer with interests in geology and botany. He collected extensively in India before his premature death at the age of thirty-one. Wallace likely consulted the posthumously published collection of Jacquemont's letters: *Correspondance de Victor Jacquemont à sa famille et plusieurs de ses amis, tome premier, par Victor Jacquemont* (H. Dumont, 1836). In 1936 Macmillan published *Letters from India, 1829–1832: Being a Selection from the Correspondence of Victor Jacquemont*, translated by Catherine Alison Phillips.

## [1]British Coleoptera . No. of species

3575

| [The column below is written verticallly] | | | [The columns below are written verticallly] | | |
|---|---|---|---|---|---|
| Cicindelidae — | 6 | | Buprestidae — | 42 | |
| Carabidae— | 479 | 485 | Eucnemidae — | 5 | 116 |
| Dyticidae — | 142 | | Elateridae— | 69 | |
| Gyrinidae — | 9 | 151 .. | Cebrionidae — | 1 | |
| Heteroceridae — | 7 | | Cyphonidae — | 19 | |
| Parnidae | 19 | | Lampyridae | 3 | |
| Helophoridae — | 42 | | Telephoridae | 57 | |
| Hydrophilidae — | 37 | Philhy- | Melyridae | 27 | Malaco- |
| Sphaeridiidae — | 69 | drida 270 | Cleridae— | 13 | derms 188.. |
| Agathidiidae — | 96 | | Ptinidae— | 31 | |
| Scaphidiidae — | 31 | | Lymexylonidae | 2 | |
| Silphidae — | 24 | | Bostrichidae — | 19 | |
| Nitidulidae — | 70 | Necro- | Scydmenidae | 16 | |
| Eugidae — | 55 | phaga 288.. | Notoxidae | 10 | |
| Cucyidae — | 11 | | anthieus — | | |
| Paussidae — | 0 | | Pyrochroidae— | 2 | |
| Myetophagidae — | 84 | | Lagriidae— | 1 | |
| Dermestidae — | 13 | | Mordellidae | 28 | |
| Staphylinidae — | 801 | | Cantharidae | 12 | |
| Pselaphidae.20. | | | Salpingidae | 12 | Hetero- |
| Byrrhidae — | 20 | | Oedomeridae | 14 | mera 142.. |
| Histeridae — | 47 | 67 .. | Melandryidae | 16 | |
| Lucanidae — | 4 | | Cistelidae— | 11 | |
| Geotrupidae — | 14 | | Helopidae— | 4 | Lacor- |
| Scarabaeidae — | 11 | | Diaperidae— | 12 | daire vol. 4 |
| Aphodiidae — | 64 | Lamelli- | Tenebrionidae— | 17 | ? |
| Trogidae — | 5 | cornes 122.. | Blapsidae— | 3 | |
| Dynastidae — | 3 | | Bruchidae — | 17 | |
| Melolonthidae — | 15 | | Attelabidae— | 121 | 507.. <5> |
| Cetoniidae — | 6 | | Curculionidae | 369 | |
| | | | Scolytidae— | 36 | |
| | | | Prionidae— | 3 | |
| | | | Cerambycidae. | 26 | 77.. <6> |
| | | | Lamiidae | 19 | |
| | | | Lepturidae— | 29 | |
| | | | Crioceridae— | 35 | |
| | | | Cassididae— | 20 | 276.. Phytoph. |
| | | | Galerucidae— | 141 | |
| | | | Chrysomelidae | 80 | |
| | | | Erotylidae— | 7 | <7> |
| | | | Endomychidae | 2 | 67.. |
| | | | Coccinellidae | 58 | |

1. Wallace's passion for natural history was sparked by British Coleoptera, when Henry Walter Bates introduced him to beetle collecting in Leicester, England. What is likely his first "scientific" publication, appearing in the April 1847 issue of the *Zoologist*, was on this subject. While Wallace was delighted to see his name in print in a naturalists' journal, judging from the editor's inserted comment it was an inauspicious start to an illustrious career:

> *Capture of Trichius fasciatus near Neath*—I took a single specimen of this beautiful insect on a blossom of Carduus heterophyllus near the falls at the top of Neath Vale. —*Alfred R. Wallace, Neath.* [The other insects in my correspondent's list are scarcely worth publishing. —*E. Newman*].

The editor, Edward Newman, was to become a friend and supporter, and the *Zoologist* would publish many of Wallace's letters and papers from the field in the years to come.

*Trichius fasciatus* is one of the "bee beetles," genus *Trichius* (family Scarabaeidae, subfamily Trichiinae). These 2-centimeter-long beetles have a fuzzy orange thorax and yellow-orange elytra with black spots, thought to resemble a bee—a form of protective mimicry that Wallace was to write about many years later.

[1]Cramer. 1775–82.
Fab.[ricius] Sys.[tema] Gloss.[atorum] *[1807 written and then wiped away]* 1805?
Esper . . 1777 &c. . .
Herbst. 1783. 95. .

---

[2]Lake Superior—Coleoptera collected by Agassiz & by Leconte in several visits.

---

| | | | |
|---|---|---|---|
| Geodephaga | 177 | species | |
| Hydradephaga | 72 | | |
| Hydrophili— | 31 | Cicindela | In 7 weeks on N.[orth] shore |
| Necrophori &c. | 14 | 6 | of Lake Superior, were |
| Brachyelytra— | 118 ? | Carabus | observed 69 sp.[ecies] of Birds |
| Necrophori &c. | 85 | 3 | viz. Passeres—52 } over |
| Lamellicornia | 26 | Calosoma | Raptores— 6 } 300 |
| Buprestidae— | 18 | 2 | Gallinaceae— 3 } miles |
| Elateridae— | 53 | Cleridae | Water Birds 8 } of |
| Malacodermia | 44 | 9 | ground |
| Heteromera | 37 | Donacea | |
| Rhyncophora— | 25 | 17 | |
| Longicornes— | 39 | | |
| Phytophaga— | 53 | | |
| Coccinellida | 26 | | |
| Total Species. | 818 | | |

1. Here Wallace lists authors and years for several entomological treatises:

Pieter Cramer (1721–1776), Dutch merchant and entomologist, published *De Uitlandische Kapellen (Papillons exotiques de trois partes de monde)*, 1775–1782.

Johann Christian Fabricius (1745–1808), Danish entomologist, published *Systema Entomologica* in 1775 and *Systema Glossatorum* in 1807 (Wallace's first date entry was correct).

Eugenius Johann Christoph Esper (1742–1810), German entomologist from Bavaria, produced a series of volumes entitled *Europaischen Schmetterlinge* (European Butterflies) commencing in 1777.

Last but not least, Johann Friedrich Wilhelm Herbst (1743–1807), another German entomologist, but from Minden-Ravensberg, undertook an ambitious project in 1785 that built on Buffon's *Histoire naturelle* with his *Natursystem aller bekannten in- und ausländischen Insecten, als eine Fortzetsung der von Büffonschen Naturgeschichte: Nach dem System des Ritters Carl von Linné bearbeitet* (Natural System of All [European] and Foreign Insects, as a Continuation of Buffon's Natural History). This was primarily a comprehensive treatment of European and foreign beetles, a subject of great interest to Wallace.

2. In the summer of 1848 the famed Swiss-born Harvard naturalist Louis Agassiz (1807–1873) led a scientific expedition to explore Lake Superior and environs. It was a great success, with findings yielding more than a dozen scientific papers and the monograph *Lake Superior: Its Physical Character, Vegetation, and Animals, Compared with Those of Other and Similar Regions*, published in 1850. *Lake Superior* was authored by Agassiz with American James Elliot Cabot (1821–1903). Cabot studied law at Harvard, but his great interest in natural history led him to accompany Agassiz on the expedition to assist with the collecting and documentation. (Agassiz's accomplished wife, Elizabeth Cabot Cary [1822–1907], a founder of Radcliffe College and its first president, was also a member of the illustrious Cabot family of Boston—they married in the same year that *Lake Superior* came out.)

*Lake Superior* included "contributions by other scientific gentlemen," as its title page indicated. Among them is a chapter entitled "General Remarks upon the Coleoptera of Lake Superior," by American beetle authority John Lawrence Leconte (1825–1883), but the beetle groups Wallace lists on this page are not derived solely from Leconte's species list. Likewise Cabot's chapter, "Report of the Birds Collected and Observed at Lake Superior," was consulted by Wallace. His interest in these data can be seen as part of his broader interest in species distribution and relationships. (See also recto 140–147.)

Wed. <u>1855</u>

| | | | |
|---|---|---|---|
| [1]March 14th. | carried things to Hill — | | Wet. |
| " 15th. - | 80 Coleoptera _____ | | Fine |
| " 16th. | 60 | | Cloudy |
| " 17th. | 50 | | Cloudy |
| Sun " 18th. | 25 " To Dyak house. | | Fine |
| " 19th. | 100 Beetles. | | Showers, Fine. |
| " 20th. | 46 " | | Showery — |
| " 21st. | 100 " 4 new Long. 3 new Buprest. | | Fine . . . |
| " 22nd | 30 | | Wet. |
| " 23rd. | 120 " 2 new Long. 1 new Buprest (22 n. s.) | | fine |
| " 24th _____ | 30 " 2 Long .. 12. n. s. | | wet. |
| Sun. 25th .. | 100 .. 55 sp. 23 new. _____ | | fine |
| 26th. | 120 . . . 76 sp. 34 new..to me. Long. 18 sp. 8 new. | | fine |
| 27th. | 90 — 20 n. s. 6 new Long. | | dull. fine — |
| 28th. | 90 — 20 n.s. (5 long. 2 Bup. 2 Car. 5 Cur.) new. | | dull — |
| 29th. | 50. 6 n. s. | | wet |
| 30th | 70. 17 n. s. (2 Long.) | | fine.. |
| 31st | 72. 20 n. s. 2 Long. | | fine.. |
| April 1st. Sun. | 45. 13 n s. 3 Long. _____ | | dull |
| 2nd | 70. 4. n. s. 0 Long | | dull |
| 3rd, | 80. 12 n.s. 3 Long | | Showery. |
| | 20 / <u>1428</u> | | |
| | 71 per day. | | |

1. On page 1 of the recto Species Notebook Wallace reported that on March 12, 1855, he "arrived at the landing place in the Si Munjon river," in Sarawak. He soon "found a thatch house erected by Mr. Coulson the superintendent of the Coal Mines at the neighbouring mountain. A rough jungle path on logs & fallen trees leads from this point to the hill through a swampy forest." The coal mine and dwellings of the superintendent and workers were located at the base of Santubong Mountain, and this is likely the "hill" referred to in this entry for March 16 (repeated in *Malay Archipelago*, 46–47): he and his assistant Charles Allen had to carry their belongings and equipment to the small settlement at the mountain's foot. He first stayed with Coulson but soon had a small house built, in which he lived for the next nine months.

1855 April

specimens

| | | | |
|---|---|---|---|
| 4th. | 40. | 12 new. [SP] 2 Long. 1 Paussus ? | Wet... |
| 5th. | 150 | 24 new 4 Long 3 Buprest | Fine |
| [1]6th | 40. | 12 new. 3 Long — (Skinning Mias.) ~ | Fine. |
| 7th. | 60 | 12 new. 3 Long..... | Fine. |
| Sun. 8th. | 12 | 6 new.. (Helius came) | Wet. |
| [2]9th. | 70 | 17 new. 4 Long. ( " went.) (wrote to Stephens) | Cloudy. [pm. wet] |
| 10th. | 90 | 12 new. 2 Long. | fine. [showers] |
| 11th. | 140 | 16 new. 2 Long. 60 of a new Carabidae ~ | fine |
| 12th. | 80 | 5 new. 1 Long. 1 paussus.. | fine |
| 13th. | 30 | 8 new. 2 Long. 1 fine Catuscopus (Dyak) | fine |
| 14th. | 18. | 2 new... 1 Long... | Wet. |
| Sun. 15th. | 30 | 5 new ... (Malays idle. | fine. |
| 16th. | 20 | 3... | wet. |
| 17th. | 115 | 6.. | fine. |
| 18th. | 90 | 10 new. (2 fine Elater, Dyak.) | fine. |
| 19th. | 120. | ~~24~~ 8 new. 5 Long. | Showers. fine |
| 20th. | 160 | 24 new. 5 Long. | showers... |
| 21st. | 110 | 18 new. 5 Long.. | |
| Sun. 22nd. | 60 | 8 new. . (Malays idle.) | |
| 23 | | 5 " | |
| 24 | | 6 " | |
| 25 | | 10 new | |

[illeg.] Dyak & Chinese coll[ns] [Written vertically in lower left margin of page]

1.  April 6, 1855, finds Wallace skinning a luckless *mias*, or orang-utan. This is likely the specimen shot the day before, as recorded on page 10 of the recto Species Notebook.

2.  Wallace notes that he wrote to his agent, Samuel Stevens, on April 9, though the letter itself is dated the eighth. Stevens published this as "Letter from Sarawak" in the August 1855 issue of the *Zoologist* (Wallace 1855a). The "Helius" mentioned on April 8 is a mail boat. As a side note, it is interesting that in this letter Wallace makes approving reference to Charles Allen, "who," he writes, "is now a rather expert collector." He was to revise that opinion dramatically in the coming months (see annotations for the following page, and for recto 2).

## 3

July & August
very hot & dry—
8th sept.[embe]r: heavy rain
dry & rains ab[ou]t once a week
*[Above wording written in top left corner of page]*
Sun.[day] 29. . . very few to this
Thurs.[day] 3 May . . very few, no new Longicorns. .
May continued a fine month.
June—very cloudy damp & wet— [1]mites attacked
insects & occasioned much trouble. . . . . . .

[2]Charles collected - Insects of all orders.

| Aug. | No. of Insects | Aug. | No. of Insects | Sept.[r] | No. of Insects | Sept. | No Insects |
|---|---|---|---|---|---|---|---|
| — | | | 1985 | | 3701 | | 4959 |
| 4th. | 88 | Tu. 21st. | 125 + 10 | Th. 6 | 72 | Mon. 24 | 88 |
| Mon. 6 | 155 | W. 22 | 167 | Fr. 7 | 73 | Tu 25 | 100 |
| Tu. 7 86 | 202 | Th. 23 58 | 202 +15 | Sat. 8 | 57 | W. 26 | 142 |
| W. 8 | 85 + 15 | F. 24. | 145 + 10 | Mon. 10 | 102 | Th. 27 | 75 |
| Th. 9 72 158 | 230 | S. 25. | 134 | Tu. 11 | 58 | Fr. 28 | 77 |
| Fr. 10 | 100 | Mon. 27 | 186 | W 12 | 82 | S. 29 | 70 |
| Sat. 11 | 134 + 10 | Tu 28 | 155 | Th. 13 | 88 | Oct. Mon. 1 | 53 + 10 |
| Mon 13 | 155 | W. 29 | 120 | Fr. 14 | 122 | Tu. 2 | 35 wet |
| Tu. 14 | 146 | Th. 30 | 68 | S. 15 | 22 | W. 3 | 86 |
| W. 15 | 85 | Fr. 31 | 82 | M. 17 | 140 | Th. 4 | 48 |
| Th. 16 | 178 | Sat. 1. | 98 | Tu. 18 | 50 | Fr. 5 | 48 |
| Fr. 17 | 65 | Mon. 3 | 110 | W. 19 | 76 | Sat. 6 | 55 |
| S. 18 | 152 | Tu. 4 | 68 | Th. 20 | 115 | M 8 | 85 |
| Mon 20 73 | 210 +10 | W. 5 | 56 | Fr. 21 | 76 | Tu. 9 | 55 |
| | 1985 | | 3701 | S. 22 | 125 | 10&11 | 98 |
| | | | | | 4959 | F. 12 | 55 |

Sept 12 deduct 70 lost by Charles  20th — 5 spoiled *[illeg]*

18   66

6198

1. Pests such as ants, flies, and mites caused endless headaches for Wallace, and he often had to take elaborate measures to safeguard his hard-won specimens from marauding arthropods. On recto 134 he commented on pesky mites that seemed to be able to withstand months of submersion in arrack, a preservative. He also inserted a comment that he had the same problem with mites on specimens collected in Borneo, after a week's soaking in preservative. That reference is likely connected to the comment on mites here, from June 1855.

2. Here Wallace summarizes Charles Allen's collecting activity for August and September 1855, when Allen was assisting Wallace collecting in Sarawak. The tally of collections is impressive, but there is evidence of mishaps too: note the comment near the bottom of the page that reads "Sept 12 deduct 70 lost by Charles." According to the tabulated data, Charles took eighty-two insect specimens that day, so the subsequent loss of seventy of them amounts to quite a large proportion. Wallace's frustrations with Allen were well recorded in his letters (see annotations for recto 2).

[4]

[1]Notes—for extent of Tropics.
Darwin's Journal see for <u>Mendoza</u> & Monte Video
Plants coll[ected] insects of Pt.[Porto] Alegre Tropical
Insects of Port Natal, Tropical—
The <u>Punjamb</u> ? tropical at Cashmere
Llaryae subtropical—
Richmond R.[iver] S.[outh] of Juneton Bay.   <u>Ornithoptera</u>
————

R.[io de] Janeiro & Calcutta both close to the
tropical—get absolutely tropical.  Exhibit
no sign of approach to temperate zone.
The base of the Himalayas at Darjeeling to
Lat.[itude] 27 & Sarula _ 31 are still greater[?]
Tropical in their productions—

1. Wallace's interest in global biogeography was first piqued by his conviction that geographical distribution held the key to species origins, and later, as he wrote in the preface to his 1876 book *The Geographical Distribution of Animals*, by the idea that "the most remarkable and interesting of the facts" of distribution could be illuminated "by means of established laws of physical and organic change." Biogeography was a record of inheritance, but in addition it shed light on "the islands and continents of a former epoch."

Wallace drew on his notes here in writing various talks and papers, perhaps the first being his "Lecture on Animal Life in the Tropics," given in 1867:

> Limits and Extent of Tropics. Almost all central and N. Asia is of this character, as well as N. America, N. of Lat. 50°; so that the real area of the temperate regions, where a varied animal life during the winter months is possible, will not be much more than half that of the tropics. But we have to consider further, that the tropics of the geographer and of the naturalist, do not coincide. The tropical regions as indicated by the preponderance of tropical forms of animal and vegetable life, almost always extends beyond the tropic of L. 23° 28K. Almost the whole of the valleys of the Ganges and the Bramaputra, are beyond this line, but their productions are purely tropical as far as 27°. N. Lat. In Brazil tropical forms of insects and birds abound as far as 30° S. Lat.; and in Australia the large tropical butterflies of the genus Ornithoptera extend to Richmond R. in Lat. 29° S.

This unpublished lecture was the second in a series of three lectures delivered at the Literary and Philosophical Society in Newcastle in November 1867. The lecture manuscripts are in the Wallace Collection at the Natural History Museum (London), manuscript number WP9/2/2.

## 5

[1]Insects  Wings of Buprestidae do not fold
at the end being no longer than the elytra.
Wings of Elateridae are longer than elytra &
do fold at the tip. A[fred] Wallace . June 55/

---

[Sketch]

[2]Cone of brown papery substance
attached by base to thatch
covering 3 or 4 small open
earthen cells containing
eggs of a wasp.
Sar.[awak]  (1

---

(Sar.[awak] 2) a [3]Black wasp makes cells of a very sticky wax
like a mixture of wax & pitch—came into my room
& deposited wax among my books under the edges
sticking the leaves together—in the grooves of my
insect setting boards &c. fly very quick hum
& buzz loudly.

---

[Sketches]
above     beneath mag.[nifie]d

[4]Euploea dehaanii—
tufts form a yellow fleshy tube
protruded in each side of
abdomen just above the
vent.

1. This note is dated June 1855, three months after Wallace's arrival in Sarawak, Borneo. Buprestidae, the metallic wood-boring beetles, are often strikingly colorful, with their iridescent blues, greens, yellows, and oranges. Like the metallic shimmering of various butterflies, including the *Morpho* and *Ornithoptera* that Wallace sought, the beetles' coloration stems from ultrastructure and light diffraction rather than pigmentation.

2. The wasp that built this nest is very likely a member of the vespid subfamily Stenogastrinae, the hover wasps, a group characterized by nests constructed of a combination of thin paper and earth. These diverse wasps range from India to New Guinea. They get their subfamilial name (which means "trumpet-gaster") from their long, thin basal abdominal segment, which gives them the appearance of having a trumpet-shaped abdomen. From Wallace's description one might conclude that the nest was hanging from the slender pedicel end when he found it. In all likelihood, however, it was cemented at the wide end of the cone to the underside of a leaf or similar shelter. This is characteristic of wasps of the genus *Eustenogaster*. Typical for social wasps, the nest Wallace found has open cells; adults must be able to access the developing larvae to fee them.

3. By this time Wallace no doubt would have come to see glued book pages as the least of his pesky insect worries. He was constantly beset by insects looking to make a meal of his specimens.

4. *Euploea dehaanii* belongs to the tribe Danaini (milkweed butterflies and relatives) of the brush-footed butterfly family Nymphalidae. This species was described in 1853 by French entomologist Pierre-Hippolyte Lucas (1814–1899), in his *Description de nouvelles espèces de Lépidoptères appartenant aux collections entomologiques du Musée de Paris* (*Revue et magasin de zoologie pure et appliquée* 5 [7]: 310–322).
   Like many danaiines, these butterflies, and their caterpillars, tend to be quite toxic to vertebrate predators, as they sequester poisonous secondary chemicals from their host plants. *Euploea* species are called "crows" for their dark background color, on which they have contrasting white spots. The caterpillars are often brightly colored and sport long, tentacular lappets and tubercles. What caught Wallace's eye on the *E. dehaanii* caterpillar were the curious tufts protruding from fleshy tubes at the tip of the abdomen. He does not mention if the tufts are retractable, but they likely are, and are deployed defensively like the osmeterium of papilionids (see verso 34, 37).

## 6

[1]Borneo

When trees are felled, the leaves while drying
are frequented by Longicorn beetles Nos. [numbers]
321, 322, 347, 396, 348, 321, 941 345 *[Five illeg. numbers struck out by ARW]*
   and by Buprestidae which
settle in the bark in the sunshine. A
little later, species of Clytus & Hesthesis are
found crawling about the bark and Xylinades
& several Anthriibidae beneath it.
The longicorns Nos [numbers] 344, 346      next come
and in the morning may often be found
on the under side of the tree.  In the
sunshine species of Mecopus now frequent
it.  After about a month the Longicorns
& Curculionidae generally leave the trees.
Numerous woodborers pierce it & may be
seen at work in the little cylindrical holes
Now numerous small Carabidae may be
found beneath it, and a few Heteromera
are attracted to it.    Still later when
the bark begins to rot away species of
Catascopus frequent its crevices &
run actively about the bark.

1. It did not take Wallace long to discover that areas with lots of felled trees were beetle hotspots. The beetle bonanza recorded here is also reported in his April 1855 letter to Samuel Stevens ("Letter from Sarawak"; Wallace 1855a): "I have now 135 species of Bornean Longicorns," he wrote, "and I do not despair of getting 200 before I leave this place, which I mean to work thoroughly." (Longicorns, as discussed on recto 156 and the annotations to that page, are long-horned beetles, family Cerambycidae, a highly diverse group of often large and spectacularly colored beetles.)

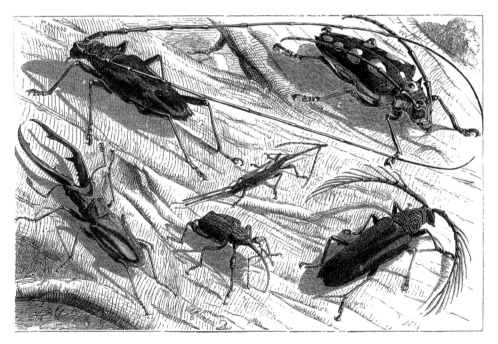

In *Malay Archipelago* this illustration is captioned "Remarkable beetles found at Simunjon, Borneo." Two of these new species were named for Wallace: Ectatorhinus Wallacei (bottom center) and Cyriopalpus Wallacei (bottom right). From Alfred Russel Wallace, *The Malay Archipelago*, 4th ed. (London: Macmillan & Co., 1872), 37.

under the decaying leaves on the ground, small
Carabidae, <u>Lebia & Demetrias</u> may be found.
When a trunk is completely rotten & falling
to pieces the fine Cetoruada, <u>Macronota Diardi</u>
& <u>Chalcothoea</u> <u>smaragdula</u> often settle upon it
either to feed or to deposit their eggs.  The former
frequent shady jungle the latter more
often sunny places flying with excessive
rapidity & strength.

—————————————

[1]In Lombock Dragon flies are eaten, they
are caught by a kind of bird line on the
end of a stick, their wings pulled off,
and then fried either alone or chopped up
with onions and preserved shrimps; they are
taken by thousands in the time of flowering of
the paddy, and are esteemed a great delicacy

—————————————

In Borneo & also in Celebes the larvae &
pupae of bees & also of wasps are eaten
often raw & alive as pulled out of the cells
Also fried when they are no doubt very good.
In Ternate, Gilolo, & Batchian, the larva of <u>Calandra</u>
are sold for food packed in bamboos.   <u>(over)</u>

1. Among Wallace's many ethnological observations, he noted instances of entomophagy (insect eating) among native peoples. He reported such observations from South America both in a paper read to the *Entomological Society* (1852a) and in his memoir *Narrative of Travels on the Amazon* (1853a). The capture and preparation of dragonflies in the vicinity of Ampanam in Lombock (Lombok) is memorably recounted in *Malay Archipelago*:

> Every day boys were to be seen walking along the roads and by the hedges and ditches, catching dragon-flies with birdlime. They carry a slender stick, with a few twigs at the end well anointed, so that the least touch captures the insect, whose wings are pulled off before it is consigned to a small basket. The dragon-flies are so abundant at the time of the rice flowering that thousands are soon caught in this way. The bodies are fried in oil with onions and preserved shrimps, or sometimes alone, and are considered a great delicacy. In Borneo, Celebes, and many other islands, the larvae of bees and wasps are eaten, either alive as pulled out of the cells, or fried like the dragon-flies. In the Moluccas the grubs of the palm-beetles (Calandra) are regularly brought to market in bamboos, and sold for food; and many of the great horned Lamellicorn beetles are slightly roasted on the embers and eaten whenever met with. The superabundance of insect life is therefore turned to some account by these islanders. (*Malay Archipelago*, 164–165)

The sago grub, *ular sagu*, which is the larva of the striking beetle *Euchirus longimanus*, is another example of an insect eaten by humans that Wallace reported from the Archipelago (see verso 9 for an account of the adult beetle).

## 8

Aru.
[1]Brenthidae. . . . pointed rostrum ♀ dilated ♂. males longest.
females bore deep hole in fallen trunks often burying
their rostrum up to its base.
*[Sketch]*

two males seen fighting
each had a fore leg
across the neck of
the other & rostrum bent in an attitude of
defiance looking most ridiculous.

———————

Another time two fighting for a female, pushed at
each other with their rostra, clawed & thumped, with
apparently great rage but most harmlessly on
their coats of mail.  The smaller one soon ran away.

———————

[2]Instinct at fault.
The little cylindrical wood boring beetles [(Platypus & Tesserocerus sp[ecies])] attacked a
newly felled tree, but most rashly as it exuded
where the bark was pierced . thick white sap
which hardened on exposure to the air,—glueing
the little insects into the holes they had bored.  Dozens
were to be seen [dead] stuck fast in the graves they had
dug for themselves their hind legs & elytra only protruding.

———————

The larva of the oryctes australis ? of Boisduval
is eaten at Dorey.  It feeds on the cocoanut.
The perfect insect is also eaten; pulling off the elytra
& wings & roasting it slightly.
Euchirus longimanus. & Dynastes gideon d[itt]o. at
Amboyna

1. These dueling brentid beetles are described and figured in chapter 32 of *Malay Archipelago*, about Wallace's visit to the Aru Islands. In writing the chapter, he drew from these notes almost verbatim, from the "ridiculous" look of the males to their seeming "rage" and the protection afforded by their "coats of mail." In the published account, post–*Origin of Species*, their behavior is recognized as being of relevance to sexual selection: "In most Coleoptera the female is larger than the male, and it is therefore interesting, as bearing on the question of sexual selection, that in this case, as in the stag-beetles, where the males fight together, they should be not only better armed, but also much larger than the female" (*Malay Archipelago*, 482).

2. These hapless wood-boring beetles were observed at Dorey (Manokwari), in western New Guinea, which Wallace visited from April through June 1858. Wallace incorporated this observation of supposed "failure of instinct" into *Malay Archipelago* (481), where he opined that "if, as is very probable, these trees have an attractive odor to certain species of borers, it might very likely lead to their becoming extinct; while other species, to whom the same odor was disagreeable, and who therefore avoided the dangerous trees, would survive." He uses this as a means to attack the idea that the beetles avoid the dangerous tree instinctively, suggesting that they would, rather, "really be guided by a simple sensation." We have seen that the distinction between instinctive and non-instinctive behaviors was of great interest to Wallace (see Species Notebook recto 112, 116–119, 143–144, 149, 155, 166–177).

However, from a modern point of view his take on the entombed beetles reveals an imperfect understanding of how natural selection operates. He does not appear to appreciate that the kind of variation in attraction to the tree that he invokes among species might exist among individuals of the same species—that is, that selection would act on individual variation in attraction, selecting for avoidance of the trees at the individual level if that attraction is lethal. The species would not be doomed to extinction if there were heritable variation for reduced sensitivity to the odor of the tree's attractant; as such individuals differentially sur-

vived from generation to generation, eventually the population would be dominated by beetles less responsive to this dangerous tree. Wholesale extinction of the species would thus be unlikely. Wallace's evolutionary reasoning in this instance is flawed, for selection acts on individual variations, not at the level of varieties or populations.

His error is an example of Wallace's tendency at times to think in terms of variety- and species-level selection, rather than selection on the level of individuals or individual variations. This has led some historians to maintain that Wallace failed to appreciate populational thinking altogether, or had no concept of selection acting on individual variation. I agree with the conclusions drawn by Kottler (1985) in his review of the question: Wallace may have been inconsistent in his use of terms, but he did have a firm grasp of the selection dynamic.

"Male Brenthidae (*Leptorhynchus angustatus*) fighting" is the caption for this illustration in *Malay Archipelago*. The Latin name *Leptorhynchus* translates as "thin-snout," and *angustatus* as "narrow" or "slender," making this the "slender thin-snouted beetle" of Aru. From Alfred Russel Wallace, *The Malay Archipelago*, 4th ed. (London: Macmillan & Co., 1872), 479.

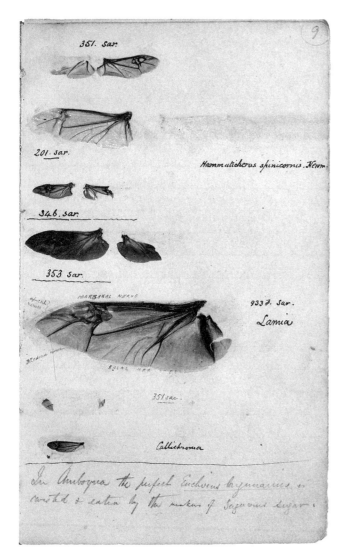

## 9

351. sar.[awak]
*[Pasted wings with portion missing]*
*[Pasted wings—one wing missing]*
201. sar.[awak]
                Hammaticherus spinicornis. Newm.[an]
[1]*[Pasted wings with portion missing]*
346. sar.[awak]

_____

*[Pasted wings with portion missing]*
353 sar.[awak]

_____

*[Pasted wing—labeled]*           933 ♂. sar.[awak]
apical nerves}              Lamia
marginal nerve
Median nerve
Basal nerve
*[Pasted wing fragments]*  351 sar.[awak]
*[Pasted wing]*    Callichroma

_____

[2]In Amboyna the perfect <u>Euchirus longimanus</u> is
roasted & eaten by the makers of Sago *[illeg.]* sugar.

*[The next notebook page was left blank by ARW and is omitted]*

1. This is the first of several pages in the verso Species Notebook with beetle wings pasted on them. These are the flight wings of the insects—the beetle order Coleoptera, "sheath-winged" insects, have greatly thickened forewings that close hatchlike over the membranous hind wings. The forewings "sheath" and protect the hind wings, but cannot be used for flight.

Note the variable size of the wings on this and subsequent pages. This is a small fraction of the actual size variation of beetles, a family that outnumbers all other species in the animal kingdom combined. Identifying his specimens in the species-rich tropics was a daunting but essential task for Wallace. A host of anatomical details were (and are) used as identifying characters. His purpose in preserving the wings of these species was to document the wing venation, the thickened network you see in the wings. These are not veins in the sense of blood conduction, but they are hollow tubular structures through which the insects' blood, hemolymph, is pumped after metamorphosis from pupa to adult, in order to "inflate" and expand the wings. Soon afterward the veins are closed off, and hemolymph does not circulate within the wings of adults.

The largest specimen on this page, number 933 from Sarawak, labeled male *Lamia*, is a long-horned beetle (family Cerambycidae), subfamily Lamiinae. This large subfamily, named by the prolific French coleopterist Pierre André Latreille (1762–1833) in 1825, is the second largest in the family, with more than 750 genera.

2. The adult stage of *Euchirus longimanus* is called the long-armed chafer—a large scarabaeid beetle well named for the extraordinarily long front legs of the males (a sexually selected trait). Wallace collected this species in Amboyna (modern Ambon).

In *Malay Archipelago* Wallace wrote of this spectacular beetle:

I was much pleased to get here the fine long-armed chafer, *Euchirus longimanus*. This extraordinary insect is rarely or never captured except when it comes to drink the sap of the sugar palms, where it is found by the natives when they go early in the morning to take away the bamboos which have been filled during the night. For some time one or two were brought me every day, generally alive. They are sluggish insects, and pull themselves lazily along by means of their immense forelegs. (309)

It is the plump larva, the sago grub, or *ular sagu*, that is roasted and eaten. (See verso 7 for a discussion of entomophagy.)

*Euchirus longimanus* dominates this figure from *Malay Archipelago*. The specific epithet *longimanus* translates as "long-handed." From Alfred Russel Wallace, *The Malay Archipelago*, 4th ed. (London: Macmillan & Co., 1872), 401. A specimen of *E. longimanus*, likely the very one that this illustration is based on, can be seen in the "Specimens of Asian beetles" collection (Drawer 25) in the online Wallace Collection, Natural History Museum, London. The other species in this illustration entitled "Moluccan beetles" include two fungus beetles (Anthribidae)—*Xenocerus semiluctuosus* (upper left) and a then-unnamed *Xenocerus* (lower left)—and two unnamed weevils (Curculionidae)—*Arachnobus* (lower center) and *Eulophus* (upper right).

[1] *[Pasted wing]* 1331. sar.[awak] Coccinella

*[Pasted wing]* 447 sar.[awak] Triniera fungus

*[The next notebook page was left blank by ARW and is omitted]*

1. Wallace left verso 10 blank, so the facsimile is not reproduced here. He used only the recto side of pages that had wings pasted on them, probably to ensure that a wing-free page both faced and closed against the wings, preventing damage that could occur if wings pasted on facing pages came into contact with each other.

(Note that the following blank page, verso 12, is omitted.)

[13]

[1]Curculionidae.
*[Pasted wing]* 381 sar.[awak] [2]Mecocerus gazella.

*[Pasted wing]* 196. sar.[awak]

1. This page was to be dedicated to Curculionidae (weevils and their relatives), but wings of only two specimens made it into the notebook. This is likely owing to the sheer numbers of insects Wallace was collecting from this area; he noted in his daily beetle collecting record for March and April 1855 (verso 1–3) that he averaged seventy-one specimens per day between March 14 and April 3. Among the ninety specimens collected on March 28 were twenty new species, five of which were curculionids.

2. Wallace mentioned a "handsome" *Mecocerus gazella* near his house on the Sadong River in Sarawak (see recto 1 and also verso 26). This specimen, number 381 from Sarawak, was recorded in his insect register (WCP4766, 25) as found "about old fallen timber," which abounded at the Simunjon coal works. The other specimen on this page, Sarawak specimen 196, is unidentified beyond family in the insect register.

## 14

[1]Sketch of Mr. Darwin's "Natural Selection"

| | | |
|---|---|---|
| [2]Chap.[ter] | I. | On variation of animals & plants under domestication, treated generally. |
| " | II. | d[itt]o. d[itt]o. treated specially,—external & internal structure of Pigeons & history of changes in them. |
| " | III. | On interc[r]ossing,—principally founded on original observations on plants. |
| " | IV. | Varieties under Nature. |
| " | V. | Struggle for existence, Malthusian doctrine; rate of increase,—checks to increase &c. |
| " | VI. | "Natural Selection" manner of its working |
| " | VII. | Laws of variation. Use & disuse reversion to ancestral types &c. &c |
| " | VIII. | Difficulties in theory. Gradation of characters |
| " | IX. | Hybridity. |
| " | X. | Instinct. |
| " | XI. | Palaeontology & Geology. |
| " | XII. and XIII. | Geog.[raphical] distribution. |
| " | XIV. | Classification, Affinities Embryology. |

1. Wallace likely acquired this table of contents for Darwin's *Natural Selection* from Darwin himself. He copied it into the Species Notebook (unfortunately without comment) from a letter, but it is not clear why or when Darwin sent it. Perhaps, in an effort to show Wallace how far along he had gotten on the subject beyond the extracts read at the Linnean, Darwin decided to send Wallace the contents soon after the Linnean event. One possibility, then, is that Darwin sent this as part of a now-missing letter written on or about July 13, 1858. That letter came on the heels of the joint reading of the Wallace and Darwin papers at the Linnean Society on July 1, 1858, and accompanied a letter from Joseph Hooker explaining the circumstances of the reading of the papers. Both letters (now lost) were sent on July 13, and their receipt by Wallace is confirmed by a letter he wrote in October: "I have received letters from Mr. Darwin & Dr. Hooker . . . which have highly gratified me" (Wallace Correspondence Project, WCP369; see also Darwin Correspondence Project, letters 2303 and 2306).

2. Details of the list of chapters suggest that it was sent to Wallace in the fall of 1858, however, rather than in mid-July. This table of contents differs in interesting ways from Darwin's working table of contents as of June 1858, when his progress was interrupted by the arrival of Wallace's manuscript. That version, discussed by Stauffer (1975, 21–24), parallels the one here up to Chapter X. In the *Natural Selection* manuscript, Stauffer designated the unfinished geographical distribution section as Chapter XI, and also discussed notes Darwin made for a continuation of his treatment of the subject in a following chapter, which would have been Chapter XII (Stauffer 1975, 528, 534).

Here in the Species Notebook, in contrast, Chapter XI is given as "Paleontology & Geology," and geographical distribution is shifted to Chapters XII and XIII. Note, further, that Darwin had added a fourteenth chapter, on classification, affinities, and embryology. The sequence and content of these last four chapters are closer to those of *On the Origin of Species* than *Natural Selection*: in the *Origin* there are two geological chapters (nine and ten), consisting of material largely absent from *Natural Selection*;

two geographical distribution chapters (eleven and twelve), only the first of which contains significant amounts of material correlated with *Natural Selection*; and a final chapter on morphology, embryology, and rudimentary organs (thirteen), almost none of which is found in *Natural Selection*.

Through the summer and fall of 1858 Darwin was planning to prepare a long paper for the Linnean Society. In a letter to Hooker he said that he would soon "set to work at abstract, though how on earth I shall make anything of an abstract in 30 pages of Journal I know not" (Darwin Correspondence Project, letter 2306). Through Hooker's efforts the Linnean Society agreed to have Darwin submit a longer abstract consisting of several sections, each to be read separately although they would be published together (letter 2313). He still intended to finish *Natural Selection*, but in the meantime his "abstract" grew in scope until, sometime that fall, he merged the two projects and opted for a book-length work (the *Origin*, far shorter than *Natural Selection*). Darwin must have sent the fleshed-out table of contents to Wallace when he still had the *Natural Selection* project in mind; otherwise he would not have used that title. There are no letters known to have been sent by Darwin to Wallace between July 13, 1858, and January 25, 1859, and by the latter date it is clear he had abandoned *Natural Selection* in favor of a "small vol. of 400 or 500 pages"—the *Origin* (Darwin Correspondence Project, letter 2405). It is likely that Darwin sent the table of contents in the fall of 1858, in a now-lost letter.

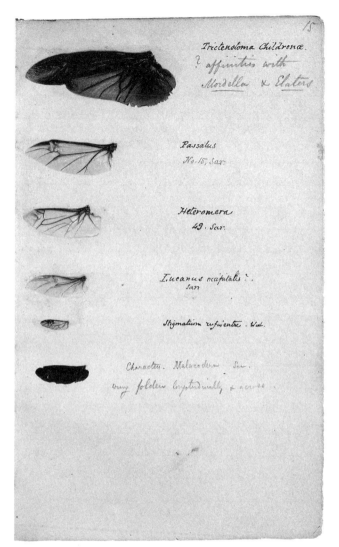

## 15

[1] *[Pasted wing]*     Trictenotoma Childrenae.
                                 ? affinities with
                                 <u>Mordela</u> & <u>Elaters</u>

*[Pasted wing]*     [2]Passalus
                                 No. [number] 15, Sar.[awak]

*[Pasted wing]*     Heteromera
                                 49. Sar.[awak]

*[Pasted wing]*     [3]Lucanus occipitalis ? .
                                 sar.[awak]

*[Pasted wing]*     Stigmatium rufiventre. West.[wood]

*[Pasted wing]*     Charactus. Malacoderm Sar.[awak]
                                 wing folden longitudinally & across.

*[The next notebook page was left blank by ARW and is omitted]*

1. The size of this wing signals a hefty beetle. *Trictenotoma childreni*, named by G. R. Gray in 1832, is one of the largest members of the family Trictenotomidae, relatives of the Tenebrionidae, or darkling beetles. Little known in Wallace's lifetime, essentially nothing is known of *Trictenotoma* biology even today—remarkable for so large an insect. See verso 26 for a sketch by Wallace of this impressive beetle.

2. In Wallace's insect collection register (WCP4766, 6), he includes an exquisite small drawing of the odd head of this specimen—beetle number 15 from Sarawak, *Passalus crucifer* (family Passalidae). Passalid beetle heads are often knobbed or horned, likely a sexually selected trait. These beetles are social, typically living in family-colonies in rotting logs, and are known to communicate with a remarkable range of sounds (see Costa 2006, chapter 14).

3. This *Lucanus* specimen may be number 55 in the insect collection register (WCP4766, 8), with a drawing of spectacular antlerlike mandibles. Indeed, stag beetles (family Lucanidae) are so called for the often stunningly elongated and branched mandibles that the males sport. These mandibles are useless for feeding but function in much the same way as do stag antlers: they are a sexually selected trait used in displays and contests by the males in competition for mates.

(Note that the following blank page, verso 16, is omitted.)

[Pasted wings]
²[Pasted wing—missing]
[Pasted wing]

[1]Lamellicornes.
Chalesthea smaragdula.
Trichogomphus milo.
[3]Protaetia obscurella ?
Sar.[awak] 221

*[The next notebook page was left blank by ARW and is omitted]*

1. "Lamellicornes" is an obsolete name for a superfamily of scarab beetles, Lamellicornia, the name of which refers to their club-shaped, or lamellate, antennae. These antennae end in a set of plate-like segments that the beetle can spread out like fingers or pull together into a compact mass. As the antennae are the organs of scent, spreading the lamellae increases the exposed surface area and likely heightens sensitivity. These beetles are now classified as family Scarabaeidae, subfamily Dynastinae.

2. The outline beside the name *Trichogomphus* indicates a missing wing of remarkable size. *Trichogomphus* species, numbering only about a dozen in total, are very large, stout scarab beetles.

3. *Protaetia obscurella* is a beetle in a different scarab subfamily: Cetoniinae, the flower beetles. Wallace noted in his insect collection register (WCP4766, 20) that he received specimens of this species from the Dyaks at Sadong, and caught them "about posts &c... Santubong."

(Note that the following blank page, verso 18, is omitted.)

[19]

[*Pasted wing*]        [1]Brachinus . . 233 . sar.[awak]
[*Pasted wing*]        [2]Cicindela aurulenta.

*[The next notebook page was left blank by ARW and is omitted]*

1. Besides collecting on a rigorous schedule, Wallace was also opportunistic. He notes by the entry for this *Brachinus* ground beetle (Carabidae) in his insect collection register (WCP4766, 20), "houses at night."

2. The Southeast Asian tiger beetle *Cicindelus aurulenta*, named by Fabricius in 1801, is breathtaking in color: it has three large orange-yellow spots lining each elytron and a metallic blue-green body with edges lined in metallic red.

(Note that the following blank page, verso 20, is omitted.)

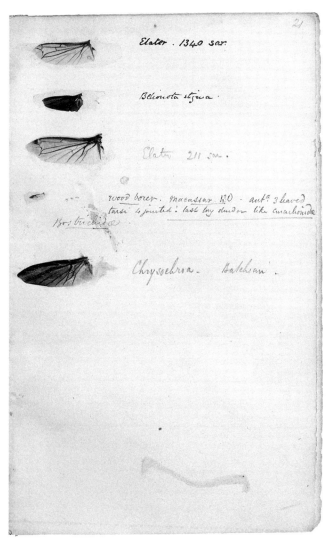

## 21

*[Pasted wing]* Elater . 1340 sar.[awak]
*[Pasted wing]* [1]Belionota stigma
*[Pasted wing]* Elater 211 sar.[awak]
*[Pasted wing]* [2]<u>wood borer</u>. <u>Macassar 130</u>. ant.[enn]^ae 3 leaved
tarsi 4 <u>jointed! last long slender like curculionidae</u>

<u>Bostrichidae</u>
*[pasted wing]*  [3]Chrysochroa.  Batchian.

*[The next notebook page was left blank by ARW and is omitted]*

1. *Belionota* are metallic wood-boring beetles, family Buprestidae. Wallace collected several *Belionota* species; in one of his published letters to Samuel Stevens from Sarawak he commented on an especially noteworthy species: "The Buprestidae I am happy to say are very fine; not that the species run very large, but they are tolerably abundant. One of the most beautiful I make out to be *Belionota sumptuosa*" (Wallace 1855a, 4804).

2. Wallace noted that this little bostrichid bark beetle "bore[s] holes in bamboo & newly fallen timber" (WCP4766, 132). This specimen surprised Wallace in that its tarsal segments number four. Tarsi (singular: tarsus) are the distal leg segments of insects. The number and shape of tarsi are useful features for identifying species, genera, even whole families. Coleopterists use a *tarsal formula*—a three-number sequence that gives the number of tarsal segments of the front, middle, and hind legs, respectively—in taxonomic keys for some groups. (See the drawings on verso 27 and 29 for examples of tarsal segments.) Bostrichidae are typically characterized by a 5-5-5 tarsal formula, but in rare cases like this one Wallace found in Macassar, the formula is 4-4-4. Wallace also notes that the distalmost tarsal segment of his specimen is long and slender, "like curculionidae." There may be something interesting about the biology of these bark beetles that relates to their having legs that are atypical for the family. Other bark beetles from this region sport different sorts of anatomical anomalies indicative of unusual biology, such as the bark beetle *Crossotarsus* with its highly modified head (Costa 2006, plate 4).

3. *Chrysochroa* is another member of the family Buprestidae. This genus is one of the largest and most widespread buprestids in Southeast Asia, and also one of the most beautiful. *Chrysochroa wallacei*, a large, striking, metallic green buprestid, was named in Wallace's honor by French entomologist Achille Deyrolle (1813–1865) in 1864. This beetle was not the only treasure from Batchian (Bacan) to be named for him, as this is the island where he found his most prized specimen, the bird of paradise that proved to be a new genus: *Semioptera wallacei* (see recto 135).

(Note that the following blank page, verso 22, is omitted.)

[23]

¹(23

1. This blank but numbered page is one of many illustrations of Wallace's working method with his notebooks. In his analog world there was no easy cutting-and-pasting method he could use to make room for insertions or close gaps created by deletions. Wallace entered different kinds of information "in parallel" in different parts of his notebooks, leaving space between sets of entries so that he could continue to add to the entries over time. On occasion he ran out of room and had to continue notes begun in one place on another, sometimes distant page (see, e.g., recto 7 and 9; recto 36 and 42). This page may have been intended as a space where he could eventually paste in more wings, but he abandoned the practice before getting this far.

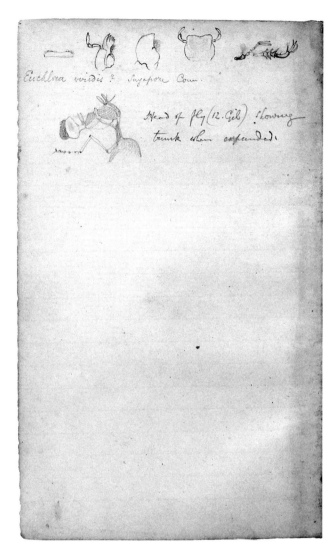

## 24

*[Pencil sketches]*
Euchlora viridis ? Singapore Com.*[?]*
*[Pencil sketch]*      [1]Head of fly (12. Gil[olo]) showing
                         trunk when expanded.

1. Wallace collected this bizarre fly on Gilolo (modern Halmahera). Frances Walker (1809–1874) cataloged fifty-nine fly species collected by Wallace on that island, most of them new (Walker 1861). Only the most spectacular of flies tend to be illustrated; in the case of Wallace's many fly collections, the most alluring are probably the antlered flies, comprising a subfamily of the fruit fly family Tephritidae. Wallace collected several species in Dorey (Manokwari), New Guinea, where they were "settling on fallen trees and decaying trunks." He illustrated them with a woodcut in *Malay Archipelago*, shown to the right.

Wallace's "Gilolo fly" may be specimen number 12 described on page 145 of his insect register (WCP4766): "On foliage in damp places, walks about slowly wings expanded & rather raised, shaking[.] head transparent pale amber." This sounds like typical tephritid courtship behavior, though Wallace does not indicate if there was more than one fly present when he observed this behavior.

William Wilson Saunders (1809–1879), who had acquired much of Wallace's insect collection, erected the genus *Elaphomyia* for Wallace's antlered flies in 1861. Today that genus is synonymized with the genus *Phytalmia*, named by Carl Eduard Adolph Gerstaecker (1828–1895) in 1860 and so has priority. Antlered flies exhibit what in behavioral ecology terms is a resource-defense mating system, in which males guard, and defend against other males, resources such as food or oviposition sites that females will find attractive. If successful, they garner most of the matings, and give their mates access to the resource.

Four species of antlered flies (now genus *Phytalmia*), in an illustration from *Malay Archipelago*. The species in the upper left of the image is a male stag-horned deerfly (*Phytalmia cervicornis*), collected by Wallace on Waigeo (Irian Jaya). Wallace's remarkable specimen now resides in Drawer 28 of the Wallace Collection at the Natural History Museum in London. It is one of many of Wallace's insect specimens that have been lovingly and meticulously restored by museum entomologist George Beccaloni, after years of neglect resulted in their disarticulation. From Alfred Russel Wallace, *The Malay Archipelago*, 4th ed. (London: Macmillan & Co., 1872), 502.

## 25

*[Pencil sketches]* [1]Bill of <u>Collocalia</u> with a
large membrane at gape

1. This is a genus of swifts (family Apodidae), described by George Robert Gray, keeper of birds at the British Museum. *Collocalia* swifts include the Asian species from which bird's-nest soup is made. The large membrane that Wallace mentions helps the bird scoop up its insect prey on the wing, something accomplished in other swift, swallow, and nightjar groups by rictal bristles. Wallace read a paper at the Zoological Society of London in 1863 showing how his *Collocalia* collections helped clarify the identity of Linnaeus's *Hirundo esculenta* (the name of which means "edible swallow," of bird's-nest soup fame); they in fact belong to the same genus (Wallace 1863b).

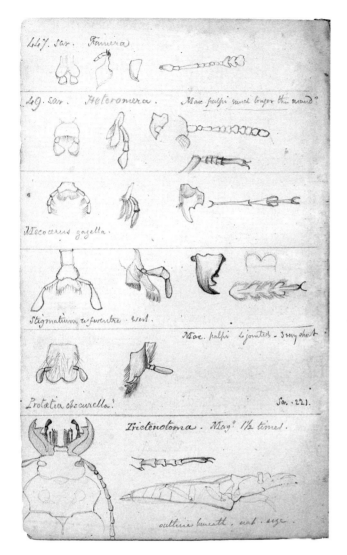

## 26

**[1]** 447. sar.[awak] Triniera
*[Pencil sketches]*

---

49. sar.[awak] Heteromera. Max[illary] <u>palpi much longer
than</u> mand.[ibula]ʳ
*[Pencil sketches]*

---

*[Pencil sketches]*
Mecocerus gazella.

---

*[Pencil sketches]*
Stigmatium rufiventre West.[wood]

---

*[Pencil sketches]*          Max.[illary] palpi 4 jointed—3 very short
Protaetia obscurella?          Sar.[awak] 221.

---

*[Pencil sketches]*          **[2]**Trictenotoma. Mag.[nifie]ᵈ 1½ times.
outline beneath. nat.[ural] size.

1. This is one of several pages of detailed sketches of beetle mouthparts and in some cases antennae and the segments of the tarsi (the distalmost leg segment), all of which Wallace used as aids in identification. In general, insects with chewing mouthparts have four easily distinguished segments: the labium, mandibles, maxillae, and labrum. In these sketches the labium, which tends to be a single but two-lobed structure, is drawn first on the far left. The second sketch from the left is one of the two paired maxillae. (Note that the maxillae and labium have palps.) Next over comes one of the mandibles, which in the case of cerambycids are often sharp, as they bore into wood. Wallace generally does not sketch the labrum, as it is not as informative in identification as the other mouthparts.

The leg tarsi often vary in segmentation (tarsomeres), and in many cases genera, tribes, subfamilies, and even sometimes whole families can be distinguished by what is termed the "tarsal formula"—given as three numbers, the formula represents the number of tarsomeres found in the fore-, mid-, and hind legs. (See also annotations for verso 21.)

2. Among the sketches of beetle mouthparts, Wallace's sketch of the imposing beetle *Trictenotoma* grabs our attention on this page. A relative of the darkling beetles (Tenebrionidae), this genus (described by G. R. Gray in 1832) is now put in its own family, the Trictenotomidae. Wallace made reference to "the handsome but anomalous *Trictenotoma* forming a distinct family" in *The Geographical Distribution of Animals* (Wallace 1876, 1: 320). There are but fourteen or so species in two genera described from this family (*Trictenotoma* and *Autocrates*), nearly all of them occurring in Southeast Asia. Wallace collected *Trictenotoma* beneath tree bark, where they assume the flattened position he sketches here. The larvae are likely to be wood-borers, feeding on the heartwood of the host tree. (See also verso 15.)

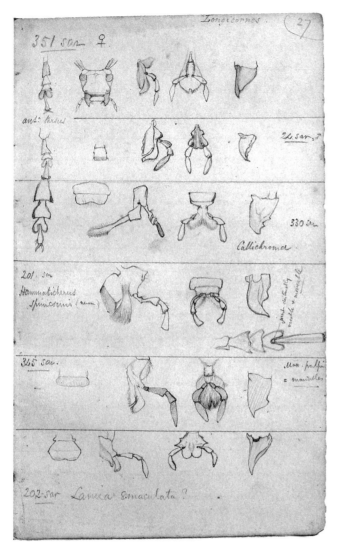

Longicornes.

351 sar[awak] ♀
*[Pencil sketches]*

---

ant.[erio]ʳ tarsus
*[Pencil sketches]*                    24 sar.[awak] ♂

---

*[Pencil sketches]*                    330 sar[awak]
                                       Calichroma.

---

201. sar[awak]                         *[Pencil sketches]*
[1]Hammaticherus
spinicornis (Newm.[an])
*[The following two lines are written vertically]*
                                       joint distinctly
                                       visible & <u>moveable</u>.

---

<u>345 sar.</u>[awak] *[Pencil sketches]*    Max.[illary] palpi
                                       = <u>mandibles</u>

---

*[Pencil sketches]*
[2]<u>202 Sar</u>[awak]  Lamia 8-maculata ?

*[The next notebook page was left blank by ARW and is omitted]*

1. *Hammaticherus spinicornis* (family Cerambycidae), a wing of which can be found at the top of verso 9, is specimen number 201 collected in Sarawak. This species was named by Edward Newman in 1842. The genus *Hammaticherus* is now synonymized with *Hoplocerambyx*, and the specific epithet *spinicornis* refers to a curious feature of the long antennae of this species: each segment bears a sizable thornlike spine near the distal joint. As the oblong beetle is almost uniformly brown and the antennae resemble slender brown thorny twigs, this may be a case of mimicry.

2. Arabic numerals are no longer legitimately used in official taxonomic names. Formerly they were commonly used to reference numbers of an identifying feature such as spots (in this beetle, "8-maculata" references its eight spots). Use of numerals was taken to an extreme by some entomologists; the ladybird beetle genus *Coccinella*, for example, included species named for numbers of small spots ranging from two (*C. 2-punctata*) to twenty-five (*C. 25-punctata*). According to the International Code of Zoological Nomenclature, such names are now rendered with latinized expressions of the numbers: *8-maculata* becomes *octomaculata*.

(Note that the following blank page, verso 28, is omitted.)

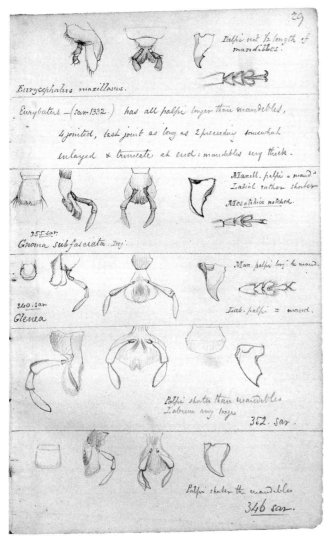

*[Pencil sketches]* Palpi not ½ length of
mandibles.

[1]Eurycephalus maxillosus.

---

Eurybatus— [2](sar.[awak] 1332.) has all palpi longer than
    mandibles,
    4 jointed, last joint as long as 2 preceeding some-
    what
    enlarged & truncate at end: mandibles very thick.

---

*[Pencil sketches]* Maxill.[ary] palpi = mand[ible][s]
Labial rather shorter
<u>Mesotibiae notched.</u>

355 sar.[awak]
Gnoma subfasciata Dej.[ean]

---

*[Pencil sketches]* Max.[illary] palpi long[e][r] than
    mand.[ibles]

340. sar.[awak]
Glenea Lab.[ial] palpi = mand.[ibles]

---

*[Pencil sketches]*

Palpi shorter than mandibles
Labrum very large
    352 Sar.[awak]

---

*[Pencil sketches]*

Palpi shorter than mandibles
    346 sar.[awak]

*[The next notebook page was left blank by ARW and is omitted]*

1. In his published letter from Sarawak dated April 1855 (Wallace 1855a), Wallace commented that "numbers of the handsome red *Eurycephalus maxillosus* are here constantly flying about and crawling on the timber." *Eurycephalus* (now *Euryphagus*) are large long-horned beetles (Cerambycidae), all the more striking for their vivid red coloration as mentioned by Wallace. The red color is characteristic of the entire genus, not just *E. maxillosus*. Incidentally, this is the species that Wallace used in experimenting with specimen label layouts on recto 36 and 42.

2. This specimen number, 1332, reflects the rich insect collecting that Wallace enjoyed in Sarawak, especially the Simunjon coal-mining area (see recto 1–4) where he resided from mid-March 1855 though November 1856. In *Malay Archipelago* he reckoned he had collected about 2,000 distinct species in Borneo, "of which all but about a hundred were collected [at Simunjon], and on scarcely more than a square mile of ground" (48). And all this despite a month-long setback: in his insect collection register Wallace noted that this *Eurybatus* specimen was a new species, found in "swamp jungle—under a leaf." Between specimens 1331 and 1332 he noted the date, "July 1, 1855." Before he could reach specimen 1335, however, he recorded "July. all month in house with sore ancle [*sic*]" (WCP4766, 70–71).

(Note that the following blank page, verso 30, is omitted.)

*[Pencil sketches]*

¹359 sar.[awak]

palpi
ab[ou]ᵗ = mand[ible]
tarsi short broad

1. This specimen was first tentatively identified as a *Lamia*, a large-bodied cerambycid species, and Wallace noted that it "walks slowly," and is "often found crawling among dead leaves" (WCP4766, 24). He later changed the identification to *Phryneta*, yet another large-bodied cerambycid. Drawings such as these helped him with such taxonomic identifications, since for highly diverse groups like long-horned beetles some species are distinguishable only by minute study of seemingly trivial characteristics of their anatomy (in this case, mouthparts).

## 32

[1]Pelopoeus sp.[ecies] ? (3) Sar.[awak]
Makes a small earthen cell in houses &c., rests on
its 4 hind legs two fore d[itt]o being kept elevated,—
constantly cleans its antennae & palpi with its fore legs—,
often holds one antenna down & the other up—moves
its antennae rapidly over surface of any small object
it approaches. Stops up the mouth of the cell
at times & reopens it.  One opened had 5 small
spiders & a small grub inside. spiders torpid as
mentioned by Abbott.  A cell in my room
was first stopped up then completely & irregularly
plastered over so as to quite conceal its shape.
After it had remained thus a fortnight apparently
unvisited by the parents wanting to move the books
to which it was attached I opened it & found
a brown transparent delicate cocoon 7/10 th inch
long containing a white grub about 2/3 the length
of the cocoon.  The outer clay cell was
very hard & solid & how the insect was to
get out does not appear.    Sar.[awak] Sept.[ember] 1855.

1. *Pelopoeus* is a genus of solitary wasp of the family Sphecidae, the species of which are now generally assigned to the genera *Sceliphron* and *Chalybion*. The Sphecidae include spider wasps, digger wasps, mud daubers, and their relatives. All are predatory wasps that stock their nest with paralyzed prey that serve to feed their young. Different sphecid species specialize on prey type, most commonly spiders or caterpillars. These wasps do not exhibit parental care but seal their prey, on which an egg is deposited, into a nest cell. On hatching, the wasp grub feeds on the paralyzed prey.

These observations are dated September 1855, from Sarawak. He collected two *Pelopoeus* species in Sarawak, according to Frederick Smith's "Catalogue of the Hymenopterous Insects Collected at Sarawak, Borneo; Mount Ophir, Malacca; and at Singapore, by A. R. Wallace" (F. Smith 1857–1858). The reference to Abbott is likely John Abbot (1751–1840), the early American naturalist celebrated for his entomological and botanical observations and paintings from Georgia in the United States.

## 33

The rows of punctures in the elytra of
beetles, are the centres of cells formed
by longitudinal & cross nervures . In a
large elater this was distinctly [seen] every puncture
being in the centre of a square cell, &
apparently formed by the nervures all round
becoming thickened from below upwards leaving
the original wing membrane only level with
the lower surface of the elytron which
forms the bottom of each puncture.
In this Elater (Sar.[awak]    ) there were 9 rows of
punctures in each elytron, equal to the chief
longitudal nervures in the under wings.
The acute shoulder angle is formed in a small
longitudinal basal cell.

---

[1]Lomaptera, fly with the elytra scarcely raised above the
abdomen,—hum & wheel about like bees, difficult
to capture. Aru—Jan.[uary]/57.
Ornithoptera male follows the female near the ground
among thickets, at times mounting over them. ♂ generally
flies high, ♀ low—Aru—

1. *Lomaptera* are scarab beetles of the subfamily Cetoniinae. This large genus is widely distributed in the Australasian region. The often large-bodied (2–3 centimeter) metallic-green beetles hum in rather beelike fashion in flight, as Wallace notes here. Large scarabs are often clumsy on the wing, but it seems these beetles were able to elude Wallace's insect net.

## 34

*Ornithoptera*

*Larva, rich purple black, 2nd. 3rd. & 4th. segments with 8 rows of tentacles, rest with 6, which are rather long acutely pointed & with a rich crimson basal band, 2 dorsal rows ½ inch long, those below the spiracles shorter, on the seventh seg. is a white stripe on each side oblique from the spiracle to & enclosing the base of the tentacle; form like that of O. Heliacon, fig. by Horsfield but tentacles longer & much more acute. Feeds on a climber, ? an Aristolochia.*

*The head tubercles are y shaped ⋎ transparent blood red, exserted as in Papilio, with no separate sheath.*

*Larva does not suspend itself from the plant on which it feeds but seeks a more rigid leaf ⅌ a stronger stalk & suspends itself from the under surface in an inclined position. The suspending loop passes round the insect as in other Papilionidæ, not fastened to each side as stated by Boisduval ? from Horsfield to be the case in O. Heliacon.. As the larva makes the loop & then escapes from its skin that mode seems impossible. At the sides the sharp thread is buried in the soft skin of the newly born pupa wh. has probably given rise to the mistake..*

*Pupa; Feb. 8th. am. alivs March. 12.*
♀. *Imago Mar. 9th. am.* Aru Islands.

[1]Ornithoptera
Larva, rich purple black, 2nd. 3rd. & 4th . segments
with 8 rows of tentacles, rest with 6, which are
rather long acutely pointed & with a rich crimson basal
band, 2 dorsal rows ½ inch long, those below the
spiracles shorter, on the seventh seg.[ment] is a white stripe
on each side oblique from the spiracle to & enclosing
the base of the tentacle; form like that of O.[rnithoptera]
    Heliacon. fig.[ure][d]
by Horsfield but tentacles longer & much more acute.
Feeds on a climber, ? an Aristolochia.
The head tubercles are y shaped *[Sketch]* transparent blood
red, exserted as in Papilio, with no separate sheath.
Larva does not suspend itself from the plant on
which it feeds but seeks a more rigid leaf *[illeg.]* a
stronger stalk & suspends itself from the under
surface in an inclined position.  The suspending loop
passes round the insect as in other Papilionidae, not
fastened to each side as stated by Boisduval ? from
Horsfield to be the case in O.[rnithoptera] Heliacon. .  As
    the larva
makes the loop & then escapes from its skin that mode
seems impossible.  At the sides the sharp thread is
buried in the soft skin of the newly born pupa wh.[ich] has
probably given rise to the mistake. . [by Boisduval]
Pupa; Feb.[ruary] 8th. am. *[illeg.]*. March. 12.
♀. Imago Mar[ch] 9th. am.                    Aru Islands.

1. The observations here were reported in a paper entitled "On the Habits and Transformations of a Species of *Ornithoptera*, Allied to *O. priamus*, Inhabiting the Aru Islands, Near New Guinea," which was read December 7, 1857, at the Entomological Society of London and published in the society's *Transactions* series the next year (Wallace 1858e).

The birdwing butterflies, consisting of thirty-five or so species in the family Papilionidae, were all placed in the genus *Ornithoptera* in Wallace's day. Modern treatments divide them into the genera *Ornithoptera*, *Trogonoptera*, and *Troides*. These largest and most spectacular of butterflies may not resemble other papilionids—the swallowtails—owing to their long, angular wings and lack of swallowtail-like extensions of the hind wings. Their caterpillars reveal the relationship, however, with their eversible, defensive osmeterium, which Wallace describes here as "head tubercles." This paired organ, arising just behind the head, appears as horns and its coloration and odor can startle and repel would-be predators. An everted osmeterium is visible in Wallace's caterpillar sketch on verso 36.

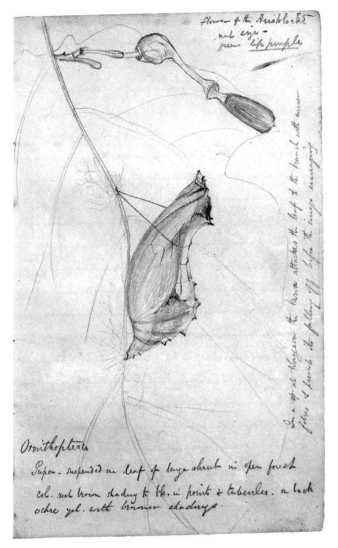

## 35

[Sketch of pupa]

[1]flower of the Aristoloch[ia]
nat[ural] size—
green   <u>lip purple</u>

*[The following two lines are written vertically on the right hand margin]*
In a sp.[ecies] at Waigiou the larva attaches the leaf to the
    branch with *[illeg.]*
fibres to prevent its falling off before the imago emerging

Ornithoptera
[2]Pupa—suspended in leaf of large shrub in open forest
col[our] rich brown shading to b[lac]k. in points & tubercles.
    on back
ochre yel.[low] with browner shadings.

1. Wallace came upon this pupa on the underside of an *Aristolochia* (this one most likely now a *Parastolochia*) leaf. This viny species is a member of the Aristolochiaceae, the Birthwort family, which consists of some 400 species in seven genera worldwide. The compound aristolochic acid, a toxic alkaloid that serves as an effective feeding deterrent against many potential herbivores, characterizes this plant group. Insects feeding on Aristolochiaceae tend to be specialists, metabolically capable of detoxifying the alkaloid. Some insects, such as the aposematic pipevine swallowtail of North America, can even sequester the compound and use it in their own defense. Several of the birdwing butterflies of the Australasian region feed on Aristolochiaceae and similarly advertise their unpalatability with spectacular coloration.

2. Look closely at Wallace's lovely—and accurate—drawing of this *Ornithoptera* chrysalis. (Another is figured on verso 37, in a drawing pasted onto the page.) It is secured to the leaf in two ways: at the tip of the abdomen is the *cremaster*, a tiny support hook (or a cluster of hooks) securely embedded in the small mat of silk that Wallace has rendered with light pencil lines radiating from the tip. The band of silk about the midsection of the pupa is called the *girdle*, also secured to a mat of silk on the leaf. Most butterfly pupae are secured with both cremaster and girdle, but in some groups the chrysalis dangles by just the cremaster. (Moth caterpillars, incidentally, pupate ensconced in silken cocoons, not exposed, and their pupae are technically not called chrysalids.)

The coloration Wallace reports here is repeated almost verbatim in his 1857/1858 *Transactions* paper: "The pupa is of a rich brown colour, on the back ochre-yellow, with the points and tubercles nearly black. It is very bulky, and nearly 3 inches long . . . Its duration in the pupa state is exactly a month (twenty-nine or thirty days), a very long period for a diurnal *Lepidoptera* in the tropics" (Wallace 1858e, 273).

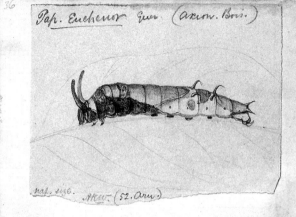

## 36

[1]Pap.[ilio] Euchenor Guer. [Guérin ] (axion. Bois.[duval])
*[Sketch on paper pasted in]*
nat.[ural] size.   A[lfred]R[ussel]W.[allace] (52. Aru.)
the above feeds on a forest shrub with large
^alternate pinnate leaves having a terminal leaflet.

———————

Another specimen at Togou has the green spot
connected with the ground colour of the back
above it to the right, & feed on a shrub
with opposite leaves having 3 leaflets.   The spots
on the 1st dark band pure violet in this specimen.
At the base of the horns is a *[tang.?]* dark brown spot divided
by a long.[itudina][1] whitish line.
Pupa. March. 14—Imago. April 6th.
Pupa, greatly bent, sharply keeled above, head dilated laterally
pinkish grey, with tinges of green & brick-red, more
on abdomen, sucker very long <u>free from abdomen</u> at
<u>end</u>.

1.  Here Wallace has rendered the larva of the swallowtail butterfly *Papilio euchenor* (see verso 37). He evidently collected the caterpillar and let it complete its development. He drew it smaller
than natural size initially (on verso 37), and then life-size on the
bluish paper pasted on this page.

The notes at the bottom of this page indicate that the caterpillar pupated on March 14 and the adult (imago) emerged on
April 6. It was a female, and was duly added to his collection as
specimen number 52 from the Aru Islands. The entry "axion Bois."
is a reference to synonymy with *Papilio axion,* named in 1832 by
Jean Baptiste Boisduval (1799–1879), the French physician and
butterfly specialist. He named the Lepidoptera and other insects
from the Malay Archipelago collected on Dumont d'Urville's *Astrolabe* and *Coquille* voyages (see recto 105).

## 37

*[Sketch of caterpillar]*

[1]Fine grass green, sides to the 6th seg.[ment] & broad dorsal
bands on the 3rd  & 4th. dark olive brown tinged, the
dorsal bands with violet whitish spots  in form like arabic letters  sides
  of the remaining
seg.[men][ts] & obliquely meeting on back of 9th. & 10th seg.
  [men][ts] fine
whitish violet, with a green wh.[ite] bordered round spot on the
sides of the 9th which with the [10th] bear each two dorsal
tubercles which are so long & slender as to hang down
at the ends, on the anal are two similar but shorter.
Above the head are two long curved diverging horns which
are rigid somewhat tuberculose & rather thickened
& rounded at the ends, they are hard, of a pale olive
col[our] at base shading to b[lac]k. at apex. From immediately
beneath these spring the forked retractile tentacles, equal in
length slender & of a transparent blood red colour.
Head partially retractile, like Larva of P.[apilio] Cresphontes
to which it is allied but has all tentacles & processes
much more developed. feeds at intervals day & night

1. *Papilio euchenor* is widespread and variable in the Australasian/Indomalayan region, with about a dozen subspecies recognized. The larvae feed on various species of Rutaceae (Citrus family). This species was described in 1829 by the noted French entomologist Félix Édouard Guérin-Méneville (1799–1874). The Aru Islands subspecies Wallace illustrates here (with forklike osmeterium, which he terms "tentacles," extended just behind the head) is today called *P. euchenor obsolescens*, a subspecies named in 1895 by Walter (Lord) Rothschild (1868–1937). Wallace initially drew the caterpillar on this page, and after it pupated he drew the chrysalis and pasted that drawing in.

[Sketch of pupa pasted by upper edge over the caterpillar sketch on verso 37 so that it can be lifted]

## 38

[1]Butterflies which settle on margins of mountain
streams N.[orth] of Macassar.

| Pap.[ilio] rhesus . . . | abundant | Terias hecabe |
| " androcles . . . | scarce | T.[erias] tilaha |
| " eucelades . . | scarce | T.[erias] 128 . . . all |
| " deucalion . . | very scarce | rather com . . . |
| " 92 . . | scarce | |
| " 2 . . . . . . | abundant | |
| " eurypelus . . | abundant. | |
| " codrus . . | very scarce | |
| " peranthus . . . | rather scarce | |
| " pamnion[?] . . | scarce | |
| " cresphontes . . | scarce | |
| Leptocircus curius . . . | scarce, flies over water. | |
| Pieris zarinda . . | abundant | |
| P.[ieris] albina . . . | abundant | |
| P. .[ieris] . . 2 sp. allied . . . | abundant | |
| P.[ieris] eperia . . . . | rather common | |
| P.[ieris] n[ew] s[pecies] (125) | rather com.[mon] | |
| Iphias glaucippe . . | rather com.[mon] | |
| Callidryas alcmene . . | —com.[mon] | |

1. As a collector and naturalist, Wallace had to know the habits of myriad species intimately—their food plants, season and time of day of flight, preference for forest interior or edge, and so on. The behavior he mentions here, settling on the margins of streams, is related to a phenomenon observed in butterflies termed "puddling" in which many individuals will gather around the margins of a puddle. The butterflies are thereby imbibing both water and, often, salts and minerals. Although Wallace does not note the sex of the various species listed here, puddling behavior is most often seen in males.

| | [1]Bowring. Hong Kong &c. | Bates 4.5 years. |
|---|---|---|
| Cicindelidae _____ | 23. | 38. |
| Carabidae _____ | 144⅑ | 272¹⁄₁₃ |
| Brachyelytra _____ | 92 | 120 |
| Melolonthidae ____ | 89 | 75 |
| Cetoniidae _____ | 30 | 19 |
| Buprestidae _____ | 31 | 95 |
| Elateridae _____ | 59 | 150 |
| Cleridae _____ | 12 | 84 |
| Rhyncophora _____ | 145 | 730 |
| Longicornes _____ | 109¹⁄₁₂ | 473 ¹⁄₇.₅ |
| Cyclica _____ | 200 | 620 |
| Hydrocanthari. . | 34. | 30. |
| | | |
| Total Coleoptera | 1300 | 3570 |

1. Wallace was interested in global patterns of species richness and diversity. We have seen many examples in the Species Notebook where he has recorded regional species numbers for different groups of organisms, drawing on a diversity of authorities. Here he appears to be comparing the richness of the beetle fauna of Hong Kong and South America, as reflected in the collections of Hong Kong–based political economist and traveler Sir John Bowring and his own long-time friend Henry Walter Bates. Bowring (1792–1872), the fourth governor of Hong Kong, was a beetle enthusiast among his many other interests, and amassed a sizable collection. This was bequeathed to the British Museum by his son John Charles Bowring (1820–1893), also an avid collector, on the latter's death in 1893.

It is unclear where Wallace got the numbers he lists here for Bowring's collections, but most likely they came from one of the Bowrings themselves, since Wallace had been in communication with both. On page 5 of his Bird and Insect Register (WCP4767) Wallace lists specimens that he "sent to Bowring." An 1855 letter extracted in the *Zoologist*, written by the son, John Charles Bowring, reported on progress in collecting in Hong Kong, his specimens numbering precisely the total reported by Wallace here: "In Hong Kong Coleoptera I cannot be expected to progress very fast; it is a rare thing now for me to fall in with a novelty, but still I do so occasionally . . . My Chinese Coleoptera now number some 1300 species" (Bowring 1855, 4910).

[Annotations for the following six facing-page tables (verso pp. 40–51) are consecutively numbered, with notes commencing on p. 482]

| Date. | Number of Species collected at Makassar. | | | | | | | | |
|---|---|---|---|---|---|---|---|---|---|
| 1856. Oct.r | Geode- phaga | Lamel- licornes | Stern- oxi. | Longi- cornes | Rhynco phora. | Alia col.d | Coleop. | Lepid. | Alia Insecta |
| 13. Mon. | 1. | 3 | 1 | 0 | 3 | 8 | 16 | 5 | 2=23 |
| 14. Tu. | 0 | 6 | 2 | 4 | 5 | 18 | 35 | | 3=38 |
| 15. W. | unwell_____ | | | | | | | | |
| 16 Th. | unwell_____ | | | | | | | | |
| 17. Fr. | 0 | 8 | 0 | 2 | 3 | 8 | 21 | 12 | 3=36 |
| 18. Sat. | 3 | 7 | 0 | 2 | 2 | 12 | 26 | 10 | 6=42 |
| 19. Sun. | 1 | 7 | 0 | 4 | 2 | 8 | 22 | 12 | 4=36 |
| 20. Mon. | | | | | | | | | |
| 21. Tu | 3 | 4 | 0 | 1 | 1 | 5 | 14 | 6 | 3=23 |
| 22. W. | not | | | | | | | | |
| 23. Th. | well | | | | | | | | |
| 26. Sun. | | | | | | | | | 20 sp. hymen |

1857
July —    66° am

| Remarks. | Thermometer | | Wind | Hours |
| --- | --- | --- | --- | --- |
| | High[st] | Low[st] | | Rain |
| arrived at Samata late out only in afternoon | 92 | 73 | E. | drops. |
| out short time, unwell.. | 91 | 73 | E. | drops |
| out a short time p.m. wet... | 88 | 72 | E. | 12 |
| not very well—cloudy rain p.m. | | | E. | 2. |
| | | | | 10 |
| unwell—— | | | | 8 |
| Here Cicindelas abundant in pathway. | | | | 12 |
| | | | | 3 |
| Sat. 2 more Cicindelas — one large | | | | rain |
| a fine new Cetonia — from nature | | | | most |
| often took 15 to 20 species of | | | | every |
| | | | | p.m. or |
| | | | | night. |
| Hymenoptera — Beetles & Lepidoptera scarce— [2]too ill to collect the other orders. | | | | |

1857.  Species of Insects collected at Dobbo.

| [3]Jan.. 8 rainy | Geode-phaga. | Lamel-licornes. | Stern-oxi. | Longi-cornes. | Rhynco-phora. | Alia Coleop. | Coleop. | Lepid. | Alia Insecta. |
|---|---|---|---|---|---|---|---|---|---|
| 9. | 1 | 1 | 1 | 2 | 5 | 7 | 17 | 30! | 20=67 |
| 10 | - | - | - | - | - | - | - | - | |
| 11 | - | - | - | - | - | - | - | - | |
| 12 | 1 | 1 | 0 | 4 | 9 | 5 | 20 | 25. | 13=58 |
| 13 | 1 | 0 | 0 | 4 | 6 | 3 | 14 | 20 | 8=42 |
| 14 | 1 | 0 | 0 | 0 | 2 | 1 | 4 | 8 | 13=25 |
| 15 | 0 | 0 | 1 | 3 | 5 | 2 | 11 | 12 | 2=25 |
| 16 | 1 | 2 | 0 | 2 | 6 | 6 | 17 | 17 | 7=41 |
| 17 | 0 | 1 | 2 | 1 | 8 | 5 | 17 | 26 | 20=63 |
| 18. | 2 | 1 | 1 | 5 | 10 | 7 | 26 | 21 | 20=67 |
| 19. | 0 | 2 | 1 | 2 | 5 | 8 | 18 | 25 | 16=59 |
| 20 | - | - | - | - | - | - | - | - | |
| 21 | - | - | - | - | - | - | - | - | |
| 22 | - | - | - | - | - | - | - | - | |
| 23 | - | - | - | - | - | - | - | - | |
| 24 | 2 | 1 | 2 | 14 | 13 | 13 | 45 | 26 | 17=90 |
| 25 | 2 | 0 | 0 | 9 | 10 | 17 | 38 | 17 | 20=75 |
| 26 | 3 | 0 | 2 | 13 | 12 | 14 | 44 | 15. | 12=71 |
| 27 | 4 | 1 | 2 | 11 | 16 | 9 | 43 | 16 | 15=74 |
| 28 | 0 | 1 | 1 | 11 | 13 | 8 | 34 | 19 | 18=71 |

| Aru Islands . . | Thermometer | | | |
| --- | --- | --- | --- | --- |
| | | | | Hours |
| Remarks. | Highest | Lowest. | Wind | Rain. |
| Idea d'Urvillei, <u>Locytia d'Urvillei,</u> | | | | |
| O. prianius fine ♀ | | | W.N.W. | 0 |
| & 4–5 beautiful Eryciniidae & Lycenidae. . | | | W. | 20 |
| *[illeg.]* <u>Hyades</u> near Horsfeldii. . fine | | | | |
| Limenitis &c. &c. | | | | |
| 2 last days strong winds & almost | | | | |
| continual rain | | | W. | 14. |
| fine, a ♂ & ♀ [4]<u>Ornithoptera</u> near | | | | |
| <u>Poseidon</u> ♀8½ in ♂6½ in | | | W. | 1. |
| strong wind & rain all day, | | | | |
| tremendous evening squalls | | | W.N.W | 6 |
| very windy & heavy showers of rain. | | | | |
| no sunshine. | | | W.N.W | 12. |
| tremendous squalls of wind & rain, | | | | |
| sunshine (1–3pm) | | | W.N.W. | 10 |
| fine after morning but little sun | | | | |
| till p.m. | | | W.N.W | 2 |
| fine & hot . . . a fine new <u>Lomaptera.</u> | | | W.N.W | 0 |
| walk to <u>Wamma/?]</u>. . cloudy— | | | W.N.W | 0 |
| fine & hot  6 new sp. of Butterflies. . | | | W.N.W | 0 |
| wet almost all day | | | | 12 |
| do. cloudy showery & windy | | | | 15 |
| do. cloudy — showery . . moved | | | | |
| to new house. . | | | | 10 |
| . . . " very wet ——— | | 78° | W.N.W. | |
| pretty fine | | 76 | " | 4 |
| " " cloudy a.m. & some showers. | 86 | 76 | " | 4 |
| " " ——— | | 80 | | 6 |
| fine morning. . rain at night | | 78 | N. | 6 |
| . . fine & hot am. cloudy pm. | | 78 | N. | 0 |

| 1857 | | | | **Species of Insects collected at** | | | | |
|---|---|---|---|---|---|---|---|---|
| Jan^y Geode-phaga | Lamel-licornes | Stern-oxi | Longi-cornes | Rhynco-phora | Alia Coleop^a | Coleop. | Lepid. | Alia Insecta. |
| 29th. | 1 | 0 | 1 | 8 | 10 | 8 | 28 | 23 | 15=66 |
| 30th. | 3 | 0 | 1 | 9 | 16 | 11 | 40 | 23 | 30=93 |
| 31st. | | | | | | | | | |
| Feb. 1st. | 0 | 1 | 2 | 8 | 10 | 7 | 28 | 21. | 23=72 |
| 2nd | 1 | 0 | 0 | 5 | 9 | 6 | 21 | 6 | 10=37 |
| 3rd | 0 | 2 | 0 | 6 | 9 | 9 | 26 | 20 | 26=72 |
| 4th. | 2 | 0 | 1 | 8 | 6 | 6 | 23 | 19 | 15=57 |
| 5th. | 0 | 1 | 3 | 5 | 6 | 3 | 18 | 24. | 30=72 |
| 6th. | 0 | 0 | 3 | 5 | 7 | 11 | 26 | 15. | 32=73 |
| 7th. | 1 | 2 | 0 | 2 | 11 | 2 | 18 | 13. | 25=56 |
| 8th.. | 1 | 1 | 2 | 2 | 8 | 4 | 18 | 16 | 8=42 |
| In one month.. | 10 | 8 | 20 | 35 | 72 | 90= | 235 | 121 | 245 |
| | | | | | | | | 601 | |
| 2 months | 13 | 9 | 28 | 54 | 102 | 140= | 346 | 144 | 360 |
| | | | | | | | | 850 | |
| 9th.. | 0 | 0 | 0 | 6 | 7 | 4= | 17 | 25 | 22 |
| 10 | 3 | 0 | 0 | 8 | 6 | 7= | 24 | 14. | 14 |
| 11 | 2 | 2 | 0 | 7 | 5 | 13= | 29 | 18 | 13 |
| 12. | | | | | | | | | |
| 13 | 1 | 0 | 1 | 3 | 4 | 4 | 13 | 12 | 8 |
| 14 | 2 | 1 | 1 | 8 | 10 | 8 | 30 | 14 | 17 |
| 15 | 0 | 1 | 1 | 1 | 3 | 5 | 14 | 17. | 11 |

[45]

## Dobbo .. Aru Islands ..    Thermometer

| Remarks. | Highest. | Lowest. | Wind | Hours Rain. |
|---|---|---|---|---|
| hot & sultry — cloudy am. wet night. . | 88 | 78. | | 6 |
| fine. . . cloudy p.m._____ | 86. | 78. | | 4 |
| wet & cloudy_____ | | | | 12 |
| sun very hot. . | 90 | 78. | | 0 |
| cloudy & rain. fine evening. | | 77 | | 6 |
| fine, hot (saw [6]P. Ulysses?) | 91 | 78 | | 0 |
| rain & cloudy with gleams of hot sunshine. | 88 | 78 | | 4 |
| fine, hot.. | 88 | | | 8 |
| cloudy_____ | 88 | 80 | | 6 |
| very windy — clouds & sunshine wet last night | 87. | 80 | | 6 |
| windy & rain. . . | 88 | 82.. | | 8 |

| Hymenop. | Diptera. | Alii.. | Eric. [&] Lycenidae. | Hesperidae | Papilios | Alii diurn. | Moths. | Total Insects |
|---|---|---|---|---|---|---|---|---|
| 105 | 80 | 60 | (33 | 12 | 5 | 38) | 32 | 600 |
| | | | | 89 | | | | |
| 150 | 120 | 90 | (40 + | 16 + | 8 + | 44) | 36 | 850 |
| | | | | 108 | | 78 | | 0 |
| | | | | | | 95 | 80 | 2 |
| (Mayai) | | | | | | 93 | 81 | 6 |
| very wet._____ | | | | | | 84 | 80 | 18 |
| wet morning.._____ | | | | | | 87 | 79 | 12. |
| wet night & morning. wind— | | | | | | 87. | 77 | 12 |
| wet a.m. (Maya.i idle.)_____ | | | | | | | 77 | |

1857    Species of Insects collected at

| Feb. | Geode-phaga | Lamel-licornes | Stern-oxi. | Longi-cornes | Rhynco-phora | Alia Coleop. | Coleop. | Lepidop. | Alia Insecta |
|---|---|---|---|---|---|---|---|---|---|
| 16th. | 1 | 1 | 0 | 1 | 6 | 2 | 11 | 7 | 5 |
| 17th. | | | | | | | | | |
| 18th. | 0 | 2 | 3 | 7 | 7 | 4 | 23 | 30! | 13=69 |
| 19th. | cloudy — | | | | Bruchidae & | pm wet | | | |
| 20th. | wet | | | Curcul. | Anthrob. | Brent | | | |
| 21st | fine-- | | | 86 | 34 | 18 | =138 | | }145 |
| 22nd. At Arru | Ke-- | | | 6 | 1 | 0 | =7 | | |
| 23rd-- | Lamidae | Pieronidae | Cerambicidae | | | | | | |
| 24th | 72 | 1 | | 32 | =105 | 114 | | | |
| Ke | 5 | 1 | | 3 | = 9 | | | | |
| 25th | | | | | | | | | |

[7]March 13 To Tojou

[8] "   20 Orang Kaya to Dobbo

In 4  [9]May 10 returned to Dobbo — bad legs—

| Months. | 15 | 20 | 40 | 100 | 120 | 175 | 480 | 180 | 440 |
|---|---|---|---|---|---|---|---|---|---|
| | | | | | Total insects — 1100 species | | | | |
| | | √ | √ | √ | √ | √ | | | |
| 6 months | 20 | 24 | 46 | 105 | 138 | 201 | 534 | | |
| Ké add[l] | 0 | 3 | 7 | 9 | 7 | 12 | 38 | | |

18

37

24

33 flies

62

174 + 10

| | |
|---|---|
| Brachyelytra — | 6 - 0 |
| Hydradephaga— | 3 - 0 |
| Heteromera | 35 - 0 |
| Mordella | 7 - 0 |
| Cleridae | 16-0 |
| Malacoderma | 43 + 2 |
| Necroph. & Xyloph. | 20 + 0 |
| Buprest.— | 17 + 6 |
| Elaterid. | 29 + 1 |
| Cyclica | 59 + 9 |
| Triniera | 12 + 1 |
| Lamell | 15 + 3 |
| Lucania | 3 - 0 |
| Passala. | 6 - 0 |

Thermometer

| Remarks. | Highest. | Lowest | Wind | Hours Rain |
|---|---|---|---|---|
| Over to Wokan. . fine | | | | |
|    forest tree ferns &c.. | | | | oo |
| . . . hot, not very well <u>Mayai ½ day</u>. | | | | oo |
| fine hot.. | | | | 4 |
| Mayai at Wamma. | | | | 6 |
| do— | | | | 15 |
| do— | | | | 4 |
| idle | | <u>April fine & hot</u> | | |
| ½ day — to Wokan | | | | |
| M[?] Idle. *[Written vertically]* | | May very damp | | |
| wet. idle. | no quantity of rain | & showery | | |
| | but continual | June do —with | | |
| | <u>drizzle</u>— | a <u>few hot</u> days | | |

| Hymen optera | Dip tera | Hemip. Homop. | <u>Orthop</u> | <u>Neurop</u> | Lycean Ericin | Hesp. | Papleo nidae | <u>Pieridae</u> | <u>Alii</u> | Satyridae |
|---|---|---|---|---|---|---|---|---|---|---|
| 170 | 160 | 90 | 15 | 5 | 50 | 16 | 14 | 12 | 32 | 9 |
| | | | | | | 13 | 3 | | + 50 moths | |
| <u>38 + 35</u> | | 143 | 85 | Sing + | | | | | | |
| 173 | | | | Malacca | 65 | 33 | 19 | 19 | 82 | 12 |
| <u>145 + 59</u> | | He.\| Ho | | Forfic. | | | | | | |
| 204 | 175 | 62 \| 44 | 18 | 10 Dermap | 55 | 16 | 15 | 12 | 40 | 9 |
| 10 | 10 | 18 \| 6 | 0 | 0 6 | 4 | 2 | 1 | 0 | 3 | 0 |
| | | | | | 59 | 18 | 16 | 12 | 43 | 9 |

Diurnes. <u>157 + 65</u> *[written over 72]* moths

<u>229.</u>

Species of Insects collected at

| 1857 Sept^r | Geode-phaga. | Lamel-licornes. | Stern-oxi. | Longi-cornes. | Rhynco-phora | Alia Coleop. | Coleop | Lepidop | Alia Insecta |
|---|---|---|---|---|---|---|---|---|---|
| 11 | 3 | 1 | 8 | 1 | 12 | 24 | 49 | 16 | 50 = 115 |
| 12 | 1 | 0 | 5 | 1 | 10 | 14 | 31 | 9 | 37=77 |
| 13 | 1 | 0 | 4 | 2 | 7 | 14 | 28 | 12 | 21=61 |
| 14 | 1 | 0 | 2 | 1 | 7 | 14 | 25 | 14 | 12=50 |
| 15 | 0 | 0 | 0 | 0 | 8 | 10 | 18 | 18 | 15.=51. |
| 16} 17} | 0 | 3 | 2 | 1 | 9 | 9 | 24 ⁹ 15 | 25 { ²⁰ 15 | ³=37 7=47 |
| 18 | 0 | 4 | 1 | 1 | 4 | 6 | 16 | 14 | 14=44 |
| 19 to 22 } | 1 | 0 | 1 | 6 | 7 | 7 | 22 | 25 | 45=92 |
| 24 | 1 | 2 | 3 | 2 | 2 | 7 | 17 | 8 | 28=51 |
| 26 | 0 | 1 | 1 | 1 | 6 | 8 | 17 | 5 | 25=47 |
| 27 | 0 | 1 | 0 | 0 | 2 | 5 | 8 | 6 | 12=26 |
| 28 | 0 | 1 | 1 | 1 | 6 | 6 | 15 | 5 | 42=62 |
| 29 | 2 | 1 | 0 | 1 | 4 | 8 | 16 | 8 | 40=64 |
| Oct. 1st. | 7 | 1 | 0 | 1 | 0 | 8 | 17 | 4 | 50=7 |
| 2nd. | | | | | | | | | |
| 10 | | | | | | | 65 | 5 | 10=8 |
| 15th. | 1 | 2 | 2 | 0 | 6 | 45 | 56 | 4 | 12=72 |
| 20 | 15 | 1 | 2 | 1 | 3 | 30 | 52 | | |
| | | | | | | 17 | 476 | | 438 |
| | | | | | | | 28 | 188 | 25=64 |
| | | | | | | | | 11 | |

[10]Amasanga W. Macassar          from Aug. 16th./57.

Thermometer

| Remarks. | Highest. | Lowest. | Wind. | Hours Rain. |
|---|---|---|---|---|
| Hot & dry.. 25 Dip. 17 Hymen. Col.a all minute | | | | 0 |
| do. 18 dip. (4 new) 16 Hymen. | | | | 0 |
| do. 8 dip. (2 new) 5 [11]ornithoptera haliphron ! | | | | 0 |
| do. 8 dip. (1 new) | | | | 0 |
| Heavy shower p.m. 5 O. hal. ♂ 2 O. remus ♂. !! (5 dip.) | | | | 2 |
| fine      2. O. hal. 2 O. rem 1.. n.s.. | | | | 0 |
| fine..      4 " ♀  1 "   2 " | | | | 0 |
| fine.. shower pm.      3 "__  3 "__  0 __ | | | | 1. |
| at Bontang [illeg.] Cascade — 8 dip. 5 new.. | | | | 0 |
| fine — Pap. androcles, new Terius & c.. | | | | 0 |
| dry.. unwell.. 4 ornithoptera .. | | 25th. at home .. | | 0 |
| hot & dry. not very well 3 or. | | | | 0 |
| hot ..  "  do....     4 or. | | | | 0 |
| hot ..  "  2 or. | | | | 0 |
| hot ...  "  3 or. 30 dip. | | | | 0 |
| hot... 36 dip.  10 n.s.! After taken 150 sp. | | | | 0 |
| a shower pm. | | | | ½ |
| 200 specimens of Col.a .. Staphylinidae & c.. | | | | 0 |
| | | | | 1 |
| 24 n.s. of Col.a .. search in buff.o dung, rotten fruit & stag.t pool. | | | | 0 |
| about the 17th _____ | | | | 6 |
| 16 n.s. of n.s. of Carab[dae] under dead leaves, under timber | | | | |
| [12]& in rotten artocarpus | | | | |

Species of Insects collected at

| 1857 Oct. | Geode-phaga | Lamelli-cornes | Stern-oxi. | Longi-cornes. | Rhynco-phora | Alia Coleop. | Coleop. | Lepidop. | Alia Insecta. |
|---|---|---|---|---|---|---|---|---|---|
| ?26 | 18 | 2 | 5 | 1 | 10 | 40 | <u>76</u> | 5 | 10=91 |
| ?27 | 20 | 7 | 4 | 2 | 15 | 26 | 74 | 2 | 4=80 |
| 28 | 15 | 3 | 4 | 2 | 6 | 11 | 40 | 0 | 5=45 |
| 29 | 10 | 8 | 2 | 4 | 6 | 10 | 40 | 2 | 6=48 |
| 30 | 10 | 4 | 4 | 4 | 18 | 21 | 61 | 3 | 13=77 |
| 31 | 6 | 0 | 0 | 1 | 3 | 9 | 19 | 4 | 1=24 |
| Nov. 1st. | | | | | | | | | |
| 2 | 7 | 0 | 0 | 3 | 6 | 10 | 26 | 0 | 3=29 |
| 3 | 5 | 3 | 2 | 2 | 8 | 11 | 31 | 2 | 1=34 |
| 4 | 6 | 1 | 3 | 5 | 6 | 9 | 30 | 2 | 4=36 |
| 5 | 12 | 2 | 2 | 5 | 6 | 11 | 38 | 3 | 7=48 |
| 6 | 7 | 4 | 3 | 4 | 8 | 10 | 36 | 6 | 8=50 |
| 7.&8. | 10 | 2 | 8 | 6 | 11 | 33 | 70 | 0 | 12=82 |
| Total | 118 | 60 | 61 | 32 | 130 | | <u>840</u> | <u>232</u> | <u>640</u> |

1857

3 months collecting at end of dry season      1710 sp.$^s$ of Insects.

Amasanga. n. of Macassar

| Remarks. | Highest | Lowest. | Wind | Rain. |
|---|---|---|---|---|

[13] {Search under rotten fruit, dead leaves & c.  2 fine days after rain          0
{22 n.s. of Col.ᵃ (3 car. 4 Staph. 1 Elat.)

16 n.s. of Col.ᵃ (2 cicin. 4 carab.)  Search under dead leaves & in dung

                                                                                      pm rain          4

9 n.s. of col. (2 carab.) — very damp & cloudy, unwell.

18 n.s. of Col.ᵃ (1 carab. 4 lamell.)  fine morning, rain pm.
collecting near house abt Dead tree & c.

17 n.s. of col. (2 carab 3. Elater 1 cetonia 1 fine brenthes &c.)  fine rain p.m.

2 n.s. of col. (no Carab).  unwell — round house, <u>fever pm</u>

in house very unwell — fine & hot. .

round house — unwell — (3 n.s. 1 Long.)

9 ns. of col. (1 carab.). weak,—storm pm.—

9 n.s. of col. (3 Long. 2 Bup. 1 Staph.).  fine— <u>on foliage dead leaves &c.</u>

8 n.s. of col: (2 carab. 1 Bup.)  fine.

18 n.s. of Col.ᵃ (3 Carab. 1 Long. 1 El.  c.)  fine storm pm.— walk often *[illeg.]*
& search on foliage & dead leaves.

25 n.s. 2 Cincindas 1 Colobothea &c. . .

---

[14]<u>Oct.ʳ 1858</u>. <u>Kaioa Islands</u>          abt 100 sp of Coleoptera  ?

| | | |
|---|---|---|
| Oct. 15. Col.ᵃ          33 sp .. 5–6 new... | | Intensely hot & dry, — |
| | | limestone (coralline) & |
| Oct. 16. Col.ᵃ          70 sp. abt 12 new. . | | crystalline rock. |
| others 10 sp.. | | several small streams |
| Oct. 17. Col.ᵃ          47 sp. abt 6 new. . | | |
| others 8 sp. | | <u>no water</u> — coral reefs. |
| Oct. 18. Col.ᵃ          40 sp. abt 8 new . others 9 sp. . | a shower last night | |
| "  19  "          56 sp      "  8 new.. | | 28 species of Longˢ . . 5 new!! |

---

Long.ⁿˢ 44 sp.  In new plantations. <u>Longicorns</u> & <u>Buprestidae</u> & <u>Ceto.</u>
mix by thousands; — a continued hum & whirr!!. most
astonishing — many fine insects.

1. This is a record of Wallace's daily collecting in Macassar, near the southern tip of the southwestern peninsula of Celebes, in October of 1856. He had arrived in September with his young collecting assistant Ali, but soon noted in his journal, "This is not a very healthy season to arrive in Macassar. My boy Ali had hardly been a day on shore before he was attacked by fever" (first Malay Journal, LINSOC-MS178a, entry 34). It wasn't long before he became ill as well, likely with malaria: "I fell ill myself with strong intermittent fever every other day. In about a week I got over it by a liberal use of quinine, when scarcely I was on my legs that Ali again became worse than ever every other day." Note the records of his bouts of illness Wednesday and Thursday October 15 and 16, and again October 22 and 23.

2. On October 26 Wallace was able to collect some Hymenoptera but noted that he was "too ill to collect the other orders."

3. "It was with considerable anxiety that, on January 8th, 1857, I took my first walk into the forest," Wallace commented in his paper "On the Entomology of the Aru Islands," continuing:

> The first insect I saw was not a very encouraging one: it was the common Diadema Auge, found over the whole Archipelago. A little further, however, and I was rewarded by Idea d'Urvillei, a beautiful Hyades, the lovely Damis Coritus, *Guér.*, and that superb insect Cocytia d'Urvillei. Two or three pretty Lycenidae of genera unknown in the western parts of the Archipelago, Tricondyla aptera, and two species of the longicorn genus Tmesisternus, with several smaller insects, composed my first day's sport; and a very satisfactory one it was, for it assured me there was work to be done, and that I was really in the midst of a New Guinea Fauna. (Wallace 1858d, 5889)

Note that two of the Lepidoptera species that Wallace collected on his first day are named for the explorer and naturalist Jules Dumont d'Urville (see recto 105). *Idea d'urvillei* is one of the beautiful tree nymph or paper butterflies (Nymphalidae: Danainae),

while *Cocytia d'urvillei* is a spectacular day-flying member of the largely nocturnal moth family Noctuidae.

4. This simple record of collecting, on January 13, 1857, an *Ornithoptera* butterfly species "near Poseidon" belies the eventual significance of this discovery to Wallace's thinking. Wallace published two speculative papers inspired by his Aru collecting: the brief "Note on the Theory of Permanent and Geographical Varieties" (published January 1858; Wallace 1858b) and the slightly longer "On the Habits and Transformations of a Species of *Ornithoptera*..." (read December 7, 1857, and published the following April; Wallace 1858e). Although the birdwing butterfly Wallace found in Aru is explicitly named only in the latter paper, it may have been the inspiration for the former too.

Wallace had previously discovered a birdwing butterfly in Borneo that he named for James Brooke (*Ornithoptera brookeana*). In the eastern part of the archipelago there were two known species: *O. priamus* from Amboyna and *O. poseidon* from New Guinea. Wallace's Aru birdwing was precisely intermediate in coloration between *O. priamus* and *O. poseidon*. In the *Note*, he put his finger on the dilemma of distinguishing species and varieties and, more specifically, the untenable view held by many naturalists that species are divinely created whereas varieties arise through secondary laws. "If an amount of permanent difference, represented by any number up to 10, may be produced by the ordinary course of nature, it is surely most illogical to suppose, and very hard to believe, that an amount of difference represented by 11 required a special act of creation to call it into existence" (Wallace 1858b, 5888). Wallace next led the reader through an exercise:

> Let A and B be two species having the smallest amount of difference a species can have. These you say are certainly distinct; where a smaller amount of difference exists we will call it a variety. You afterwards discover a group of individuals C, which differ from A less than B does, but in an opposite direction; the amount of difference between A and C is only half that between A and B: you therefore

say C is a variety of A. Again you discover another group D, exactly intermediate between A and B. If you keep to your rule you are now forced to make B a variety, or if you are positive B is a species, then C and D must also become species, as well as all other permanent varieties which differ as much as these do: yet you say some of these groups are special creations, others not. Strange that such widely different origins should produce such identical results. (Wallace 1858b, 5888)

We might read *O. priamus* for "A" and *O. poseidon* for "B"—species with a very small difference between them. The Aru species is "exactly intermediate," or "D." The precise definition of species and varieties, which speaks to their origin, is at the heart of these remarks, as Brooks (1984, 91) first pointed out.

5. These beetle groups Wallace collected in late January and February 1857 at Dobbo, Aru, are familiar "usual suspects" by now: *Geodephaga* are the carnivorous ground-dwelling beetles, especially Carabidae (ground beetles) and Cicindelidae (tiger beetles). *Lamellicornes* are long-horned beetles (Cerambycidae). *Rhynchophora* is an obsolete name for the snout beetles (hence *rhyncho-*), now Curculionidae. One name that may be unfamiliar is the *Sternoxi*. This name, also obsolete, was applied to the beetle tribe Sternoxia, containing the families Buprestidae (metallic wood-boring beetles) and Elateridae (click beetles), roughly equivalent to the superfamily Elateroidea today.

6. The Ulysses butterfly (*Papilio ulysses*, Papilionidae), named by Linnaeus for the wily Homeric hero, ranges broadly across the island-dotted Australasian region—invoking the far-ranging travels of Odysseus across the island-dotted Mediterranean. Also known as the blue mountain swallowtail, this large butterfly exhibits striking sexual dimorphism: the dorsal surface of the wings of the males gleam with a lustrous, metallic blue-green color, while those of the females are a rich chestnut brown, and both also display a row of small purple crescents.

Wallace encountered this species at several locales. In the Aru interior he didn't have to look far to find it: "Close to the house I saw the splendid butterfly *Papilio Ulysses* of which I had at length got one specimen & several other fine species were also seen flapping along in the woods" (second Malay Journal, LINSOC-MS178b, entry 76).

7. Wallace visited the Aru Islands from January to July 1857. He collected in and around the village of Dobbo for some months, but eventually succeeded in exploring the odd archipelago more fully. Note the ink entry of his March 13 journey "to Tojou." This was the start of his long-awaited journey to the Aru interior in quest of birds of paradise (see recto 71–74 for Wallace's accounts of *Paradisea apoda* and *Cicinnurus regia*, the two bird of paradise species that he collected in Aru)." Wallace opened a fresh journal, and the first entry, for March 13, reads: "My boat being at last ready & having, with as much difficulty as usual obtained two men besides my own Malay & Macassar boys, we left Dobbo for the main land of Aru" (second Malay Journal, LINSOC-MS178b, entry 69).

8. Although the next ink entry in the notebook, dated March 20, reads "Orang kaya to Dobbo," in his journal Wallace reports that it was on March 28 that "the Orang-kaya having an attack of ague had begged to go home" (LINSOC-MS178b, entry 73). *Orang-kaya* translates from Malay literally as "rich man" but is also used to mean "headman," a member of the ruling elite; see annotation for recto 9.) The departure of the *Orang-kaya* on the twenty-eighth is also reported in *Malay Archipelago*, so perhaps the entry here for the twentieth was when the *Orang-kaya* first told Wallace of his wish to depart for Dobbo. In any case, Wallace stayed some months longer at Wanumbai, successfully procuring many rare specimens, including black cockatoos and birds of paradise.

9. The pencil entry below the ink one about the *Orang-kaya* reads: "May 10 returned to Dobbo—bad legs." In entry 85 of his second Malay Journal (LINSOC-MS178b) Wallace reports:

"Ever since leaving Dobbo I have suffered terribly from insects who have seemed determined on revenging my persecution of their race." Eventually the bites turned from itchy annoyances to inflamed ulcers: "In the plantations where my daily walks led me the day biting mosquitoes swarmed & seemed especially to delight in attacking my unfortunate legs. These after a months incessant punishment rebelled against such treatment, & broke into open insurrection, throwing out numerous inflamed ulcers, which are very painful & prevent walking. So here I am for more than a week in the house & with no immediate prospect of leaving it."

By May the insects had very much taken their toll. "At length our rice being again out, my bird box nearly full, & no prospect of getting the use of my legs again very soon, I determined on returning to Dobbo" (second Malay Journal, entry 90). These trials and tribulations did not mean that Aru did not yield rare and precious specimens, for this is where Wallace succeeded at last in procuring the coveted great bird of paradise, *Paradisea apoda*. He appended this postscript to his letter to Samuel Stevens from Dobbo (Wallace 1857a, 93): "Postscript.—Dobbo, May 15.—I have returned from my visit to the interior, and the brig is not gone yet; so I add a postscript. Rejoice with me, for I have found what I sought; one grand hope in my visit to Arru is realized: I have got the birds of Paradise (that announcement deserves a line of itself); one is the common species of commerce, the Paradisea apoda." (See also recto 71–73.) He remained in Aru until July, when, on the second, he sailed west with a flotilla of native praus, heading for Macassar.

10. After returning to Macassar in mid-July 1857, Wallace soon departed to collect in an area identified here as Amasanga, Máros in *The Malay Archipelago*, to which he traveled on horseback. This site was located "about 30 miles North of Macassar where a brother of my friend Mr. M. resided & had kindly offered to find me house room & give me assistance should I feel inclined to visit him" (second Malay Journal, LINSOC-MS178b, entry 107). Mr. "M" is Mr. Mesman, a "Dutch gentleman" who befriended Wallace

and assisted him in various ways. Wallace gave a fuller account of the area and his collecting progress in a letter published in the *Zoologist*:

The mountains are of limestone or basalt, the former rising from the plain in immense perpendicular walls quite inaccessible, except where a few streams break through them; the basalt hills are more rounded, and at the foot of one of them is a forest of palms and jack fruit. I had a small bamboo house built; when I arrived in August there had not been rain for two months and it was fearfully hot and parched; dead leaves strewed the ground, and a beetle of any kind was sought for in vain. After some time I found a rocky river-bed issuing from a cleft in the mountains, and though dry it still contained a few pools and damp hollows; these were the resort of numerous butterflies . . . Here, therefore, I made daily excursions and procured good series of many of these insects; the paths in the forest adjoining this stream were pretty abundant in Ornithoptera; of two species, O. Remus and the very rare O. Haliphron. (Wallace 1858g, 6120–6121)

His collecting in Amasanga while visiting Mr. Mesman's brother Jacob yielded entomological riches: on verso 50 he tallied 1,710 species collected between September and November 1857! Note, on this and the next few pages, the various places where Wallace found insects: under bark and dead leaves, on foliage, in dung and rotten fruit, and more. "A few minutes' search on the fallen trees around my house at sunrise and sunset would often," he wrote of his visit to Máros in *Malay Archipelago*, "produce me more beetles than I would meet with in a day's collecting."

11. Wallace was understandably excited to catch five *Ornithoptera haliphron* specimens, which he had called "very rare" in his *Zoologist* letter. These butterflies are jet black with a single large bright yellow patch on the upper surface of each hindwing.

12. Rotten *Artocarpus* proved to yield a beetle bonanza for Wallace. *Artocarpus*, jackfruit, got a middling ranking in his listing of the "proportionate excellence of fruits of the Eastern Archipelago" (see recto 120). Were the list a ranking of "proportionate excellence to beetles, when rotten," *Artocarpus* would no doubt have ranked highly. Just as he was despairing of the collecting at Amasanga, he wrote that

> Chance showed me a new and very rich beetle station. My lad brought me one day a fine large Nitidula which he had found in an over-ripe jack fruit (Artocarpus sp.); this set me to searching these fruits, of which there were a number about in various stages of decay, and I soon found that I had made a discovery,—Staphylinidae, large and small, Nitidulae, Histers, Onthophagi, actually swarmed on them: every morning, for some weeks, I searched these rotten fruits, and always with more or less success; I placed ripe ones [of] the fruit here and there, which I visited once a day, and from some of them got even Carabidae; in all I found not much short of one hundred species of Coleoptera on the fruit, including most that I had before found in dung, so that it seems probable that, in tropical countries, the large fleshy fruits in a state of decay and putrescence are the true stations of many of the . . . Coleoptera. (Wallace 1858g, 6121)

13. "Letter from Amboyna," published as one of the *Zoologist's* periodic "Proceedings of Natural-History Collectors in Foreign Countries" articles (Wallace 1858g), draws on these notes; the resourceful Wallace collected where he could when the going got tough—in mud holes; rotten fruit; dead trees, bark, and leaves; buffalo dung; under stones . . .

14. Wallace wrote this inked entry for October 15–19, 1858, a year later than the other entries on this page, making use of available space near the bottom of the page. Wallace first visited the Kaióa Islands on his way to Batchian (Bacan). Notice the geological notes. Here Wallace found evidence of Lyellian uplift and subsidence. As he described it in his third Malay Journal (LINSOC-MS178c, entry 155):

> I found that at an elevation of about 150 or 200 feet the coralline rock ceased & was succeeded by a very hard crystalline metamorphic rock. The last change here therefore has probably been one of elevation the height of the coral showing the depth of submergence. The very dense & crystalline nature of the central rocks shows however that it had been formed under great pressure, probably at an immense depth beneath the ocean & superincumbent strata of which it was afterwards denuded.

In a footnote on this page in the journal he added: "At the harbour the ground is sinking as shewn by the forest trees dead & dying from the rise of the salt water."

The collecting was good here too. In fact, most of the 100 or so Kaióa (Kayoa) Coleoptera species summarized here came from a single bonanza site, a particular clearing in the forest that Wallace kept returning to:

> It was a glorious spot, and one which will always live in my memory as exhibiting the insect-life of the tropics in unexampled luxuriance. For the three following days [October 17, 18, and 19] I continued to visit this locality, adding each time many new species to my collection, the following notes of which may be interesting to entomologists. October 15th [1858] 33 species of beetles; 16th, 70 species; 17th, 47 species; 18th, 40 species; 19th, 56 species—in all about a hundred species, of which forty were new to me." (*Malay Archipelago*, 330)

## 52

1857 Insects Coll[d]. at <u>Amboyna</u>

| Dec. | [1]Col. sp. | Col. n.s. | Lep. sp. | Lep. n.s. | Alia sp. | Notes. . |
|---|---|---|---|---|---|---|
| Fr. 4. | 60 | all? | 15. | | 20. | Long. 15. Rhyn. 20. |
| Sat. 5 | 38 | 16? | 22 | | 5 | 7 fine Buprest! 11 Long.. *cloudy—* |
| Sun. 6 | 50 | 20. | | | | |
| Mon. 7. | 20 | 5 | 3 | | 1 | 3 new Long. — wet — |
| Tu. 8. | 30 | 8 | 3 | | 5 | very cloudy and wet. |
| W. 9 | 45 | 12 | 5 | | 16 | cloudy and wet.. |
| Th. 10 | 35 | | 24 | 6 | | fine p.m.—wet a.m. |
| Fr. 11 | 40 | | 8 | | | rain am. . |
| Sat. 12 | 70 | 15 | | 12.. | | fine but cloudy. |
| Sun 13 | 50 | | | | | cloudy & showers. |

[2]In 10 days.    Long. 50 sp.    Rhyn. 62.    Bup 18    Col.[a] 215 sp

| Dec. | Col. sp. | Col. n.s. | Lep. sp. | Lep. n.s. | Alia sp. | Notes. . |
|---|---|---|---|---|---|---|
| Mon. 14. | 0 | <u>Heavy rain</u> | & | <u>wind</u> | – | <u>unwell.</u> |
| Tu. 15 | 58. | 25. | 2 | 0 | 5 | 2 <u>hours fine</u> only! |
| W. 16. | 40. | 5. | 2 | 0 | 6 | storm a.m p.m. fine *rice 20 catt. 1.80 R* |
| Th. 17. | 22. | 2 | 1 | 0 | 4 | very wet. . |
| Fr. 18 | 40. | 3 | 0 | 0 | 3 | fine — shower pm. |
| Sat. 19 | 36 | 5 | 1 | 0 | | fine — *cloudy* unwell *[illeg.]* left |
| Sun. 20. | 36 | 4 | | | | fine . . unwell fevers |
| Mon. 21 | 25 | 3 | 5 | 0 | | fine — do.. |
| Tu. 23 | 15 | | | | | fine |

large <u>prionus</u> & <u>lamia</u> in spirits only.

1.  Note that Wallace keeps separate tallies for species (sp.) and new species (n.s.) of beetles and butterflies in the top half of this table.

2.  This is an impressive number of beetles collected in Wallace's first 10 days in Amboyna; 215 species amounts to just over 20 per day.

[53]

[1]Collected at Amboyna in 20 days Dec. 1857.
about 550 species of Insects of all orders.

| | | | | | |
|---|---|---|---|---|---|
| Geodephaga. | 16 | | Papilionidae— | | 7 |
| Hydradephaga &c. | 0 | 1 | Pieridae— | | 14 |
| Brachyelytra. — | | 1 | Danaidae— | | 9 |
| Necrophaga, Xyloph. &c. | 6 | | Nymphalidae | | 21 |
| Lamellicornes. | 9 | 2 | Satyridae .. | | 4 |
| Lucani..... | 1 | 2 | Lycenidae | | 27 |
| Passali...... | 3 | | Hersperida | | 5 |
| Buprestidae _____ | 25 | 4 | | | 87 |
| Elateridae _____ | 14 | 2 | Nocturin. | | 35 |
| Longicornes _____ | 58 | 6 | | | |
| Rhyncophora. | 85 | 3 | Diptera (54+20..) | | 74 |
| Heteromera_____ | 20 | | Hymenoptera __ | | 48 |
| Cleridae_____ | 10 | | Alia___ | | 22 |
| Malacoderma | 12 | 1 | | | 144 |
| Cyclica_____ | 22 | 1 | From Dr. M... | | 6 |
| Trimera_____ | 8 | | | | 150 |
| | 290 +22 | | | | |

290
87
35
74
48
22
556

312  total species

1. Wallace often obtained specimens from friends and associates, such as the last six specimens in this tally courtesy of "Dr. M"—Dr. Mohnike, the chief medical officer of the Dutch Moluccas and a keen beetle collector (see verso [b]).

It was on this visit to Amboyna that Wallace discovered a massive python coiled up in the thatched roof of his abode. "One evening as I was sitting reading on the verandah I heard a noise over head as if some heavy animal crawling on the thatch." In his field journal and the *Malay Archipelago* Wallace vividly described in a section aptly titled "An Unwelcome Guest" the ensuing panic and scuffle as his assistants tried to dispatch the snake. They eventually succeeded, and the scene is memorialized by the illustration in *Malay Archipelago* entitled "Ejecting an Intruder."

"Ejecting an intruder" from Wallace's dwelling in Amboyna (Ambon), an illustration in *Malay Archipelago*. The preserved skin of this hapless snake is now on display at the Linnean Society in London. From Alfred Russel Wallace, *The Malay Archipelago*, 4th ed. (London: Macmillan & Co., 1872), 297.

## 54

<u>1858</u>  [1]Ternate &c. Insects.

| <u>January</u> | Col.[a] sp. | do. n.s. | Lep. sp. | do. n.s. | Alia Insect. | Remarks. |
|---|---|---|---|---|---|---|
| In two months | <u>200.</u> | In. Sept. | | 300 col. | | |

| Sept[r] | Djilolo village. | | | | In a 20 days trip. . 13 days coll.[g] | |
| | | | | | <u>400 species of Insects. .</u> | |
| ~~Tu~~ W 15 | 24 | 5 | 13 | 1 | 6 | Cloudy day - along Sahoe road |
| 16 | 26 | 6 | 10 | 1 | 14 | fine..to Cuboñ & do. |
| 167 | | | 5. | 1 | 3 | fine hot . . sh.[r] walk <u>unwell.</u> |
| 178 | }33 | 3 | 2 | 0 | 2 | walk to <u>Sahoe</u> & back hot |
| Sun. 189 | . . | . . | . . | . . | | . . <u>lame</u> and <u>fever.</u> hot |
| ~~Sun.~~ ~~19~~ 20 | 28 | 5 | 3. | 0. | 4. | in house [short walk] Lahaji hot |
| 21. | 6 | 1 | 2 | 0 | 2 | in house. Lahaji hot |
| 22. | . . | . . | . . | . . | | at house packing &c. . hot |
| 23. | 19 | 2 | 2 | 1 | 8 | walk to Sahoe— wet |
| <u>Sahoe village.</u> | | | | | | |
| 24. | 12 | 2 | 3 | 0 | 12 | round village, — no forest paths. |
| 25 | 25 | 6 | 4 | 1 | 25. | near village — hot. |
| Sun. 26 | 24 | 7 | 6 | 1 | 33 | do. . do. . — hot. rain night. |
| 27 | 16 | 2 | 3 | 0 | 24. | do. do. . hot rain night. |
| 28 | 40 | 6 | 4 | 1 | 35 | do. . do. hot . . . . . . |
| 29 | 24 | 3 | 5 | 2 | 30 | do — do. . hot, rain, p.m. |
| 30 | . . . . . . . . . . . . . . . . | | | | | packing . . |
| Oct. . 1 | Left at 6 am | | | | 2nd. 7. am. at Ternate. |

| Long. | Curc. | Carab. | Lamell. | | Phytoph&c. | Alia | |
|---|---|---|---|---|---|---|---|
| 27 | 40 | 14 | 12 | 40 | 55 | = 188 Coleoptera | |
| 7 | 5 | 5 | 4 | 15 | 20 | about 60 n.s. | |

Lepidoptera ..55 Other orders. . 157. || <u>Total Insects.. 400 sp.[s]</u>

1. There is maddeningly little information here regarding Wallace's two month visit to Ternate and Gilolo (modern Halmahera) in January–February 1858, the very period in which he had his insight into natural selection and penned the landmark paper now known as the Ternate essay (Wallace 1858f). Note the contrast with the previous dozen or more pages, which give daily records of collections, weather, and other things. January–February 1858 is basically a blank here; we are only told that he obtained perhaps 200 beetle species, which is consistent with the fact that he was ill at the time. The meaning of the adjacent entry, which appears to read "In Sept. 300 col." perhaps means 300 insect specimens taken in September. Adjacent to that, under "Remarks," is a conspicuously empty box: if only Wallace had jotted something about his momentous breakthrough.

In his second Malay Journal (LINSOC-MS178b, entry 122) Wallace recorded that he "left Amboyna & in the morning of [January] 8th arrived at Ternate." After describing, among other things, the topography, he reported that "after a fortnight in Ternate I determined to visit the island of Gilolo for a month." His entries for the month on Gilolo, the scene of his discovery of natural selection, describe the people and landscape. Next thing we know, Wallace writes that "on the 1st. of March I returned to Ternate" to prepare for a journey to New Guinea.

It is clear that Wallace was on Gilolo and not Ternate when he penned the "Ternate" essay, but it was mailed from the latter. No information is given in the Species Notebook regarding his discovery or precisely when he mailed off his essay. It has been assumed that it was mailed on March 9, 1858, based on Wallace's later accounts, saying that, following his discovery of the principle of natural selection, he wrote out his essay in two or three evenings and sent it off "by the next post" to Darwin, with whom he had previously corresponded. (See further discussion in comments for verso 57.)

Note, finally, that most of the entries on this page stem from Wallace's second visit to Ternate, in August–October 1858. He put these entries here for convenience, as they chronologically actually follow the New Guinean entries (Dorey) that begin on verso 55. During the second Ternate visit Wallace stayed in the villages of Djilolo and Sahoe. As he recorded in his journal for September 13, 1858, "After packing up & sending off my collections & waiting a week for the mail steamer in vain, I left for the village of Djilolo" (third Malay Journal, LINSOC-MS178c, entry 152).

## 55

[1]Dorey (N. Guinea)     Insects.

| 1858. | Col. sp. | do. n.s. | Lep. sp. | do. n.s. | Alia Insect. sp. | Remarks.. |
|---|---|---|---|---|---|---|
| April. | | | | | | |
| Sat.17. | 30 | 8 | 9 | 2 | 10. | a short walk.. *[Written vertically]* showery & cloudy every day |
| Sun. 18. | 32 | 10 | 9 | 0 | 3 | walk to Arfak village |
| Mon. 19. | 52 | 20? | 3 | 0 | 5 | search in new Cuboñ. |
| Tu. 20 | 50 | 15? | 5 | 1 | 16 | walk on hill & in old Cuboñ |
| W. 21 | 30 | 12? | 4 | 3 | 18 | Exploring walk to upper village |
| Th. 22 | 44 | 15? | 3 | 2. | 10. | a short walk. very wet. |
| Fr. 23 | — | — | — | — | — | very wet, — wet |
| Sat. 24 | 36 | 16? | 2 | 0 | 5.. | walk to new Cubon; fever wet |
| Sun. 25 | 10 | 2 | 1 | 0 | 6 | near house .. cloudy. |
| M. 26 | 40 | 17? | 12 | 4. | 10 | Exploring walk - fine! |
| T. 27 | 40 | 15? | 3 | 1 | 5 | short walk. fever. shower. |
| W. 28 | 15 | 2 | 3 | 0 | 4 | short walk .. hot. tree fell. |
| Th. 29 | 27 | 5. | 10. | 1 | 8 | Exploring walk .. hot.. |
| Fr.= 30 May | 10 | 4 | 3 | 0 | 5 | fever .. wet .. — |
| May Sat. 1 | 60 | 18 | 5 | 0 | 11 | walk to hill .. |
| Sun. 2 | 5 | 0 | 0 | 0 | 0 | at home sore foot .. |
| M. 3 | ..home.. | | | | | |

| | Geodepd. | Staph. | Necroph &c | Lamell. | Buprest. | Elat. | Long. | Rhynch. | Mal. |
|---|---|---|---|---|---|---|---|---|---|
| Tu. 5 .. do. | | | | | | | | | |
| W. 6. do. | 11 | 5 | 22 | 6 | 9 | 12 | 39 | 59 | 24 |
| Th. 7 do | 13 | 15 | 43 | 8 | 12 | 16 | 66 | 93 | 38 |

| | Cleridae. | Heterom. | Cyclica | Triru. | |
|---|---|---|---|---|---|
| Fr. 7 do | | | | | |
| Sat. 8. | 18 | 46 | 73 | 8 | =332 sp. of Coleoptera in 15 days. |
| Sun. 9th | 28 | 65 | 90 | 18 | |

| Capt & Dr.Tu. 11. | W. 12 | Sat. 22 | 26th. | 500 in 1 ½ month. |
|---|---|---|---|---|
| called   Capt. Duvᵈ called. | Dr. callᵈ still bad Jumaat but get better | Ali left night ill - | | 1 month in house, collections by Lahagi— |

1. Wallace had high hopes for New Guinea—practically the ends of the earth for him, situated as it is on the far eastern edge of the vast archipelago, a place very few Europeans had trod and that was known to hold biological wonders (and very dangerous indigenous people). He arrived at the outpost of Dorey (now Manokwari), on the eastern side of the large western peninsula of New Guinea, in April 1858, his Ternate essay slowly making its way in the opposite direction and ultimately to England.

Dorey was, he wrote in his journal, a poor place for birds and insects but the only safe place in which to venture onto mainland New Guinea. The visit was ultimately to be more than a disappointment. Collecting was generally poor, competitors (the prince of Tidore) were buying up all of the good stuff, and he and his collecting assistants were ill much of the time. Tragically, one of his assistants, Jumaat, succumbed to disease and died. On this page we see the first record of Jumaat's illness, on Wednesday May 12 (the dates run horizontally on the bottom line, as Wallace ran out of space). By the twenty-second he was "still bad but get[ting] better," Wallace recorded.

The entry for the twenty-sixth, "Ali left night," is a reference to Wallace's dispatching two of his assistants to collect specimens in distant Amberbaki:

> At length after the usual amount of talk & discussion I have succeeding in getting a boat & 6 men to take two of my boys to Amberbaki, one charged to buy & skin birds of paradise & shoot all he can besides,—the other to collect insects. They are to remain a month & I trust will have good luck & bring me back a fine collection. They left May 27th. [the Notebook says 26th] I still remaining a prisoner in the house with my unfortunate lame foot. (third Malay Journal, LINSOC MS-178c, entry 139)

His inflamed foot did not improve very readily: "After a month's close confinement I am at length able to walk out a little, though not without pain." Worse, his assistants fared no better collecting in Amberbaki, arriving back safe "but alas! brought almost nothing" (ibid., entry 142).

[56]

1858.    [1]Dorey, N. Guinea, (<u>Insects collected</u>.

| May | <u>Coleop</u> species. | n.s. | <u>Lep.</u> species. | n.s. | Other orders | <u>Remarks.</u> |
|---|---|---|---|---|---|---|
| 30. Sun | 60 | 28 | 6 | 1 | ? | Search in new plant[n] |
| 31. Mon | <u>78</u> greatest no. of sp. in a day yet. | 26 | 4. | 1 | 12 | Search in old do. dead trees & under bark. |
| June 1. | 40 | 10 | 0 | 0 | 6 | Unwell short walk to do. |
| 2 | 42 | 15. | 14 | 1 | 3 | Along <u>new Road</u> {to Steamer Evening. } |
| 3 | 12 | 3 | 1 | 0 | 3 | wet am. a short stroll. |
| 4 | 66 | 9. | 12 | 1 | 5 | walk to new plant[n] |
| 5 | 50 | 15 106 | 10 | 1 | 8 | walk to Arfak house. |
| Sun. 6 | 33 | 3 | 0 | 0 | 2 | short walk on new road, wet. |
| 7 | 47 | 8 | 3 | 0 | 5 | walk to old Cuboñ.. very hot. |
| 8. | 30 | 4 | 0 | 0 | 0 | In house, foot bad, at tree am Jumaat out p.m. |
| 9 | 50 | 16 | 6 | 0 | 4 | Walk to new Cuboñ <u>Hot</u> |
| 10 | 36 | 3 | 2 | 0 | 2 | Along new Road. fine |
| 11. | 60 | 13 | 3 | 0 | 6 | walk to old Cubon. fine. |
| 12 | 58 | 10 | 2 | 0 | 0 | walk along upper path. |
| Sun. 13 | 38 | 3 | 0 | 0 | 2 | Walk to old Cubon |
| 14 | 27 | 4 | 3 | 0 | 0 | along new Road <u>ill.</u> Ali returned evening. |
| 15 | 17 | 1 | 1 | 0 | 0 | at home fever, Lahagi out heavy rain night. |
| 16 | 16 | 0 | 4 | 0 | 0 | fever – Lahagi out |
| 17 | 40 | 6 | 2 | 0 | 2 | do .. <u>Steamer left</u> 4 pm. |
| 18. | wet all day — nothing — still unwell. | | | | | |

1. The words "wet," "hot," "fever," and "unwell" on this page say it all about the period from May 30 through June 18. Note the entry for that last day: "wet all day—nothing—still unwell." This is the very day that Darwin apparently received Wallace's Ternate manuscript, "On the Tendency of Varieties to Depart Indefinitely from the Original Type," announcing Wallace's insight into the mechanism of transmutation (Wallace 1858f).

## 57

1858    Dorey, N. Guinea.

| June. | Coleop. species | n.s. sp | Lep. n.s | Misc. | Remarks. |
|---|---|---|---|---|---|
| 19. | nothing | | | | very wet .. |
| Sun. 20. | 15 | 1 | 2 | 0 | 0 | showery, Lahagi out |
| 21 | .... | .... | .... | .... | very wet am .. M. Otto called .. |
| 22 | 55 | 12 | 3 | 0 | walk to old Cuboñ .. fine *[written vertically]* mouth very sore. |
| 23 | 24 | 4 | 5 | 1 | along new road .. hot. |
| 24 | 14 | 2. | 3 | 1 | wet am. along new road. |
| 25 | 26 | 3 | 3 | 0 | a long do. . . |
| 26 | 56 | 12 [105] | 0 | 0 | wet - [1]Jumaat died. <u>Diarrhaea.</u> walk to new Cuboñ. |
| Sun. 27 | 34 | 4 | 4 | 0 | walk along new road .. mouth little better. |
| 28 | 70 | 16 | 4 | 1 | To new Cuboñ; search under bark. |
| 29 | 52 | 11 | 9 | 0 | To old Cuboñ — unwell.. |
| 30 | 95! | 32! | 2 | 0. 17. misc | Search among dead leaves, foliage & under bark. |

all obscure but 1 Carab & 1 Staph. very pretty.

n.s. Long. 4. Carab.2. Staphlylia. 7. Curcul.7 Copridae 2. Gelica 4. El. 1. Bup. 1. Heter. 3.

July. 6 hours work to sort them out & determine species.

| | | | | | |
|---|---|---|---|---|---|
| [2]1st. | 40 | 18 | 2 - | 12 misc | Short walk - unwell. (7 new Staph!) |
| 2 | 42 | 10 | 0 | 6 | walk along new road - unwell. |
| 3 | 30 | 5 | 0 | 0 2 | Mr. Otto - sent 6th. Paradisea wet .. to brook (½ hr) unwell. 4 new Stapha |
| Sun. 4 | 55 | 8 [104] | 0. 0. | 0. 0. | wet am. short walk. unwell. |
| 5 | 28 | 10 | House beetles mostly | | took salts - wet unwell - (Otto 3 P.) |
| 6 | 50 | 15 | (6 minute) | 0 | short walk ... unwell .. cold . wet. |

1.  Tragedy struck on June 26, when young Jumaat succumbed to his illness. In the previous journal entry Wallace wrote, "This is a terrible country & after the first fortnight we have never been without two or three sick out of our small party of five." Now, on the twenty-sixth, he solemnly recorded in his journal:

> Jumaat died. He had been suffering some time with diarrhoea but did not complain & seemed pretty well. In the two days however he has been worse. I gave him a dose of calomel the only appropriate medicine I had but he was too far gone got rapidly weaker & expired this morning at 7 o'clock. He was about 17 or 18 years of age, a native I believe of Boutong & a quiet lad not very active but doing what he did as well as he was able. (third Malay Journal, LINSOC-MS178c, entry 143)

Tragedy struck just days later in the Darwin household as well, halfway around the world, when Charles and Emma's infant son Charles Waring died of scarlet fever, adding immeasurably to the turmoil Darwin felt on receipt of Wallace's letter and manuscript from Ternate.

2.  Wallace and Darwin were both grieving at the time their papers were read before the Linnean Society on July 1, 1858. The circumstance of this "delicate arrangement" coordinated by Lyell and Hooker, as it has been called, has raised ethical questions, since Wallace did not ask that his essay be published. (In the event, the reading of Wallace's essay was preceded by readings of two pieces by Darwin, with the likely intent of Lyell and Hooker being to preserve Darwin's priority in a manner that they [rather unilaterally] decided was fairest.)

In addition, although irrelevant to the event of the readings, the precise date on which Darwin received Wallace's manuscript has long been a subject of discussion. The belief that it was received on June 18, 1858, is based on a letter of that date written by the distraught Darwin to his friend Lyell, in which Darwin refers to having "received this day" the package from Wallace. It has been generally assumed that Wallace mailed his manuscript on March 9, 1858, in which case Darwin should have received it some weeks sooner than he said. The March 9 posting date is based on Wallace's later recollection that he sent his essay to Darwin "by the next post" after writing it in February.

In an alternative scenario (Van Wyhe and Rookmaaker 2012), Wallace sent his letter and essay a month later, in early April, in *reply* to a letter from Darwin dated December 22, 1857, which Wallace received from the very March 9 steamer on which he is supposed to have sent off his essay. In that letter Darwin noted Lyell's favorable impression of Wallace's Sarawak Law paper (Darwin Correspondence Project, letter 2192). In one of his later accounts Wallace said that he had asked Darwin to show his essay to Lyell, "who had thought so highly of my former paper" (Wallace 1905, 363), a request that makes sense if he knew of Lyell's approbation, which he could have learned about only from Darwin's letter received on March 9. Van Wyhe and Rookmaaker (2012) argue that, encouraged by Darwin's letter, Wallace decided to write to Darwin of his exciting new insight and did so via the *next* mail boat, in early April. They support this by showing an unbroken transport chain (by steamship, train, and even camel caravan) extending from Darwin on June 18 back to Ternate on April 4 or 5. See, however, Davies (2012) and C. H. Smith (2013)—the matter remains unresolved. In any case, even had Darwin received Wallace's manuscript weeks earlier, and fretted before writing to Lyell, abundant evidence shows that he took no ideas from Wallace (Beddall 1988), and in the end he did as requested, forwarding the manuscript to Lyell.

## 58

1858    [1]Dorey — N. Guinea.        Insects.

| July. | Coleoptera | | Leidop. | | Alia | Remarks |
|---|---|---|---|---|---|---|
| | sp. | n.s. | sp. | n.s | sp. | |
| Wed. "7th. | 50 | 9 | I | 0 | 2 | short walk. — wet Steamer returnd.. |
| 8 | 42 | 6 | I | 0 | 5 | wet — stroll along new road — Capt[n] Royes called. . |
| 9 | 40 | 10 | 7 | I | 2 | wet . Steamer left, walk to Arfak hill. |
| 10 | 60 | 16 | 3 | I | I | To Mansinam — fine! day . . 2 Long. 3 El. n.s. |
| Sun. II | 58 | 7 | 4 | 0 | 2 | Walk to new Cuboñ . . cloudy. |
| 12 | 12 | 3 | 0 | 0 | 0 | Very wet, & cloudy all day.: |
| 13 | 24 | 3 | 2 | 0 | 3 | Very wet, & cloudy all day — short walk p.m. |
| 14 | 40 | 10 | 6 | 0 | 3 | walk to old Cubon — fine. |
| 15 | 15 | 2 | 2 | 0 | 2 | Wet, we t! wet!! . . . . . . |
| 16. | 52 | 6 | 5 | 0 | 8 | Fine. along new road & to old Paddy field. |
| 17 | 0 | — | 0 | — | 0 | very wet! morn. till night. |
| Sun. 18 | 25. | 2 106 | 3 | 0 | 4 | across bay — to jungle path . . fine |
| 19 | 48 | 12 | 4 | I | 7 | walk to upper path . — fine. |
| 20 | 50 | 5. | 3 | 0 | 3 | short walk, — cloudy, wet p.m |
| 21 | 38. | 10 | 5 | I | 3 | to upper path. wet pm. |
| 22 | 68 | 16 | 10 | 0 | 4 | to do. cloudy — Schooner came. |
| 23 | 12 | 8 | | | | House beetles. unwell — at home. |
| 24 | I | I | 0 | 0 | 0 | fine Prionus from native. |
| Sun 25. | 0 | — | — | — | — | at home clearing up |
| Mon. 26 | 20 | 0 | | | | showery walk along new r[d]. p.m. |
| W. 28 | | | | | | on board schooner. |

29th 5 am. sailed.

1. In the next journal entry Wallace reported: "Terrible wet weather now when we expected it fine. Nothing to be done, nothing to eat & all of us ill. Fevers & cold succeed one another & make me long to get away from New Guinea," and then: "July 16th. Still dreadful wet weather,—few birds & nothing to eat. The schooner which is to take us back is now anxiously expected as we are all pretty tired of Dorey" (LINSOC-MS178c, entry 145). Wallace may have made a mistake in his journal, or here in the Species Notebook, as here the sixteenth looks to be a fair weather day, with "wet, wet! wet!!" noted on the fifteenth and "<u>very wet!</u>" on the seventeenth.

Note the entry for July 22 on this page: "Schooner came," then "on board schooner" on July 28 and "5 am. sailed," at long last, on July 29. In his journal he wrote:

On the 22nd of July the schooner arrived and in five days more we bade adieu to Dorey,—without much regret for in few places I have visited have I encountered more disagreeables & annoyances continual rain, continual sickness, little or nothing to eat with a plague of ants & flies surpassing everything I have yet met with, required all a naturalists ardour to encounter, & when they were uncompensated by success in his collection, became all the more insupportable. (third Malay Journal, LINSOC-MS178c, entry 147)

## 59

Coleoptera collected at Dorey (N. Guinea)
from April 15 to July 25. 1858.

| | | | |
|---|---|---|---|
| Cicindelidae___ | 8 | Heteromera | 10 |
| Carabidae_____ | 26 | Lagriidae | |
| Hydrophili_____ | 3 | Mordelldae }&c. | 38 |
| Necrophaga___ | 56 | Helopidae | |
| Staphylinidae__ | 85 | Tenebriondae}&c. | 63 |
| Histeridae_____ | 15 | Rhynocophora | 194 |
| Passali_____ | 7 | Brenthidae | 32 |
| Lamellicornes__ | 14 | Anthribidae.. | 41 |
| Scarab. . dae. | 3 | Longicornes____ | 118 |
| Trogidae. . | 1 | Prioni . . . . | 3 |
| Dynastdae. | 2 | Cerambycdae.. | 30 |
| Ruteli^dae } Melolon^dae } | 6 | Lamii . . . | 85 |
| Cetoniidae. | 2. | Phytophaga __ | 150 |
| Buprestidae__ | 21 | Crioceride. | 6 |
| Elateridae____ | 40 | Cassididae.. | 12 |
| Malacoderma _ | 181 | Galerucidae. | 102 |
| Lampyr^dae } Telephoridae }&c | 72 | Chrysomeli .. | 30 |
| Cleridae perhaps. | 50..60.. | Erotylidae_____ | 4 |
| Bostrichidae } Scolytidae } | 38 | Endomychidae___ | 3 |
| Scyclmenidae } Ptinidae.. &c.. | 21 | Coccinellidae____ | 14 |

Total Coleoptera
[1]1040 species!

1.  Despite all the hardships, the bottom line, literally and figuratively, on this page looks on the bright side: just over 1,000 species collected between mid-April and late July 1858.

## 60

1858.  [1]Register of Insects at Batchian

| Oct. | Col. sp. | do. n.s. | Lep. sp. | do. n.s. | Alia sp. | Remarks.. |
|---|---|---|---|---|---|---|
| Su. 24 | 48 | 12 | 7 | I | 16 | Exploring walk to coal mines.. |
| M. 25 | — | — | — | — | — | wet all day .. soaked .. |
| Tu. 26 | 50 | 6 | 3 | 0 | 5 | wet .. short walk, pm. |
| W. 27 | 42 | 7 | 22 | 4. | 6 | fine but windy & little sun.. |
| Th. 28 | 39 | 6 | 4 | 0 | 4 | walk to mines — fine .. |
| Fr. 29 | 28 | 5 | 11 | I | 7 | Short walk in forest.. fine |
| S. 30 | 38 | 4 | 20 | 2 | 5 | To T[illeg.] Cuboñ .. windy. |
| Sun. 31 | 18 | I | 3 | 0 | I. | Lahagi out. Mr. Schinder[?] called. windy & showery Secretary left |
| Nov. 1 | 30 | I | 4 | 0 | 5 | Lahagi out .. I, one foot sh[t] walk [hot] |
| 2 | 22 | 2 | 3 | 0 | 0 | Lahagi to mines.. hot.. |
| 3 | 15 | I | 0 | .. | 0.. | bad foot .. Lahgi out. rain. pm. |
| 4 | 30 | 2 | 2 | 0 | 0 | Lahagi out .. hot .. |
| 5 | 28 | 0 | I | 0 | 0 | Lahagi out pm hot.. |
| 6 | 24 | I | I | 0 | 0 | Lahagi out. hot.. rain p.m. |
| Sun. 7 | 20 | 0 | 0 | 0 | I | do  do  .. fine.. [house] robbed .. |
| 8 | 15 | 0 | 0 | 0 | — | do. do. fine [box &] keys stolen |
| 9 | 18. | I | 0 | 0 | 0 | Lahi — out — fine. |
| 10 | 20 | 0 | 2 | 0 | 2 | Lahi out.. I sh[t] walk.. |
| 11 | 24 | 4 | 10 | I | 2 | Lahi — I two short walks.— |

1. Wallace spent October 1858 to April 1859 on Batchian (Bacan). His collecting records, summarized here, show steady progress, especially with beetle collecting (see verso 21, note 3). Note, however, the mishaps of November 7 and 8: his bungalow was first burglarized on the seventh and then the thief, not finding much of value yet apparently aware there must be more, returned the following day. Wallace had put his money in a strongbox but carelessly left his keys out. The thief got away with the strongbox containing all of his money, plus all of his keys, themselves "a great loss ten times over than all the rest put together," as Wallace put it in his journal (third Malay Journal, LINSOC-MS178c, entry 161).

## 61

1858  [1]Register of Insects taken at Batchian

| Novr.. | Col.ª | do. | Lep.ª | do. | Alia | |
|---|---|---|---|---|---|---|
| | sp. | n.s. | sp. | n.s. | sp. | Remarks . |
| 12 | 20 | 5 | 10 | 0 | 2 | walk to a new clearing.. [am.] P. telem! |
| 13 | 27 | 2 | 5 | 0 | 5 | search for P. Telem. took ♀.. |
| Sun. 14 | 5 | 0 | — | — | — | Lahi — I to village 3 times |
| 15 | 10 | 1 | 5 | 0 | 2 | Short walk. Boxes from Ternate. |
| 16 | 35 | 3 | 9 | 0 | 6 | heavy rain pm. wet! To [illeg.] Cuboñ. |
| 17 | 10 | 0 | 4 | 0 | 10 | short walk — Coal mines Rᵈ. |
| 18 | 12 | 1 | 5 | 1 | 2 | short walk .. do. |
| 19 | | | | | | foot bad — short walks, — wet every day |
| 20 | | | | | | in the afternoon — few insects. |

Sun. 21  a schizorhina black 1 ½ inch! from native..
a gleucea the handsomest sp. I have taken.

Sat. 27.  Sent for part of rice — whole bag. 128 catties ...

Dec.??  Jan. 10th finished eat once a day.

Sun. 5.  not out much this week. very wet .. few insects ..
all week wet every day — short walks. get little.

Mon. 13.  at 1/4 to 5 pm. a series of sharp Earthquake shocks
making house shake & all small articles on table vibrate
for several minutes. all week wet ..

Sun. 19.  Walk to mines — M. Hugueunis books — Saw a ♂.of
20th. to  large new Ornithoptera — appeared like "Tithonus." new?
village  From 25th. fine hot. weather, before very wet.

1859.

Sat. Jan 1st.  At 4 am. a single strong Earthquake shock.

—— 6  Took 1st ♂ of grand Ornithoptera!!!

Mon. 10. 9 pm. a small earthquake

Sat. 15  Aurachi!  Mon. 20  generally rain pm.

Sun 30  Ali to Ternate.

1. The mundane entries on this page—weather ups and downs, a bad foot, and a fair amount of good collecting—are all punctuated by earthquakes and prize specimens. Note the entries on December 20 and then January 6: a new birdwing butterfly, which he "at once knew to be a new species of the giant *ornithoptera* the bird-winged butterflies the pride of the Indian Archipelago." His description of this exciting find contains this oft-quoted passage from Wallace's Malay Journal:

> None but a naturalist can appreciate the intense excitement I experienced on at length capturing it. On taking it from my net & opening the glorious wings my heart beat violently the blood rushed to my head & I have been never so near fainting when in apprehension of instant death, as from the excitement produced by what will to most people appear a very absurd & inadequate cause. I escaped however with a headache for the rest of the day, so that even butterfly catching is not without its perils to the enthusiastic naturalist. (third Malay Journal, LINSOC-MS178c, entry 164)

This was later recounted on page 342 of *Malay Archipelago*.

## 62

### Insect Register at [1]Ruruka Menado.

| 1859 | | | | | fl. 3500 elevation |
| June July. | Coleop. sp. n.s. | Leidop. sp. n.s | Alia sp. | Remarks.. |

| June July. | sp. | n.s. | sp. | n.s | sp. | Remarks.. | |
|---|---|---|---|---|---|---|---|
| W.23 | | | | | | To Ruruka — wet. 69° at noon. | |
| Th. 24. | 42 | 16? | 3 | 2 | 10 | 63° very wet after 11 am . . . | 75°.. |
| Fr. 25 | 26 | 2? | 5 | 1 | 2 | 62° very wet! — . . . . | 73° |
| Sat. 26. | 48 | 6? | 12 | 2 | 5 | 62° fine, — cloudy, wet. p.m. | 74° |
| Sun. 27. | 12 | .. | 3 | 1 | 2 | 63° rain & cloudy . . . | 73°. |
| Mon 28. | 80 | 20 | 4 | 1 | 10 | 64 rather fine _ Mr. Neys came | 75° |
| [2]Tu 29 | 3 | 0 | 3 | 0 | 0 | 62°.. wet pm. 8 ¼ p.m strong earthquake. at Ratahand 9 pm. | 75° |
| Wed. 30 | 67 | 12 | 4 | 0 | 6 | 64° pretty fine — clouds &c. Earthquake shocks continually. | 74° |
| Th. 1 [July.] | 30 | 2 | 3 | 1 | 2 | 66° fine—to Kembas .. | 75° |
| Fr. 2 | — | — | — | — | — | 67° rain, am. To Tondano. | |
| Sat. 3 | 48 | 8.. | 2 | 0 | 5. | 66° — pretty fine . . . unwell. | |
| Sun. 4 | wet. Ja.[?] To Tondano & to Kakas & Laryowan & Pangu — all wet!! | | | | | | |
| Pangu | 1500 feet elevation . . 40 miles from Menado. | | | | | | |
| Tu. 13. | 20 | 1 | 4 | 2 | 12 | 70° exploring walk .. finish .. | 76° |
| W. 14 | 18 | 3 | 2 | 0 | 10 | 69—do pm [am] & storm. . | 78° |
| Th. 15 | 15 | 1 | 3 | 0 | 12 | 69° — am. fine — rest very wet. | 79° |
| Fr. 16 | 8 | 2 | 1 | 0 | 5 | 72° soaking wet all day. 2 new Cicindelas!!! | 78° |
| [3]Sat. 17 | 40 | 6 | 2 | 0 | 10 | 70° pretty fine, [cloudy..] 3 new Cicindelas!!!! | 78° |
| Sun. 18. | 33 | 4 | 3 | 1 | 10 | 71° finish — in brook.. heavy rain night. | 79° |
| Mon. 19. | 25 | 2 | 2 | 1 | 4 | 70° . . . cloudy & drizzle . . | 78° |
| Tu 20. | — | — | — | — | — | 70° soaking day. | 75° |
| W. 21. | — | — | — | — | — | 70° clouds & drizzle— mist. | 74° |

1. Wallace reached the town of Menado in Celebes (modern Sulawesi) in early June 1859, a town that he declared "by far the prettiest I have yet seen in the east." In the company of some Dutch gentlemen he visited coffee plantations and the village of Tondano, situated on a high plateau between 3,000 and 4,000 feet in elevation.

2. Note the entry in the remarks for Tuesday July 29: "strong earthquake at Rathahand." In his journal Wallace wrote, "At length I have felt what I have been wishing for,—a pretty strong earthquake." He got a bit more shaking than he bargained for, though:

> On the morning of the 29th of June at ¼ after 8, as I was sitting reading, the whole house began shaking very gently at first but rapidly increasing I sat still enjoying the sensation for some seconds but in less then ½ a minute, it became strong enough to shake me in my chair & made the house visibly rock about & creak & crack as if it would fall to pieces. Then began a cry of "Tanna goyang!" "tanna goyang!!" (Earthquake earthquake) the whole village rushed out of their houses women screamed & children cried & I thought it prudent to follow. In doing so I felt my head giddy & my steps unsteady & was pretty near falling. In about a minute the shock was over. My head felt as if I had been turned rapidly round & round, & I was altogether unsteady & seasick. (third Malay Journal, LINSOC-MS178c, entry 180)

This is recounted in *Malay Archipelago* as well. Fortunately the structural damage caused by this quake was not great. His philosophical response is reminiscent of Darwin's upon experiencing a strong earthquake at Concepción, in Chile, while on the *Beagle* voyage. In Wallace's case:

> The sensation produced by an earthquake is unique, & never to be forgotten. We feel ourselves in the grasp of a power to which the wildest fury of the winds & waves are as nothing in comparison, yet the effect is more a thrill of awe, than the terror which the war of the elements produces. There is a mystery & an uncertainty in the amount of danger we incur, which gives greater play to the imagination, leaving more to hope as well as to fear. (*Malay Archipelago*, 259)

3. Note the remark "3 new Cicindelas!!!!" entered for Saturday the seventeenth. *Cincindela* is the type genus of the tiger beetle, family Cicindelidae. These are likely the beetles he mentions collecting in *Malay Archipelago* (268): "I was . . . very successful in one beautiful group of insets, the tiger-beetles, which seem more abundant and varied here than anywhere else in the Archipelago."

Wallace goes on in that account to describe "the handsome Cicindela heros," his "finest discovery . . . the Cicindela gloriosa," and the very find that seems to be reported here in the notebook: "It was in the mountain torrent of the ravine itself that I got my finest things. On dead trunks overhanging the water, and on the banks and foliage, I obtained three very pretty species of Cicindela, quite distinct in size, form, and color, but having an almost identical pattern of pale spots" (*Malay Archipelago*, 269).

[63]

Insect. Register.   [1]N. Celebes.

1859

| July. | Coleop. sp. | n.s. | Leidop. sp. | n.s | Alia sp. | Remarks. | |
|---|---|---|---|---|---|---|---|
| Th. 22 | – | – | – | – | – | 70° Rain & mist. _____ | 76° |
| Fr. 23 | 18 | 3 | 4 | 1 | 3 | 67° . . showers mist . . . . . | 75° |
| Sat. 24 | 30 | 4 | 3 | 0 | 12 | 69° . . pretty fine . . . . . . . | 76° |
| Sun. 25. | – | – | – | – | – | 70° gloomy & showers all day | 74° |
| M. 26 | 30 | 4 | 5 | 1 | 4 | 68° tolerible _____ | 75° |
| Tu. 27. | 20 | 5 | 4 | 0 | 10 | 70° — finish . . | 76° |
| W. 28 | 18 | 3 | 5 | 1 | 8 | 71° fine . . . . warm. | 79° |
| Th. 29 | 27 | 9 | 3 | 0 | 7 | 71° — ¼ to 2 [p.m Rain]— Strong Earthquake— | |
| Fr. 30. | | | | | | 71° — 8¼ am. sharp earthquake . . | |
| August. | | | | | | | |
| Sun. 1st. | | | | | | | |
| Mon. 2. | Left Paryhu[?] — | | | | | | |
| W. 4 | At Menado — | | | | | | |
| W. 18. | Left Menado — At Lempias — night. | | | | | | |
| M. 30. | To Licoupang — arrived evening . . | | | | | | |
| Sat. 4 | Thomas began shooting (25 cents a day). | | | | | | |
| Mon 19. | 12¼ pm. Earthquake — | | | | | | |

1. Wallace made his way to the earthquake-rocked northern peninsula of Celebes (Sulawesi) in August 1859, a region celebrated for its birds as well as rare mammals. He was able to procure several skulls and skins of the remarkable babirusa, the Malayan "deer-pig" (*Babyrousa babyrussa),* family Suidae. The sapiutan is referred to in *Malay Archipelago* as the "forest-ox" but is more commonly known today as the midget buffalo, or anoa, of which there are two species in the genus *Bubalus.* Although Wallace procured many bird and mammal specimens in northern Celebes, here he records only his insect collections.

### 1859     Insect Register — [1] W. Ceram

| Nov.[r] | Coleoptera | | Lepidop. | | Alia | Remarks — |
|---|---|---|---|---|---|---|
| | sp. | n.s | sp. | n.s | | |
| Th. 2 | 57 | 6 | 3 | 0 | 0 | An hour in new Plantation — |
| Fr. 3 | 75 | 3 | 13 | 0 | 2 | Exploring walks. fine |
| Sat. 4 | — | — | 5 | 1 | — | walk to Cairáto. |
| Sun 5 | 70 | 2 | 4 | 5 | 5 | ab.[r] Cuboñ. |
| Sat. 11 | — | | | | | Mr. V. der Beck left evening — |
| Tu. 21. | | | | | | Left in prow for Elpaputih. |

At <u>Mr. V. der Beck's Plantation   20 days.</u>     Coleoptera <u>species</u>

Elaters 15 Buprest. 18 Cleridae 12. Lamell. 12 Lucan — 2 ⎱
Geodeph 8. Sundries 64. Longicorns 75 Curcul.[dae] 58 ⎰ <u>264.</u>

Dec.

| | Coleoptera | | Lepidop. | | Alia | Remarks |
|---|---|---|---|---|---|---|
| 12th. | 66 | 5? | 5 | 0 | 8 | [2]Awaiya Cuboñ. fine. . |

Dec.[r]   <u>Totals collected.</u>     Col[a] <u>species</u>

27th.    Elaters & 26.   Buprest. 20.    Cleridae 13    Lamell. 23 ⎱
       Lucan &c. 6   Geodeph. 11.   Sund. 132   Long. 100   Curcul. 95 ⎰ <u>426</u>

<u>Note.</u> Up to this date probably <u>450</u> sp. of Amboyna & Ceram
        perhaps <u>500</u>       Coleoptera.

1. Wallace and his assistants spent much of November 1859 as guests at the plantation of Captain van der Beck, near Amboyna (Ambon) on the island of Ceram (Seram):

> We staid a day at two villages near the W end of Amboyna at one of which we had to discharge some wood for the missionaries house, & on the third afternoon reached Mr. Van der Becks plantation. This was a clearing in a rather swampy forest, of about 20 acres extent & mostly planted with cacoa & tobacco. Mr. V. offered me the use of part of his tobacco drying house which I accepted thinking I should find good collecting ground both for birds & insects in the vicinity. (fourth Malay Journal, LINSOC-MS178d, entry 193)

The notebook entry following Tuesday November 21 records that Wallace and his assistants spent the next twenty days as a guest of Mr. van der Beck. Note the low numbers in the columns headed "n.s."—new species. This explains Wallace's anxious record in his journal from this locale. It turned out that

> Coleoptera indeed were tolerably plentiful where the timber had been recently felled but the species were rather few & mostly the same as those which I had obtained during my first visit to Amboyna. Butterflies & birds were however very scarce, day after day brought nothing worth notice & I soon saw that I must change my locality if I wished to do any thing worthy of the almost unknown & unexplored island of Ceram. (Ibid.)

2. He left Captain van der Beck's plantation before the end of the month, and the December entries in this table correspond to greener collecting pastures, in Awaiya (Awaiyia).

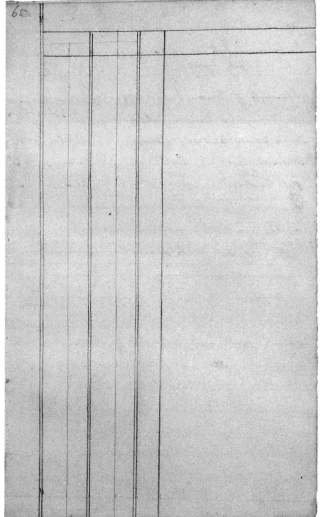

1. This page, set up for a continuation of Wallace's insect register tables but unused, signals the end of the detailed collecting records of the verso Species Notebook.

[1]Cacatua equatorialis. Tern.[ate] ? Menado, N.[orth] of Halmahera N.[ew] Guinea?

Microglossum alecto . . small & large sp.[ecies] with 2 mid.[dle] feathers
of tail longest.

Mammals of Borneo—(Verhand) 59 sp.[ecies] 14 peculiar. 26 in Java

45 in Java & Sum.[atra]
only 6 of these in Celebes.

| Timor Mammals. 21 sp.[ecies] | Amboyna | | Celebes 16? | |
|---|---|---|---|---|
| Mac.[accu][s] cynomolgus !! | Bats. 14 | | Quadrumana 3? | |
| Felis megalotis. | Alia 7 | 21 | Bats— | 5 |
| Phal.[angista] cavifrons | cuscus 3 | | Alia— | 8 |
| Sus Timoriensis— | Sorex. 1 | | | |
| Cervus moluccensis— | Mus— 1 | | | |
| | Paradox.[urus] 1 | | | |
| | Cervus. 1 | | | |

Mus decumanus—
Par.[adoxurus] musanga    Java 85 mammals..
Sorex tenuis
Bats . . . 13 sp.[ecies]    N.[ew] Guinea. 8 . . (7 marsup.[ials]
1 pig.)

*[The next notebook page was left blank by ARW and is omitted]*

1. The entries on this page present another of the several examples of Wallace's comparative biogeographical notes found throughout the Species Notebook. He considers the mammals of Borneo, Timor, Amboyna, Celebes, Java, and New Guinea, likely consulting Salomon Müller's *Verhandelingen over de Natuurlijke Geschiedenis der Nederlandsch Overzeesche Bezittingen door de leden natuurkundige commissie in Indie andere Schrijvers* (Treatises on the Natural History of the Dutch Overseas Territories...) (Müller 1839–1840). This is the "Verhand" that he mentions on this page after "Mammals of Borneo."

Several of these species are discussed in the chapter "The Natural History of the Moluccas" in *Malay Archipelago*. To name a few rarities from the island of Timor alone: *Phalangista cavifrons*, a Timorese opossum; *Sus timoriensis*, the Timor warty pig; and *Cervus moluccensis* (now *Cervus timorensis moluccensis*), the Timorese dwarf deer.

Others no longer exist, not necessarily owing to extinction but to taxonomic changes. Thus *Felis megalotis*, named by Müller in 1839 and dubbed the feral striped domestic cat of Timor, is now simply recognized as an ordinary domestic cat, *Felis catus*.

(Note that the following blank page, verso 67, is omitted.)

## 68

[1]Notes for address
It is a great pleasure to me to find that what
I have been able to do for Natural History is
appreciated by you.

My path has not always been among roses,—
but if one thing more than another has caused me
to persevere & to overcome difficulties,—it was
the knowledge that my labours were exciting some
attention at home.

[2]Among the disagreeables of my position during
the last 8 years I may mention as the chief that of being
in a state of almost perpetual motion.
"A rolling stone gathers no moss" is a proverb
that in my case was quite untrue, for it was
only by continual rolling from island to island
in that vast archipelago, that my little A[lfred] W[allace] stores of
"moss" have been accumulated.

I have made a little account of the number
of "movings" I suffered, to go to & return
from my various collecting localities & I find
they amount to nearly 100 or just on the average

1. This is a draft of the opening of an address that, according to historian Wilma George (1964, 48) Wallace prepared for one of the scientific societies soon after his return to England in early 1862. There is no evidence that he gave the address, however. As Charles Smith has pointed out to me, the wording suggests that the remarks could not have been delivered much later than a few months after his return, and the only two societies whose meetings Wallace attended that year were the Zoological Society of London and the Entomological Society of London. In fact, on April 8, 1862, just one week after Wallace got home, there was a meeting of the Zoological Society and in its *Proceedings* the donation of Wallace's live birds of paradise was announced. Wallace was likely at the meeting and may have spoken, though no birds of any kind are mentioned in the address drafted here.

2. Wallace was indeed in a "state of almost perpetual motion." By his reckoning he traveled about 14,000 miles over his eight years in the archipelago, but he revised the number of separate journeys to between sixty and seventy (Malay Archipelago, vii)—still a prodigious number in eight years, averaging nearly nine journeys annually. What is difficult for the modern reader to comprehend, however, is the enormity of the logistical issues behind most of those journeys, which were often complicated by sociocultural differences and very real dangers, from pirates and less-than-friendly natives to powerful currents and storms.

one a month. / Now the packing & contriving[?] & forethought required was just as much for a journey of ~~2 weeks~~ ¹ month as for ~~2—3~~ ⁶ months.. In both cases every thing has to be thought of, & the same problem has to be worked over & over again of how to take every thing that will be wanted or may be wanted in the smallest possible space.—

———————

And in those countries every thing was wanted, *[illeg.]* kettles frying pans & crockery, soap salt & lamp-oil as well as every other necessary civilized man requires, have to be seriously considered by the unhappy traveller in ~~those countries~~ such regions .. ¹"Punch" I believe says that 3 movings are equal to one fire,—my sufferings from 96 consecutive movings may therefore be imagined.

Of real dangers I say nothing,—for I do not believe that with proper precautions there is more danger in one part of the world than another. "Familiarity breeds contempt" of dangers as well as of persons,—& a

1. Wallace may have read this aphorism in *Punch*, but it likely originated in early eighteenth-century England or the American colonies. In *Early American Proverbs and Proverbial Phrases* (Cambridge, MA: Harvard University Press, 1977), Bartlett J. Whiting expresses it as "three removes is as bad as a fire," citing Benjamin Franklin's *The Way to Wealth, or Poor Richard Improved* (1758). Emanuel Strauss's *Concise Dictionary of European Proverbs* (London: Routledge, 1998) reports an earlier origin, pushing the first appearance of the aphorism to 1739.

Wallace is making a joke here about his frequent moves—if three are equal to a fire, imagine how much he has lost with some ninety-six consecutive moves. In fact he had firsthand experience of how one fire could obliterate the labors of years of incessant moving: the burning of the ship *Helen* mid-Atlantic on his voyage home from South America in 1852 sent his treasure trove of specimens, notebooks, and sketches from the past two years of his Amazonian journeys up in smoke (see Wallace 1852c). After ten days adrift in a lifeboat, Wallace and the *Helen*'s crew and other passengers had the good fortune to be picked up by a passing ship, aboard which Wallace wrote a long letter to his friend the botanist Richard Spruce detailing the tragedy ("Brig Jordeson, Lat. N. 49.30, Long. W. 20" was Wallace's "address" in the letter; remarkably, the letter reached Spruce, though it was addressed simply to "Senr. Ricardo Spruce, Rio Negro, Brazil." The loss was devastating to Wallace, as he related in the letter (Wallace Correspondence Project, WCP349):

And now I began to think, that almost all the reward of my four years of privation & danger were lost. What I had hitherto sent home had little more than paid my expenses, & what I had with me in the "Helen" I calculated would realize near £500—But even all this might have gone with little regret had not far the richest part of my own private collection gone also. All my private collection of Insects & birds since I left Pará was with me, & contained hundreds of new & beautiful species which would have rendered (I had fondly hoped) my cabinet, as far as regards American species, one of the finest in Europe.

But fortunately for science Wallace was made of stern stuff; by the letter's postscript he could write: "Fifty times since I left Pará have I vowed if I once reached England never to trust myself more on the ocean. But good resolutions soon fade & I am already only doubtful whether the Andes or the Phillipines [sic] are to be the scene of my next wanderings." By early 1854 he was in Southeast Asia.

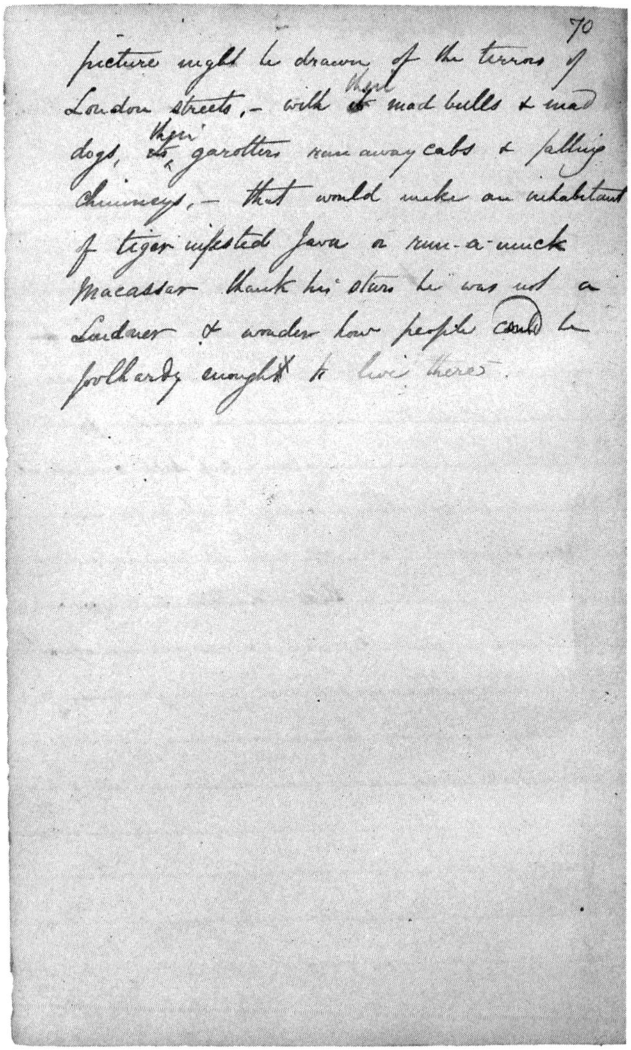

picture might be drawn of the [1]terrors of
London streets,—with its their mad bulls & mad
dogs, its their garotters run away cabs & falling
chimneys,—that would make an inhabitant
of tiger infested Java or run-a-muck
Macassar thank his stars he was not a
Londoner & wonder how people can ould be
foolhardy enought to live there.

*[The next notebook page was left blank by ARW and is omitted]*

1. Wallace's sense of humor is in evidence here, suggesting that the various "terrors" of London he names would make an inhabitant of "tiger infested Java" or "run-a-muck Macassar" wonder how people could be so foolhardy as to live there. The tendency of some Malays to "run amok" was memorably described by Wallace in one of his letters from Macassar, published by Stevens in the *Zoologist*:

> The people here [in Celebes] have some peculiar practices. "Amok," or, as we say, "running a-muck," is common here; there was one last week: a debt of a few dollars was claimed by a man of one who could not pay it, so he murdered his creditor, and then, knowing he would be found out and punished, he "run a-muck," killed four persons and wounded four more, and died what the natives consider an honourable death! (Wallace 1857b, 5560)

(Note that the following blank page, verso 71, is omitted.)

## 72

¹List of Morphidae of Malay Archipelago.

1. Clerome faunula . . Westwood. G.[enera] [of] Diur.[nal] Lep.[idoptera]
   1 spec.ᵐ [specimen] Malacca, (Singapore, West?)

2. ~~Clerome xanthotaenia~~

   Clerome chitone . Hewitson. Ex.[otic] But.[terflies] Macassar. 2
   spec.[imen]

3. Clerome menado . Hew.[itson]      "  "      Menado. 1 spec.ᵐ
   [specimen]

4. Clerome arcesilaus . Fab.[ricius] Java, Mal.[acca] Sum.[atra] 4 spec.ᵐ
   [specimen]

5. Clerome stomphax . West.[wood] Borneo. 3 spec.ᵐ [specimen]

6. Clerome besa Hew.[itson] Borneo . . (wanting)

7. Xanthotaenia busiris. West.[woo]ᵈ Malacca, Sumatra . (3 spec.ᵐ) [specimen]

8. Drusilla horsfieldii . Sw.[ainson] Java. (1 spec.ᵐ) [specimen]

9. Drusilla urania . (jairus, Cr.[amer]) L[innaeus] Amb.[oyna] Ceram &
   Bouru . 8 spec.ᵐ [specimen]

10. Drusilla macrops. Feld.[er] Aug. 1860 Batchian, Gilolo, Ternate, Morty.

11. Drusilla dioptrica . v.[an] Voll.[enhoven] ♂. 1859 New Guinea,
    Waigiou
    "   artemis . v.[an] Voll.[enhoven] ♀. ? . . . . Mysol & Aru Is.[lands]
    "   myops . Feld.[er] Ap.[ril] 1860 . . . .

12. Drusilla catops . West.[wood] Gen.[era] [of] D[iurnal] But.
    [Lepidoptera]—N[ew] G[uinea] Mysol, Waigiou, Aru

13. Drusilla domitilla Hew.[itson] Gilolo, Kaioa, Morty.

14. Drusilla divona Hew.[itson] Ceram, Aru Is.[lands]

15. Drusilla bioculatus . Guer[in] Wagiou . 1 spec.ᵐ [specimen].
    (Hyades indra Bd. [Boisduval])

16. Hyantis hodeva. Hew.[itson] Wagiou & New Guinea .

1. The butterfly family Wallace calls "Morphidae" consists of the spectacular *Morpho* butterflies and their relatives. They are now considered a subfamily (Morphinae) of the brush-footed butterfly family Nymphalidae. *Clerome, Xanthotaenia, Drusilla,* and *Hyantis* are Southeast Asian genera, though some of them were later synonymized with other names (*Clerome* is now synonymized with *Faunis,* for example).

This list was likely compiled in 1862 or perhaps later; the genus *Clerome* was named by the accomplished entomologist William Chapman Hewitson (1806–1878) in his *Exotic Butterflies,* published in that year.

## 73

See [1]Westwood's Paper on Eastern species of Morphidae
in Trans.[actions] of [the] Ent.[omological] Soc.[iety] vol.[ume] 3. New Series.

---

17. Thaumantis odana. God.[t].... [Godart] Java. 1 spec.[m] [specimen] (Horsf.[el][d])
18  Thaumantis                Sumatra. 2 spec.[imen]
19  Thaumantis                Malacca. 1 spec.[m] [specimen]
20  Thaumantis lucipor . West.[wood] ♂♀ Malacca, Sum.[atra] Borneo.
21  Thaumantis noureddin . West.[wood] ♂♀ "        d[itt]o.   d[itt]o..
     11 W.[est]              10 E.[ast]

---

[2]Satyridae

1.  Debis                    ♂. & ♀. Java.
2.  Debis Europa. Fab.[ricius]    ♂. Java
3.  Debis arete. Cr.[amer]    ♂. ♀. Makass.[ar] Sula Is.[lands] Amboy.[na]
     Bouru.
4   Debis                    ♂. Bali, Sumatra
5   Debis                    ♂. Sumatra
6   Debis arcadia. Cr.[amer] ♂. ♀.        Java, Sumatra.
7   Coelites epiminthia. West.[woo][d] ♂.? Sing.[apore] Borneo. Sumatra.
8   Cyllo amabilis. Biosd.[uval]  ♂. ♀.    Bouru, Ceram, Dorey, Mysol
9   Cyllo constantia. Cr.[amer]   ♂. ♀.    Bouru, Amb.[oyna], Mysol,
     Aru, Batch.[ian]

                              Dorey, Gilolo, Sula Is.[lands]

10  Cyllo leda. L.[innaeus]  ♂. ♀. Sumatra, Singapore, [small,] apex subtruncate
     "      "                Timor.   small, outer wing straight
                                    apex nearly acute
     "      "                Ceram, Bouru, Wagiou, Batch.[ian] Gilolo
     & Mysol

outer margin slightly hollowed, apex rather rounded.
in ♀s. the apex is subtruncate & subfalcate.

1. John Obadiah Westwood (1805–1893) was a prolific English entomologist based at the University of Oxford, where he became the founding Hope Professor of Zoology. In 2010 the Royal Entomological Society established the J. O. Westwood Medal for excellence in insect taxonomy.

Westwood described many of Wallace's insects. The paper Wallace refers to here is Westwood's 1858 "On the Oriental Species of Butterflies Related to the Genus *Morpho*," but this appeared in volume 4 of the *Transactions*, not volume 3; see *Transactions of the Entomological Society of London*, n.s., 4 (1858): 158–189.

2. The butterflies that Wallace refers to as Satyridae are now classified as Satyrinae, with Morphinae also a subfamily of the brush-footed butterfly family Nymphalidae. They are commonly called satyrs or browns, a hugely diverse group containing half of the nymphalid species thus far described. The caterpillars feed on many monocotyledonous plants, such as palms, grasses, and grass relatives such as bamboos.

Wallace's list of satyrid butterflies continues to verso 76, reflecting their rich diversity and his good collecting success. Hewitson worked on Wallace's satyrids, and in December 1864 he read before the Linnean Society a paper entitled "A List of Diurnal Lepidoptera Collected by Mr. Wallace in the Eastern Archipelago." The paper opens with these adulatory words: "The very valuable collections of Satyridae, Erycinidae, Lycaenidae, and Hesperidae amassed by the indefatigable industry of Mr. Wallace having been transferred to my keeping, I am happy to comply with his wishes by compiling a list of the species, with notice of all their varieties and localities" (Hewitson 1865, 143).

## 74

[1]Malay Satyridae          74

| | | | |
|---|---|---|---|
| 11. | Cyllo | ♂. ♀. Celebes, ♂ apex simple | |
| | | | ♀ apex produced, truncate |
| | | | & + *[illeg.]* below |
| 12 | Cyllo | ♂. Malacca | |
| 13 | Cyllo | ♂. Sumatra, Java. | |
| 14 | Cyllo lowii. Doub.[leday] | | Malacca, Borneo, Sumatra. |
| 15. | Cyllo | ♂. Sula Is.[lands] | |
| | | | |
| 16. | [2]Ragadia crisia . Hub.[ner] | | Sing.[apore] Borneo, Sum.[tra] Java. |
| 17 | Erites madura. Horsf.[ield] | | Sing.[apore] Borneo, |
| | | | Sum.[tra] |
| | | | |
| 18 | Mycalesis | ♂. Java. | |
| 19 | Mycalesis dorycus. Bois.[duval] | | Aru, N.[ew] |
| | | | Guinea, Waigiou, |
| | | | Mysol. ♂. ♀. |
| 20 | Mycalesis cyamites. Bios.[duval] | | New Guinea. ♂. |
| 21 | Mycalesis wessene. Hewits.[on] | | Batch.[ian] |
| | | | Gilolo, Ternate, Morty. |
| | | | ♂. ♀ |
| 22 | Mycalesis | | Mysol, Wagiou, N.[ew] |
| | | | Guinea . ♂. ♀. |
| 23 | Mycalesis remulia . Cr.[amer] | | Amboyna, Ceram, Bouru ♂. |
| | | | *[illeg.]* |
| 24 | Mycalesis | | N.[ew] Guinea, Wagiou, |
| | | | Mysol, Salwatty |
| 25 | Mycalesis | | Malacca, Sar.[awak] ♂. ♀. |

1. Most of these taxonomic names have changed, owing to the establishment and formalization of nomenclatural rules like priority. The genus name *Cyllo*, for example, named by Boisduval in 1832 from specimens collected on Durmont d'Urville's *Astrolabe* voyage, has given way to the earlier (1807) name *Melanitis* conferred by Fabricius.

2. A name on this list that has not changed is *Ragadia crisia*. What did change, however, is the range of this species in the decades following Wallace's visit to the region. Sydney Skertchly (1850–1926), an English geologist, writer, and avid amateur entomologist, commented on the expansion in a letter published in the May 2, 1889, issue of the journal *Nature* (Skertchly 1889):

> Mr. W. L. Distant, in his admirable "Rhopalocera Malayana,"
> calls attention to the recent appearance of Ragadia crisia
> in the Malay peninsula. As I have had opportunities of
> studying the habits of the species in what seems to be one
> of its head quarters, it may be desirable to record the facts.

Wallace and Skertchly knew one another, probably through scientific circles in London. In his letter Skertchly also related that this butterfly "has the feeblest (and wickedest) flight of any butterfly known to me. I never saw it rise six feet above the ground, and it flaps slowly along, apparently with effort, its wings not stiff but bending with each stroke. It looks like a certain capture; but this . . . is delusive and elusive" (10).

Incidentally, English entomologist William Lucas Distant (1845–1922) produced several noted entomological treatises beginning with his *Rhopalocera Malayana* (1882–1886), which treated the butterflies of the Malay Peninsula. Wallace reviewed the first volume of *Rhopalocera Malayana* for *Nature* (Wallace 1882). The review was favorable, but, ever the biogeographer, Wallace suggested that Distant consider adding information on the distribution of the butterflies among the islands of the Malay Archipelago and not just peninsular Malaysia: "We believe that such an extension of the scope of the work would double its value, and add largely to the list of subscribers; while the increased expenditure would be comparatively unimportant."

75  <u>Malay Satyridae</u>

26  [1]Mycalesis dexamenus. Hewits.[on] Tondano  ♂. ♀.

27  Mycalesis        (var.[iety] Hew.[itson]) Macassar.

28  Mycalesis dejanira. Hew.[itson]  Tondano, Sula Is.[lands] ♂. ♀.

29  Mycalesis        Gilolo, Batch.[ian] Ternate, Morty.

30  Mycalesis asophis. Hew[itson]    Salwatty, N.[ew] Guinea,
     "    mehadeva Bois.[duval]?     Aru, Wagiou, Mysol.

31  Mycalesis phidon Hewits.[on]    Aru, N.[ew] Guinea, Wagiou,
    Mysol.

32  Mycalesis        Sumatra  Borneo. ♂. ♀.

33  Mycalesis deniche . Hew.[itson]   Sing.[apore] Malacca, Sumatra,
    Borneo.

34  Mycalesis diadis . Hewits.[on]   Ceram [Amboyna] Bouru, Sula
    Is.[lands] ♂. ♀.
     "       manipa . Boisd[uval] ?.

35  Mycalesis jopas . Hew.[itson]    Makassar, Tondano, Sula
    Is.[lands]

36  Myclesis hesione Fab.[ricius]    Ceram, Bouru, Morty
    Is.[lands] Timor, Sarawak
    Sumatra, Malacca . ♂. ♀.

37  Mycalesis mineus L.[innaeus] ?   Singapore, Sumatra .

38  Mycalesis        Sum.[atra] Singapore

39  Mycalesis        Singapore, Sula, Amboyna
    Gilolo.

40  Mycalesis        Sumatra, Java .

41  Mycalesis        Java .

1. Wallace's list of *Mycalesis* species numbers thirty, beginning on the previous page and extending to the following page. Over eighty species are recognized today, ranging from Central through Southeast Asia. As with the other satyrine butterflies on these pages, note the geographical information: it is through lists like this that an underlying pattern may be glimpsed in the seeming chaos of species distributions.

*Mycalesis* butterflies go by the common name of bushbrowns, in reference to their drab gray-brown coloration, which is often offset by conspicuous spots (eyespots) along the margins of the wing undersides. The eyespots are thought to function in defense, and are interpreted as "deflective markings" that fool vertebrate predators into attacking the rear and less critical end of the body. Wallace, in "The Colours of Animals and Sexual Selection," chapter five in his 1878 essay collection *Tropical Nature and Other Essays*, discussed how this idea was first advanced by Darwin:

> In *Descent of Man,* 2nd ed., pp. 307–316, we find an elaborate account of the various modes of colouring of butterflies and moths . . . Mr. Darwin has himself referred to cases in which the brilliant colour is so placed as to serve for protection; as for example, the eye-spots on the hind wings of moths, which are pierced by birds and so save the vital parts of the insect, proving that the coloured parts are always more or less displayed. (Wallace 1878, 201–202)

This "deflection hypothesis," as it is now called, is well supported (see Ruxton, Sherratt, and Speed 2004; Stevens 2005). However, there is more to these spots than meets the eye, so to speak: spotted species often exhibit *seasonal polyphenism*, with wet- and dry-season forms differing especially in regard to the size and number of eyespots.

Wallace appears to have not noticed this in the various *Mycalesis* species he collected, but in his paper on Malayan Papilionidae he introduced the term "seasonal dimorphism" in describing an analogous phenomenon in the European map butterfly, *Araschnia levana* (Wallace 1865a, 9), a member of the same family (Nymphalidae) as *Mycalesis* but a different subfamily. *A. levana* exhibits a striking dimorphism (now more properly termed diphenism) in response to temperature during development: it produces two generations annually, a spring generation of conspicuous orange and black butterflies, and a summer generation that is mostly black with some white markings (*A. levana* form *prorsa*). The noted German evolutionary biologist August Weismann (1834–1914) discussed this at length in his *Studies in the Theory of Descent: On the Seasonal Dimorphism of Butterflies* (Weismann 1880), which Wallace reviewed glowingly for *Nature* (Wallace 1880b).

Wallace and Weismann did not, however, discuss the kind of polyphenism (wet-season, dry-season) found in *Mycalesis*. In the tropics, many butterflies, especially satyrines, appear more cryptic during the dry season and have larger eyespots in the wet season. This polyphenism is understood as the outcome of the interplay between selective pressures bearing on predation, balancing crypsis and the more conspicuous eyespot antipredator signal (e.g., Brakefield and Larsen 1984, Brakefield and Zwaan 2011).

| 42 | Mycalesis | Timor. |
|---|---|---|
| 43 | Mycalesis | Makassar. (4) |
| 44 | Mycalesis | Malacca, Sumatra, Java . |
| 45 | Mycalesis myamede . Hewits.[on] | Gilolo, Ternate Batch.[ian] Sula Is.[lands] |

[1]22 W.[est]—23 E.[ast]

1.  Here we have a clue as to Wallace's purpose in compiling these butterfly lists: their geographical distribution. Note that the numbers here, 22 and 23, reflect the total number of satyrid butterflies in his list: forty-five species in six genera. These are essentially evenly split between the eastern and western parts of the archipelago. Similarly, there are twenty-one Morphidae butterflies in his list on verso 72–73, also split evenly, with eleven species among the western islands and ten among the eastern islands. He may have compiled these lists intending to do an analysis like the one he published on the Papilionidae of the Malayan region (Wallace 1865a).

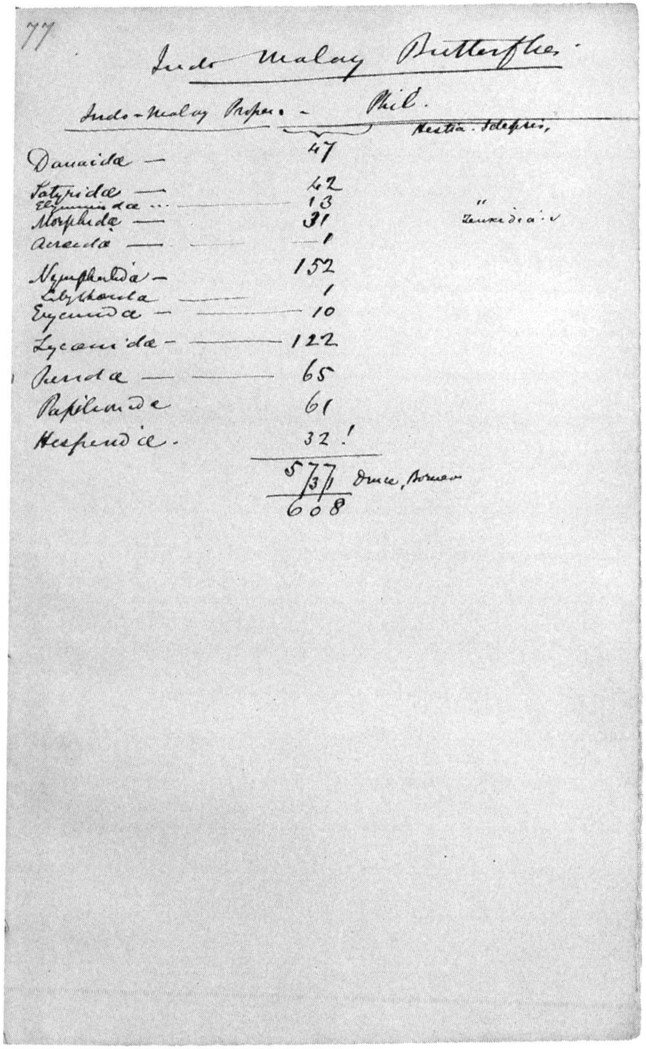

[77]

<u>Indo Malay Butterflies</u>

<u>Indo-Malay Proper — Phil.[ippines]</u>

Hestia Ideopsis*[?]*

| | | | |
|---|---|---|---|
| Danaidae— | 47 | | |
| Satyridae— | 42 | | |
| Ely*[illeg.]*dae | ...13 | | " |
| Morphidae— | 31 | | *[illeg]* |
| Acroidae— | 1 | | |
| Nymphalidae— | 152 | | |
| Liby*[illeg]*— | 1 | | |
| Ery*[illeg]*— | 10 | | |
| Lycaenidae— | 122 | | |
| [1]Pieridae— | 65 | | |
| Papilionidae | 61 | | |
| Hesperiidae. | <u>32!</u> | | |
| | 577 | | |
| | <u>31</u> | Druce, Borneo | |
| | 608 | | |

*[The next fourteen notebook pages, verso 78–91, were numbered by ARW but are otherwise blank and are omitted]*

1. Simple tallies of species by family and locality are informative in revealing broad-brush patterns of diversity. A finer scale of resolution, island by island, reveals still other patterns. Wallace continually mined his data for such insights, and one example worth noting pertains to these Pieridae (cloud and sulfur butterflies) and the following group in this list, the Papilionidae (swallowtails).

While sorting his specimens after his return home, he noticed that many butterfly species from Celebes (Sulawesi) sport an odd variant wing shape—specifically, the front wings are distinctly angular, or falcate relative to closely related species found in neighboring areas. In modern terms, these butterflies exhibit *parallel evolution*. Later, in *Island Life* (1880a), Wallace estimated that about thirty species of butterflies in Celebes, representing three families, have this modified wing shape, "by which they can be distinguished at a glance from their allies in any other island or country whatever" (Wallace 1880a, 425).

He exhibited pierid and papilionid specimens when he discussed the subject at the May 2, 1864, meeting of the Entomological Society of London, offering an adaptive explanation: he posited that the falcate wings gave those individuals greater speed and maneuverability, and the preponderance of this form on Celebes suggested to him that predation pressure was greater on that island than on the neighboring ones, or at least had been in the past. His idea met with resistance from his fellow entomologists at the meeting, but he stuck with this explanation and discussed it at length in his paper on the Malayan Papilionidae (1865a) and in *Malay Archipelago*, where he concluded that "there seems no unusual abundance of insectivorous birds to render this necessary; and as we cannot believe that such a curious peculiarity is without meaning, it seems probable that it is the result of a former condition of things" (288).

James Mallet (2009) has discussed this curious phenomenon and Wallace's interest in these butterflies as providing geographical evidence for the action of natural selection and its role in speciation. Mallet determines that it seems to be a clear case of parallel evolution, suggesting that the various species exhibiting the phenomenon responded to a similar (maybe common) selective pressure that still has not been adequately explained. Wallace's idea of predation pressure may indeed be correct, though Mallet points out that the group has yet to be reinvestigated with modern methods, and the variant wing shape may be related to other, equally interesting characteristics of these butterflies, such as the fact that they tend to be larger-bodied in comparison to related species found elsewhere.

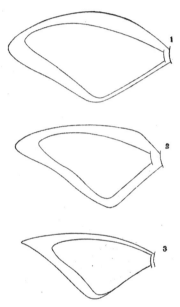

Illustration of the variant wing shape that Wallace noted in many butterfly species collected from Celebes (Sulawesi) compared with related species from nearby islands. Each drawing shows, in actual size, the superimposed forewing outlines of a Sulawesi species (outer line) and a related non-Sulawesi species (inner line). The Sulawesi butterflies have more angular or falcate forewings. From Alfred Russel Wallace, *The Malay Archipelago*, 4th ed. (London: Macmillan & Co., 1872), 281.

(Note that the remaining verso pages, 78–91, are blank and are therefore not reproduced in this volume.)

# Appendix 1

## Species Notebook Entries Bearing on Transmutation and Related Topics

As a working field notebook, the Species Notebook is neither chronological nor topically systematic. Some pages are filled with narrative while others are crowded with brief memoranda: drawings, references, ideas, lists, book extracts, specimen notes, and so on. Given the scattered nature of the entries, and the fact that a topic or theme is often commented on in multiple places in the notebook, the following index of entries grouped into evolutionarily relevant categories can be used as a quick topical guide and to get a sense of the themes that were of prevailing concern to Wallace.

The entries are grouped into six categories: design, affinity, distribution, transmutation, instinct, and humans. The subjects encompassed by each are summarized as follows: *Design* includes critiques of the concept or claims of "designedness," related critiques of claims of the supposed balance or harmony of nature, and entries that have ethical or spiritual references. *Distribution* includes entries on comparative species richness of different areas and relationships of island species. *Affinity* includes entries relating to homology, morphology, and classification. *Transmutation* encompasses widely ranging entries that can be best seen as offering evidence and arguments for the reality of transmutation. These include entries on species and varieties (natural and domestic varieties), extent of variation, and geological succession and "progressive development." The extracts of Lyell's *Principles* and attendant rebuttals constitute the bulk of this category,

with twenty-three pages dedicated to critiquing Lyell (see Appendix 2). Here, too, is where Wallace's planned book on the organic law of change is referenced. *Instinct* is explored in several contexts, including birds' nests, construction of honeycomb by bees, and a critique of claims of certain instinctual behaviors in humans (instinct versus experience is discussed). Human behavior also relates to Wallace's ethnological and racial observations, treated, finally, under *Humans*. This category includes descriptions of ethnic and "racial" features and behaviors of various peoples that Wallace encountered in his travels. Wallace's observations of the infant orangutan he adopted are included in this section because his close observations of the young primate are of a piece with his broader interest in human origins and diversity.

Page numbers listed are for recto pages in the Species Notebook except where "verso" is indicated.

## Design

| | |
|---|---|
| Critiques of supposed proofs of design | 12, 31–33, 53, 173–177 |
| Ecology / Competition | 108–109, 140–141, 146–147 |
| Harmony / Balance of nature | 49–50, 140–141, 146–147 |
| Religion / Belief / Ethics | 103–104, 154–155, 178 |
| Renan, Ernest | 154–155 |
| Spencer, Herbert | 178 |

## Distribution

| | |
|---|---|
| Canary Islands | 90, 151 |
| Distribution, species / Biogeography | 108–109, 132, 145–147, 151, 177; verso [d-1], 66, 72–77 |

# Appendix 2

## On Wallace's Critique of Charles Lyell and *Principles of Geology*

The notebook entries bearing on Lyell's arguments in *Principles of Geology* speak to Wallace's transmutational views with special clarity. Most of these entries are part of Wallace's extended critique of Lyell's anti-transmutation arguments that constitutes the central evolutionary section of the Species Notebook. In this appendix, Lyell's anti-transmutationism and Wallace's attack on Lyell's arguments in the *Principles* are discussed more fully than the annotations permit. The key points addressed by Wallace are treated under the following categories: history, non-progression of the fossil record, domestication, geographical distribution, and limits of variation.

Wallace's notebook entry titled "Note for Organic law of change" comes early in the Lyell critique, on the second page, suggesting that Wallace had the idea to undertake a book-length treatment of the transmutation question almost immediately after initiating the section he headed "Notes from Lyell's 'Principles.'" He may have then decided to go through the *Principles* systematically, highlighting the key objections that Lyell leveled against the idea of transmutation, as Wallace saw them. The main entry in support of the idea that Wallace was planning such a book is found on recto 51, where in reference to a particular assertion by Lyell he wrote, "Introduce this and disprove all Lyells arguments first at the commencement of my last chapter." It is likely that Wallace's book would have consisted of an expanded treatment of the positive evidence for transmutation presented in the Sarawak Law

paper (Wallace 1855c), ending with a review of the major objections to transmutation à la Lyell, which Wallace would then have attempted to refute.

· · ·

Lyell's monumental *Principles of Geology* went through twelve editions, the first appearing in 1830–1833 and the last published posthumously in 1875. The third through fifth editions were issued in four volumes; it was the fourth edition, issued June 1835, that Wallace took with him on his Southeast Asian travels. The table that follows summarizes the relevant volumes and pages from the *Principles* that correspond to Wallace's main critique (consisting of entries on recto 34–53 of the Species Notebook) and the additional points he made considerably later (on recto 142 and 149–150). The main narrative critique largely follows the *Principles* sequentially, with entries on recto 34–38 corresponding to statements found in volume 1, those on 39–44 corresponding to volume 2, and those on 45–53 corresponding to volume 3. The later entries return to issues raised in all three of those volumes plus volume 4.

Wallace's critique of Lyell's anti-transmutation arguments in *Principles of Geology*. Notebook entries are given with the corresponding volume and page numbers in *Principles*.

| Species Notebook pages (all recto) | *Principles of Geology*, 4th ed. (1835) (volume and pages) |
| --- | --- |
| 34 | I: 35–42, 69 |
| 35 | — |
| 36 | I: 226 |
| 37–38 | I: 231–234 |
| 39–40 | 2: 435 |
| 41–42 | 2: 437, 446–448, 452 |

| | |
|---|---|
| 43 | 2: 414, 464 |
| 44 | 2: 443 |
| 45–48 | 3: 22–28 |
| 49–50 | 3: 115–116, 154 |
| 51–52 | 3: 161–162 |
| 53 | 3: 172–173 |
| 142 | 1: 266; 2: 334; 4: 60–61 |
| 149–150 | 1: 239; 3: 21 |

## History

Wallace opens his critique by recording Lyell's historical account of the early resistance to recognizing fossils for what they are, and the comte de Buffon's supposed forced recanting of his opinions on the subject as "contrary to scripture." Lyell's rhetorical strategy in the *Principles* was to first present a narrative history of geology that identified a few lone figures, like Buffon, ahead of their time, heroically advancing ideas then considered heretical but now vindicated by Lyell and his circle. Lyell did this in a deliberate manner, borrowing heavily (and not always accurately) from the writings of Italian geologist Giovanni Battista Brocchi (1772–1826) to construct his "Historical Sketch of the Progress of Geology" (McCartney 1976; Rudwick 1970, 1998), which made up chapters 2–4 of volume 1. Lyell wrote of opposition to Buffon's claim that "secondary causes" were responsible for the mountains and valleys of the earth, setting himself up as a Buffon-like figure proclaiming uniformity despite resistance. Wallace likely saw irony in Lyell's inability (or unwillingness) to see the transmutational message written so clearly in the patterns of geological succession he eloquently details in the *Principles.*

Wallace may have thought to use Lyell's point as part of his own argument in his planned book, to anticipate and defuse opposition to the evolutionary ideas he intended to present. In this his approach anticipates Darwin's, who also made extensive use of Lyell (citing him twenty times in *On the Origin of Species*—more than any other person cited; Costa 2009, 519). Darwin did not critique Lyell but rather invoked him—and the revolution in geology he helped bring about—to further his own cause. Writing of the "lapse of time" in the *Origin* (Darwin 1859, 282), he constructed a parallel between the reception of Lyell's ideas and his own ideas: naturalists once resisted Lyell's vision of earth history but now universally accepted it, and this should give pause to readers resisting his (Darwin's) vision of organic change. Wallace was contemplating a similar rhetorical use of Lyell: on recto 35 of the Species Notebook Wallace cites Lyell's uniformitarian approach to good effect. The page is headed "Note for Organic law of change," and he applies with logic and consistency the implications of the Lyellian vision of change in the inorganic world to that in the organic world. "We must in the first place assume that the regular course of nature from early Geological Epochs to the present time has produced the present state of things & still continues to act in still further changing it," he writes. Wallace insightfully notes (again, recto 35), that "while the inorganic world has been strictly shown to be the result of a series of changes from the earliest periods produced by causes still acting, it would be most unphilosophical to conclude without the strongest evidence that the organic world so intimately connected with it, had been subject to other laws which have now ceased to act, & that the extinctions & production of species and genera had at some late period suddenly ceased." Outdoing Lyell in applying Lyell's own ideas, Wallace concludes that change in both the organic and inorganic worlds is "so perfectly gradual from the latest Geological to the modern epoch, that we cannot help believing the present condition of the Earth & its inhabitants to be the natural result of its immediately preceding state modified by causes which have always been & still continue in action" (recto 35–36).

## Nonprogression of the Fossil Record

Wallace next launches into an attack on the idea of nonprogression in the fossil record, which Lyell insisted on as part of his anti-transmutation rhetorical strategy. He copied out Lyell's anti-transmutation arguments from the *Principles,* in most cases following them with a critique of Lyell's arguments and a defense of transmutation. Wallace, like most naturalists of the day, was convinced that the fossil record was one of progressive change, so this was a logical and necessary point on which to attack Lyell. (Even many of Lyell's geological colleagues, anti-transmutationists, were incredulous at Lyell's insistence on nonprogressionism; see Rudwick 1998, 11–12). Lyell apparently saw progressive development as lending too much support to Lamarckian ideas; promulgating his steady-state theory that argued against directional change in the fossil record was another way to undermine the "transformists" (Corsi 1978). The trouble was, the fossil record appears to exhibit a strong pattern of directional change (reviewed by Burchfield 1998, Haber 1968, Rudwick 1972) and so his position was untenable. Wallace saw this, and was even gathering evidence to controvert Lyell with Lyell himself, as when he noted Lyell's statement in the *Principles* that "some of the more ancient Saurians approximated

more nearly in their organization to the types of living Mammalia than do any of our existing reptiles." "Which?" Wallace asked; "just what I want" (recto 37).

There follows another insightful passage in which Wallace critiques Lyell's claim that "the Didelphys [a marsupial mammal] of the Oolite [a Jurassic age limestone formation] is fatal to the theory of progressive development." "Not so," declares Wallace, "if low organized mammalia branched out of <u>low</u> reptiles, fishes. All that is required for the progression is that <u>some</u> reptiles should appear before Mammalia & birds or even that they should appear together. In the same manner reptiles should not appear <u>before fishes</u> but it matters not how soon after them." He continues: "All that the development theory requires is that <u>some</u> specimens of the lower organized group should appear earlier than any of the group of higher organization" (recto 37; emphasis in the original). Thus Wallace grasps the essential genealogical pattern of descent with modification, a significant departure from the primarily linear or within-lineage mode of change envisioned by Lamarck and others. His branching-tree conception of descent was likely inspired by the treelike diagram in *Vestiges of the Natural History of Creation* based on embryological development (Chambers 1844, 212; see recto 150). Lyell's comment about *Didelphys* might apply to simple linear chains of descent characteristic of Lamarck's earlier thinking, but Wallace had a ramifying pattern of relationship in mind. On the following page he makes this clear when he notes that "the supposed contradictions all arise from considering it necessary that the highest forms of one group appear before the lowest of the next succeeding, not considering that each group goes on progressing after other groups have branched from it. They then go on in parallel or diverging series & may obtain their max[imum] together" (recto 38).

## Domestication

Several of the passages from Lyell that Wallace singled out for critique pertain to domestication. In the *Principles,* Lyell cited a number of ways that domesticated varieties undermine transmutationism, among them: (1) in that no domestic variety has been transmuted into a new species, supporting the notion that species vary only within limits; (2) in the tendency of domesticated varieties to "revert" to type, which also illustrates the limited nature of their capacity for change; and (3) in that preserved remains of animals (including domestic species like cats) from ancient tombs suggest no change over thousands of years. There are two ways to address such claims: they can be attacked for their inaccuracy or dismissed as irrelevant. Wallace goes on the attack, in so doing suggesting that domesticated varieties are the products of a transmutational process, and that they hold lessons for the nature and extent of variability and change. It is noteworthy that a few years later in his Ternate essay on natural selection (Wallace 1858f), Wallace cites domestic varieties in a very different context. There he drops the idea of their putative informativeness and argues instead that domestic varieties are irrelevant to understanding transmutation; being unnatural or artificial beings, they can teach us nothing about the possibility of transmutation in nature. It is not clear why Wallace did not pursue in the Ternate essay the spirited arguments regarding domestic varieties that we find in the Species Notebook.

Here Wallace first rhetorically asks, what positive evidence do we have that species vary only within limits? He suggests a scenario in which all dog breeds but one become extinct, and the one is spread far and wide around the world and used as stock to develop new breeds. He then supposes that all breeds but one of *those* (in fact, the one "farthest removed from the original") become extinct and the process begins again. "Does it not seem probable that again new varieties would be produced," asks Wallace, "and have we any evidence to show that at length a check would be placed on any further change & ever after the species remain perfectly invariable" (recto 40)—clearly finding untenable this notion, which adherents of the idea that varieties can change only within limits must accept. Those adherents have painted themselves into a corner, Wallace believes.

Domestic varieties are themselves evidence of transmutation, Wallace insists, pointing to the diverse varieties of dog: "Is not the change of one original animal to two such different animals as the Greyhound & the bulldog a transmutation?—Is there more essential difference between the ass the giraffe & the zebra than between these two varieties of dogs" (recto 41). While two domestic varieties may differ from each other in "specific" [species] characters no more than two recognized species, he argues, we do not label the varieties as distinct species because we know they came from the same ancestral stock. Notwithstanding the criterion of reproductive compatibility, here Wallace's point is that such positions are taken without "positive knowledge" on the matter. Pressing this point, he asserts that the varieties of primrose that Lyell discusses as proof *against* transmutation are indeed proof positive *for* transmutation. "It only shows the impossibility of convincing a person against his will," Wallace scornfully comments.

"Where an instance of the transmutation is produced, he [Lyell] turns round & says 'You see they are not species they are only varieties'" (recto 42).

In other lines of attack Wallace disparages the notion that there has not been time enough for the production of new species. "Does Mr Lyell think that from any one race of dogs the greyhound bulldog & spaniel can be produced in a short space of time?" (recto 43). Examples of species or varieties being dramatically transformed interested Wallace. The case of the goat grass *Aegelops ovata*, purported (incorrectly) by Esprit Fabre in 1854 to have been transformed into wheat, caught Wallace's attention; he commented on it in both his 1854 notebook (LINSOC-MS179) and the Species Notebook (recto 44). If true this would be remarkable indeed, considering that the two species were classified in different genera. But Wallace, with a note of irritation, anticipates the response of anti-transmutationists: even should a member of one genus be shown to be transformed into another genus, this would be dismissed those by claiming that the two were merely varieties of a single species after all.

There ensues a series of entries that echo issues Wallace addresses in the Sarawak Law and Ternate papers, though at the time these entries were made, only the Sarawak Law paper had been written. Note his language at the bottom of recto 44: "Many of Lamarck's views are quite untenable & it is easy to controvert them, but not so the simple question of a species being produced in time from a closely allied distinct species which however may of course continue to exist as long or longer than its offshoot." More remarkable is the next entry, which essentially states the main thesis of the Ternate essay: "Changes which we bring about artificially in short periods may have a tendency to revert to the parent stock … This is considered a grand test of a variety. But when the Change has been produced by nature during a long series of generations, as gradual as the changes of Geology, it by no means follows that it may not be permanent & thus true species be produced" (recto 45).

## Geographical Distribution

In the subsequent passages in the Species Notebook Wallace comments on geographical distribution—why species found in "distant countries of similar climate" should differ, and the affinity of island endemics with species of the nearest mainland. "If they are special creations why should they resemble those of the nearest land? Does not that fact point to an origin from that land," he asks. He read Darwin's *Journal of Researches* closely here, as he specifically refers to the Galápagos Islands: "Again in these islands we find species peculiar to each island, & not one of them containing all the species found in the others as would be the case had one been peopled with new creations & the others left to become peopled by winds currents &c. from it." How to explain this? He anticipates the response of Lyell and others—"Here we must suppose special creations in each island of peculiar species though the islands are all exactly similar in structure soil & climate & some of them within sight of each other" (recto 46)—finding this unsupportable, of course. Such naturalists make this out to be a much greater puzzle than it is: "It may be said it is a mystery which we cannot explain, but do we not thus make unnecessary mysteries & difficulties by supposing special creations contrary to the present course of nature," he asks on the following page.

Wallace's critique has an ecological as well as evolutionary dimension. He points out that *recent* volcanic islands contain species from the nearest land, and no others—"nothing peculiar!" he emphasizes on recto 47—while islands "peculiarly inhabited" are of great antiquity ("A long succession of generations appears therefore to have been requisite, to produce those peculiar productions found no where else but allied to those of the nearest land"; recto 48–49). He argues that we can hardly accept that islands would be "stocked" in this way for ages, and then later (when the island reached a certain age?) suddenly have "new & peculiar" creations introduced. Why, this would be just when they were *not* wanted, Wallace declares, and what's more, "They would hardly be able to hold their own against the previous occupiers of the soil & there would have to be a special extermination of them to make room for the new & peculiar species" (recto 47–48). Wallace's appreciation of ecological competition is evident, as is his awareness that holding to the idea of the special creation of island endemics requires acceptance of an increasingly complicated scenario of colonization, creation, and extinction that begins to resemble the epicycles upon epicycles that were invented to preserve the Ptolemaic model of the solar system. He continues his attack at some length, putting his finger on the relationship between an island's age and the degree to which it harbors endemic species; this, too, is consistent with the idea that islands are initially colonized by species from the nearest mainland, species that, with the passing of ages, transmutate into the "peculiar productions" we see. Wallace is convinced that force of argument highlighting the inadequacies and inconsistencies of Lyell's claims and interpretations will lead to the collapse of the anti-transmutationist position.

As seen in the Sarawak Law paper, Wallace at that time understood that the evident patterns of biogeography speak to species change. In the Species Notebook he returns to this point toward the end of the long critique of Lyell. Gradual geological changes go hand in hand with change in environment, but while Lyell asserted that as a landscape changes—as the Sahara gives way to a chain of lofty mountains, for example—its species disappear and are gradually replaced by others "perfectly dissimilar in their forms habits & organization," Wallace strongly disagrees. "But have we not reason to believe they would be modified forms of the previously existing Northern African species," he protests. "The climate might then more resemble that of the W. Indies, but we know the productions would not resemble them" (recto 50). With a consistency that Lyell was then incapable of, Wallace points out that "it would be an extraordinary thing if while the modification of the surface took [place] by natural causes now in operation & the extinction of species was the natural result of the same causes, yet the reproduction & introduction of new species required special acts of creation, or some process which does not present itself in the ordinary course of nature" (recto 50–51). Lyell claimed that one species cannot change into another by a change of environment because other species already well suited to those conditions would preclude the new species, being already well established. Wallace does not think this likely if new species come about very gradually. It was regarding this point, on recto 51 of the Species Notebook, that Wallace inserted his note to himself to "introduce this and disprove all Lyells arguments first at the commencement of my last chapter."

## Limits of Variation

The remaining contexts in which Lyell is discussed include a criticism of the iceberg-transport theory championed by Lyell and, more important in terms of evolutionary thought, further passages from the *Principles* that aim to undermine the possibility of species change. In the first of these entries (recto 149) Wallace addresses the analogy Lyell drew between change from an irrational to a rational being and change from simpler to more complex organisms. Lyell saw discontinuity, not a gradual series with many intermediate steps. Wallace argued, on the other hand, that discontinuity assumes an absolute distinction between reason and instinct, and points out that Lyell's argument is semantic, depending on the terms themselves. In the other noteworthy passage (recto 149–150) he quotes Lyell on the reality of species, where Lyell argues in his anti-transmutation "recapitulation" that species can vary only so much and that varying beyond certain bounds is fatal to the individual. Wallace says that Lyell assumes that "only change of circumstances produce variety" and demands to know how, then, varieties are constantly produced in the same place under the same conditions? Furthermore, he points out that Lyell cannot prove that variation will not continue "at a rate commensurate with Geological changes." In the final comment in this passage, Wallace asks, "How can man's hasty experiments settle this—[Lyell's] 'though ever so gradually' is a gratuitous assumption. What are '<u>the defined limits</u>',—he assumes that they exist" (Wallace's emphasis). The experiments Wallace refers to are found immediately prior to Lyell's recapitulation: German physiologist and comparative anatomist Friedrich Tiedemann (1781–1861), who studied under Georges Cuvier in Paris, published an influential study of fetal brain development in 1816 in which he argued that the brain appears to pass, developmentally, through different stages: piscine, reptilian, avian, and so on. Lyell grasped the transmutational interpretation of Tiedemann's ideas (Corsi 1978, 17) and attacked Tiedemann on this point in the *Principles,* stating that, "on the contrary, were it not for the sterility imposed on monsters, as well as on hybrids in general, the argument to be derived from Tiedemann's discovery, like that deducible from experiments respecting hybridity, would be in favour of the successive degeneracy, rather than the perfectibility, in the course of ages, of certain classes of organic beings" (Lyell 1835, 3: 20). Wallace felt that the weighty question of limits on variation in the grand sweep of time could not be settled by mere "hasty experiments," implying that Lyell was too quick to marshal any such observations as ammunition in his anti-transmutation arsenal.

# References

## Electronic and Online Bibliographic Resources

The Alfred Russel Wallace Correspondence Project, wallaceletters.info/, referred to as simply the Wallace Correspondence Project (WCP), is directed by George Beccaloni at the Natural History Museum in London. This project's publically accessible electronic archive of Wallace-related documents is named Wallace Letters Online (WLO), www.nhm.ac.uk/wallacelettersonline. WLO contains metadata, digital scans, and transcripts of all known letters sent to and written by Wallace, as well as a selection of other important Wallace-related manuscripts. Note that the documents catalogued in WLO each have a unique identifier, known as a WCP number, and that these are cited in this volume. Many of the letters published in Wallace's autobiography (1905) and in Marchant (1916) were edited and often differ markedly from the originals, so for accuracy the quotations from Wallace's letters cited in this volume have been taken from the original manuscripts in WLO.

The Alfred Russel Wallace Page, people.wku.edu/charles.smith/index1.htm. Created and maintained by Charles H. Smith, Western Kentucky University, this online archive includes comprehensive Wallace bibliographic records, many with full text. Works are uniquely designated with an "S" number on this site; these numbers are included for Wallace works listed in the "Literature Cited" section that follows.

Wallace Online, wallace-online.org/. National University of Singapore–based digital archive and transcripts of the writings of Alfred Russel Wallace; directed by John van Wyhe, assisted by Kees Rookmaaker.

Darwin Correspondence Project, www.darwinproject.ac.uk/. University of Cambridge–based digital archive of the letters of Charles Darwin; directed by Jim Secord et al. Darwin letter citations in this book include the Darwin Correspondence Project letter number.

Darwin Online, darwin-online.org.uk/contents.html. Directed by John van Wyhe, this is the most extensive scholarly website on Darwin, featuring complete transcriptions and scans of manuscripts, published works, private papers, and other documents. Made possible by a consortium of universities, museums, libraries, and other institutions.

## Manuscripts

The following is a complete list of Wallace's unpublished notebooks, journals, and species registers from his eight-year journey in Southeast Asia.

## Natural History Museum, London

WCP4766: Insect, Bird and Mammal Register, 1855–1860. Description from the Wallace Letters Online website: "Manuscript notebook, opened at both ends. On the covers is written 'Insect Notes' and 'Birds and [?]animals.'"

WCP4767: Bird and Insect Register, 1858–1862. Description from the Wallace Letters Online website: "Manuscript notebook, opened at both ends. Writing on covers reads 'Register Birds. 1858' and 'Register Insects. 1858.'"

## Linnean Society of London

LINSOC-MS176: Undated Notebook ("Index to Hardy Plants").

LINSOC-MS177: North American Journal, October 1886–August 1887.

The Malay Journals consist of four parts (notebooks) with sequentially numbered entries:

LINSOC-MS178a: first Malay Journal (June 1856–March 1857); entries 1–68.

LINSOC-MS178b: second Malay Journal (March 1857–March 1858); entries 69–128.

LINSOC-MS178c: third Malay Journal (March 1858–August 1859); entries 129–192.

LINSOC-MS178d: fourth Malay Journal (October 1859–May 1861); entries 193–245.

LINSOC-MS179: Field Notebook, 1854–1861.

LINSOC-MS180: Field Notebook, 1855–1859. (Species Notebook)

LINSOC-MS181: Field Notebook, 1854–1862. (Eastern Butterflies)

LINSOC-MS182: Field Notebook, 1854–1862. (Sketches of palms of the Amazon)

Registry of Consignments. 1854. Record of material sent to Samuel Stevens.


## Literature Cited

Alberti, F. von. 1834. *Beitrag zu einer Monographie des bunten Sandsteins, Muschelkalks und Keupers, und die Verbindung dieser Gebilde zu einer Formation* [Contribution to a monograph on colored sandstone, shell limestone, and mudstone, and the joining of these structures into one formation]. Stuttgart and Tübingen: J. G. Cotta.

Agassiz, L., with J. E. Cabot. 1850. *Lake Superior: Its Physical Character, Vegetation, and Animals, Compared with Those of Other and Similar Regions.* Boston: Gould, Kendall & Lincoln.

Aldrovandi, U. 1599. *Ornithologiae.* Bologna: Bononiae.

Austin, S. 1833. *Characteristics of Goethe,* vol. 1. London: Effingham Wilson.

Baer, K. E. von. 1828. *Entwicklungsgeschichte der Theire: Beobachtung und Reflexion.* Konigsberg: Borntrager.

Baker, D. B. 2001. Alfred Russel Wallace's record of his consignments to Samuel Stevens, 1854–1861. *Zoologische Mededelingen Leiden* 75: 251–341.

Barlow, N., ed. 1958. *The Autobiography of Charles Darwin 1809–1882: With Original Omissions Restored* [edited and with appendix and notes by his grand-daughter Nora Barlow]. London: Collins.

Barrett, P. H., P. J. Gautrey, S. Herbert, D. Kohn, and S. Smith, eds. 1987. *Charles Darwin's Notebooks, 1836–1844.* Ithaca, NY: Cornell University Press.

Bartholomew, M. 1973. Lyell and evolution: An account of Lyell's response to the prospect of an evolutionary ancestry for man. *British Journal for the History of Science* 6: 261–303.

Bastian, H. C. 1872. *The Beginnings of Life: Being Some Account of the Nature, Modes of Origin, and Transformations of Lower Organisms.* 2 volumes. London: Macmillan and Co.

Bates H. W. 1862. Contributions to an insect fauna of the Amazon Valley. Lepidoptera: Heliconidae. *Transactions of the Linnean Society of London* 23: 495–566.

Beddall, B. G. 1988. Darwin and divergence: The Wallace connection. *Journal of the History of Biology* 21: 1–68.

Bentham, J. 1824. *The Book of Fallacies: From the Unfinished Papers of Jeremy Bentham,* ed. P. Bingham. London: John and H. L. Hunt.

———. 1838–1843. *The Works of Jeremy Bentham,* ed. J. Bowring. 11 volumes. Edinburgh: William Tait and Simpkin, Marshall, & Co.

Berghaus, H. 1836–1848. *Physikalischer Atlas.* Gotha: Justus Perthes.

Berry, A., ed. 2002. *Infinite Tropics: An Alfred Russel Wallace Anthology.* London & New York: Verso.

———. 2008. "Ardent beetle hunters": Natural history, collecting, and the theory of evolution. In *Natural Selection and Beyond: The Intellectual Legacy of Alfred Russel Wallace,* ed. C. H. Smith and G. Beccaloni, 47–65. Oxford: Oxford University Press.

Berry, A., and J. Browne. 2008. The other beetle hunter. *Nature* 453: 1188–1190.

Blyth, E. 1835. An attempt to classify the "varieties" of animals, with observations on the marked seasonal and other changes which naturally take place in various British species, and which do not constitute varieties. *Magazine of Natural History* 8: 40–53.

Boswell, J. 1900. *The Life of Samuel Johnson, L.L.D.,* vol. 1. London: Macmillan and Co.

Bouton, M. E. 2007. *Learning and Behavior: A Contemporary Synthesis.* Sunderland, MA: Sinauer Associates.

Bowler, P. J. 2003. *Evolution: The History of an Idea,* 3rd ed. Berkeley: University of California Press.

———. 2009. Charles Darwin and his Dublin critics: Samuel Haughton and William Henry Harvey. *Proceedings of the Royal Irish Academy* 109C: 409–420.

Bowring, J. C. 1855. Coleoptera of Siam. [Extract of a letter by

J. C. Bowring read by Edwin Shepherd.] *Zoologist* 13: 4910–4911.

Brackman, A. C. 1980. *A Delicate Arrangement: The Strange Case of Charles Darwin and Alfred Russel Wallace.* New York: Times Books.

Brakefield, P. M., and T. B. Larsen. 1984. The evolutionary significance of dry and wet season forms in some tropical butterflies. *Biological Journal of the Linnean Society* 22: 1–12.

Brakefield, P. M., and B. J. Zwaan. 2011. Seasonal polyphenisms and environmentally induced plasticity in the Lepidoptera: The coordinated evolution of many traits on multiple levels. In *Mechanisms of Life History Evolution: The Genetics and Physiology of Life History Traits and Trade-Offs,* ed. T. Flatt and A. Heyland, 243–252. Oxford: Oxford University Press.

Bronn, H. G. 1842–1843. *Handbuch einer Geschichte der Natur.* 2 volumes. Stuttgart: Schweizerbart.

Brooks, J. L. 1984. *Just Before the Origin: Alfred Russel Wallace's Theory of Evolution.* New York: Columbia University Press.

Brougham, H. P. 1839. *Dissertations on Subjects of Science Connected with Natural Theology: Being the Concluding Volumes of the New Edition of Paley's Work.* 2 volumes. London: C. Knight.

———. 1844. *Dialogues on Instinct; With Analytical View of the Researches on Fossil Osteology.* London: C. Knight and Co.

Browne, J. 1980. Darwin's botanical arithmetic and the "principle of divergence," 1854–1858. *Journal of the History of Biology* 13: 53–89.

Bryant, H. N. 1995. The threefold parallelism of Agassiz and Haeckel, and polarity determination in phylogenetic systematics. *Biology and Philosophy* 10: 197–217.

Buch, L. von. 1825–1831. *Physikalische Beschreibung der Kanarische Inseln.* Berlin: Druckerei der Königlichen Akademie der Wissenschaften.

———. 1836. *Description physiques des îsles Canaries, suivie d'une indication des principaux volcans du globe,* trans. C. Boulanger. Paris.

Burchfield, J. D. 1998. The age of the Earth and the invention of geological time. In *Lyell: The Past Is Key to the Present,* ed. D. J. Blundell and A. C. Scott, Geological Society Special Publication 143, 137–143. London: Geological Society.

Camerini, J. R. 1993. Evolution, biogeography, and maps: An early history of Wallace's Line. *Isis* 84: 700–727.

Candolle, A. P. de. 1820. Géographie botanique. In *Dictionnaire des sciences naturelles,* ed. G. Cuvier, vol. 18: 359–422. Strasbourg: F. G. Levrault; Paris: Le Normant.

Carroll, S. 2005. *Endless Forms Most Beautiful.* New York: W. W. Norton.

Chambers, R. 1844. *Vestiges of the Natural History of Creation.* London: John Churchill.

Champagne, F. A., and R. Mashoodh. 2009. Genes in context: Gene-environment interplay and the origins of individual differences in behavior. *Current Directions in Psychological Science* 18: 127–131.

Chen, J.-Y., J. Dzik, G. D. Edgecombe, L. Ramsköld, and G.-Q. Zhou. 1995. A possible early Cambrian chordate. *Nature* 377: 720–722.

Chevrolat, A. 1861. Réflexions et notes synonymiques sur le travail de M. James Thomson sur les Cérambycides, avec descriptions de quelques nouvelles espèces. *Journal of Entomology* 3(4): 185–192, 245–254.

Claeys, G. 2008. Wallace and Owenism. In *Natural Selection and Beyond: The Intellectual Legacy of Alfred Russel Wallace,* ed. C. H. Smith and G. Beccaloni, 235–262. Oxford: Oxford University Press.

Clench, M. H. 1978. Tracheal elongation in birds-of-paradise. *Condor* 80: 423–430.

Combe, G. 1828. *Constitution of Man.* Edinburgh: John Anderson Jr.

Corsi, P. 1978. The importance of French transformist ideas for the second volume of Lyell's *Principles of Geology. British Journal for the History of Science* 2: 1–25.

Costa, J. T. 2006. *The Other Insect Societies.* Cambridge, MA: Harvard University Press.

———. 2009. *The Annotated Origin: A Facsimile of the First Edition of Charles Darwin's* On the Origin of Species. Cambridge, MA: Harvard University Press.

———. 2013. Synonymy and its discontents: Alfred Russel Wallace's nomenclatural proposals from the "Species Notebook" of 1855–1859. *Bulletin of Zoological Nomenclature* 70(2): 131–148.

Cranbrook, Earl of, et al. 2005. A. R. Wallace, collector: Tracing his vertebrate specimens, part 1. In *Wallace in Sarawak—150 Years Later: An International Conference on Biogeography and Biodiversity,* ed. A. A. Tuen and I. Das, 8–34. Kota Samarahan, Sarawak: Institute of Biodiversity and Environmental Conservation, Universiti Malaysia Sarawak.

Cronin, H. 1991. *The Ant and the Peacock: Altruism and Sexual Selection from Darwin to Today.* Cambridge: Cambridge University Press.

Curtis, J. 1829. *A Guide to the Arrangement of British Insects: Being a Catalogue of All the Named Species Hitherto Discovered in Great Britain and Ireland.* London: Printed for the author; published and sold by J. Pigot.

Dampier, W. 1697–1703. *A New Voyage Round the World.* London: James Knapton.

Darwin, C. R. 1845. *Journal of Researches into the Natural History and Geology of the Countries Visited during the Voyage of H.M.S. Beagle Round the World,* 2nd ed. [*Voyage of the Beagle*]. London: John Murray.

———. 1859. *On the Origin of Species by Means of Natural Selection.* London: John Murray.

Darwin, F., ed. 1909. *The Foundations of the Origin of Species: Two Essays Written in 1842 and 1844.* Cambridge: Cambridge University Press.

Davies, R. 2012. How Charles Darwin received Wallace's Ternate paper 15 days earlier than he claimed. [A comment on Van Wyhe and Rookmaaker (2012).] *Biological Journal of the Linnean Society* 105: 472–477.

Dejean, P. F. M. A. 1802–1837. *Catalogue des Coléoptères de la collection d'Auguste Dejean.* Paris.

Didik, P., M. Ancrenaz, H. C. Morrogh-Bernard, S. S. U. Atmoko, S. A. Wich, and C. P. van Schaik. 2009. Nest building in orangutans. In *Orangutans: Geographic Variation in Behavioral Ecology and Conservation,* ed. S. A. Wich, S. S. U. Atmoko, T. M. Setia, and C. P. van Schaik, 270–275. Oxford: Oxford University Press.

Distant, W. L. 1882–1886. *Rhopalocera Malayana: A Description of the Butterflies of the Malay Peninsula.* London: West, Newman and Co.

Domjan, M. 2012. Learning and instinct. In *Encyclopedia of the Sciences of Learning,* ed. N. Seel. Berlin: Springer-Verlag. doi: 10.1007/SpringerReference_319536 2012-05-03 07:58:02 UTC; www.springerreference.com.

Dumont d'Urville, J.-S.-C. 1825. De la distribution des fougères sur la surface du globe terrestre. *Annales des Sciences Naturelles* 6: 51–73.

———. 1832. Sur les îles du grand océan. *Bulletin de la Société de Géographie* 17: 1–21.

Edwards, W. H. 1847. *A Voyage Up the River Amazon, Including a Residence at Pará.* London: John Murray.

Fabre, E. 1854. On the species of aegilops of the south of France, and their transformation into cultivated wheat. Translated from the French. *Journal of the Royal Agricultural Society of England* 15: 167–180.

Fichman, M. 2004. *An Elusive Victorian: The Evolution of Alfred Russel Wallace.* Chicago: University of Chicago Press.

Forbes, W. A. 1882a. Note on a peculiarity in the trachea of the twelve-wired bird-of-paradise (*Seleucidis nigra*). *Proceedings of the Zoological Society* of London 50: 333–335.

———. 1882b. On the convoluted trachea of two species of Manucode (*Manucodia atra* and *Phonygama gouldi*); with remarks on similar structures in other birds. *Proceedings of the Zoological Society* of London 50: 347–353.

Forrest, T. 1780. *A Voyage to New Guinea and the Moluccas from Balambangan*…London: G. Scott; repr., Kuala Lumpur: Oxford University Press, 1969, with an introduction by D. K. Bassett.

Fullom, S. W. 1852. *Marvels of Science and Other Testimonies to Holy Writ.* London: Colburn and Co.

George, W. 1964. *Biologist Philosopher: A Study of the Life and Writings of Alfred Russel Wallace.* London: Abelard-Schuman.

Gessner, C. 1551–1558, 1587. *Historiae Animalium.* 5 volumes. Zurich.

Goethe, J. W. von. 1790. *Versuch die Metamorphose der Pflanzen zu erklären.* Gotha: Ettingersche Buchhandlung.

Gould, J. 1875–1888. *The Birds of New Guinea, and the Adjacent Papuan Islands.* London: Bedford Square.

———. 1850–1883. *The Birds of Asia.* 7 volumes. London: Printed by Taylor and Francis; published by the author.

Gray, A. 1856. Statistics of the flora of the northern United States. *American Journal of Science and Arts* 12: 204–232.

Gray, G. R. 1859a. List of the birds lately sent by Mr. A. R. Wallace from Dorey or Dorery, New Guinea. *Proceedings of the Zoological Society of London* 27: 153–159.

———. 1859b. [Notes on a new bird-of-paradise discovered by Mr. Wallace.] *Proceedings of the Zoological Society of London* 27: 130.

———. 1859c. [Report of the meeting of the Zoological Society of March 22, 1859, with description of *Semioptera wallacei*]. *The Athenaeum* 1639: 425.

Gray, J. E. 1858. A list of species of Mammalia sent from the Aru Islands by Mr. A. R. Wallace. *Proceedings of the Zoological Society of London* 26: 106–113.

Greenwood, W. 1940. The food-plants or hosts of some Fijian insects, IV. *Proceedings of the Linnean Society of New South Wales* 65: 211–218.

Gross, C. 2010. Alfred Russel Wallace and the evolution of the human mind. *Neuroscientist* 16: 496–507.

Haber, F. C. 1968. Fossils and the idea of a process of time in natural history. In *Forerunners of Darwin: 1745–1859,* ed. B. Glass, O. Temkin, and W. L. Straus, Jr., 222–261. Baltimore: Johns Hopkins University Press.

Hall-Jones, J. 1995. *The Horsburgh Lighthouse.* Invercargill, New Zealand: John Hall-Jones.

Harlow, H. F. 1958. The nature of love. *American Psychologist* 13: 673–685.

Haughton, S. 1863. On the form of the cells made by various wasps and by the honey bee; with an appendix on the Origin of Species. *Annals and Magazine of Natural History,* 3rd ser., 9: 415–429.

Heer, O. de. 1855–1859. *Flora tertiaria Helvetiae: Die tertiäre flora der Schweiz.* 3 volumes. Wintertur, Switzerland: J. Würster & Co.

Henslow, J. S. 1856. On the triticoidal forms of *aegilops* and on the specific identity of *Centaurea nigra* and *C. nigrescens.* Report of the twenty-sixth meeting of the British Association for the Advancement of Science, held in Cheltenham; transactions of the sections, 87–88.

Hewitson, W. C. 1865. A list of diurnal Lepidoptera collected by Mr. Wallace in the Eastern Archipelago. *Journal of the Proceedings of the Linnean Society (Zoology)* 8: 143–149.

Huc, M. 1854. *L'Empire Chinois: Faisant suite à l'ouvrage intitulé Souvenirs d'un voyage dans la Tartarie et le Thibet,* 2nd ed. 2 volumes. Paris: Librairie de Gaume Frères.

Humboldt, A. von. 1818–1821. *Personal Narrative of Travels to the Equinoctial Regions of the New Continent during the Years 1799 to 1804,* trans. H. M. Williams. 7 volumes. London: Longman, Hurst, Rees, Orme & Browne.

Jin, Y. G., Y. Wang, W. Wang, Q. H. Shang, C. Q. Cao, and D. H. Erwin. 2000. Pattern of marine mass extinction near the Permian-Triassic boundary in south China. *Science* 289: 432–436.

Johnston, G. 1853. *The Botany of the Eastern Borders, with the Popular Names and Uses of the Plants, and of the Customs and Beliefs Which Have Been Associated with Them.* London: J. Van Voorst.

Jones, G. 2002. Alfred Russel Wallace, Robert Owen and the theory of natural selection. *British Journal of the History of Science* 35: 73–96.

Joron, M., and J. L. B. Mallet. 1998. Diversity in mimicry: Paradox or paradigm? *Trends in Ecology and Evolution* 13: 461–466.

Knighton, W. 1855. *Tropical Sketches, or, Reminiscences of an Indian Journalist.* London: Hurst & Blackette.

Kottler, M. J. 1974. Alfred Russel Wallace, the origin of man, and spiritualism. *Isis* 65: 144–192.

———. 1985. Charles Darwin and Alfred Russel Wallace: Two decades of debate over natural selection. In *The Darwinian Heritage,* ed. D. Kohn, 367–432. Princeton, NJ: Princeton University Press.

Kunte, K. 2008. Mimetic butterflies support Wallace's model of sexual dimorphism. *Proceedings of the Royal Society B* 275: 1617–1624.

———. 2009. The diversity and evolution of Batesian mimicry in *Papilio* swallowtail butterflies. *Evolution* 63: 2707–2716.

LaCroix, P. [P. Dufour]. 1851–1856. *Histoire de la prostitution.* Paris.

Lawrence, W. 1819. *Lectures on Physiology, Zoology and the Natural History of Man.* London: J. Callow.

Lesson, R.-P. 1829. *Voyage médical autour du monde, exécuté sur la corvette du roi La Coquille commandée par M. L. I. Duperrey pendant les années 1822, 1823, 1824 et 1825* [*Mémoire sur les races humaines répandues sur les îles du Grand-Océan, et considérées sous les divers rapports physiologiques, naturels et moraux,* 153–228]. Paris: Roret.

———. 1838–1839. *Voyage autour du monde entrepris par ordre du gouvernement sur la corvette "La Coquille."* 2 volumes. Paris: P. Pourrat Frères.

Lindley, J. 1832. *An Introduction to Botany.* London: Longman, Rees, Orme, Brown, Green, & Longman.

Lindley, J., and W. Hutton. 1831–1835. *Fossil Flora of Great Britain.* 3 volumes. London: James Ridgeway.

Locke, D. P., et al. 2011. Comparative and demographic analysis of orangutan genomes. *Nature* 469: 529–533.

Lyell, C. 1830. *Principles of Geology; Being an Attempt to Explain the Former Changes of the Earth's Surface, by Reference to Causes Now in Operation,* vol. 1. London: J. Murray.

———. 1835. *Principles of Geology,* 4th ed. 4 volumes. London: J. Murray.

———. 1847. The structure and possible agency of the coal field of the James River, North Richmond, Virginia. *Proceedings of the Geological Society of London* 3(April 14): 261–282.

———. 1849. *A Second Visit to the United States of North America.* 2 volumes. London: John Murray.

Lyle, I. F. 1978. John Hunter, Gilbert White, and the migration of swallows. *Annals of the Royal College of Surgeons of England* 60: 485–491.

Macleay, W. S. 1819–1821. *Horae Entomologicae: Or Essays on the Annulose Animals, &c.,* vol. 1, parts 1 and 2. London: S. Bagster.

Mallet, J. 2008. Wallace and the species concept of the early Darwinians. In *Natural Selection and Beyond: The Intellectual Legacy of Alfred Russel Wallace,* ed. C. H. Smith and G. Beccaloni, 102–113. Oxford: Oxford University Press.

———. 2009. Alfred Russel Wallace and the Darwinian species concept: His paper on the swallowtail butterflies (Papilionidae) of 1865. *Gayana* 73(2)(supp.): 42–54.

Malthus, T. R. 1798. *First Essay on Population.* London [much reprinted].

Marchant, J., ed. 1916. *Alfred Russel Wallace: Letters and Reminiscences.* 2 volumes. London: Cassel and Company.

Maw, G. 1861. [Review of *On the Origin of Species* and other works.] *Zoologist* 19: 7577–7611.

Mayr, E. 1963. *Animal Species and Evolution.* Cambridge, MA: Belknap Press of Harvard University Press.

McCartney, P. J. 1976. Charles Lyell and G. B. Brocchi: A study in comparative historiography. *British Journal for the History of Science* 9: 175–189.

McGuire, J., R. M. Brown, Mumpini, A. Riyanto, and N. Andayani. 2007. The flying lizards of the *Draco lineatus* group (Squamata: Iguania: Agamidae): A taxonomic revision with descriptions of new species. *Herpetological Monographs* 21: 179–212.

McKinney, H. L. 1966. Alfred Russel Wallace and the discovery of natural selection. *Journal of the History of Medicine and Allied Sciences* 21: 333–357.

———. 1972. *Wallace and Natural Selection.* New Haven, CT: Yale University Press.

Milne-Edwards, H. 1851. *Introduction à la zoologie générale.* Paris: Victor Masson.

———. 1856. *A Manual of Zoology,* trans. R. Knox. London: H. Renshaw.

Mivart, St. G. 1871. *On the Genesis of Species.* London: Macmillan and Co.

Moore, J. 1997. Wallace's Malthusian moment: The common context revisited. In *Victorian Science in Context,* ed. B. Lightman, 290–311. Chicago: University of Chicago Press.

Müller, S. 1839–1840. Over de zoogdieren van den Indischen Archipel. In *Verhandelingen over de Natuurlijke Geschiedenis der Nederlandsch Overzeesche Bezittingen door de leden natuurkundige commissie in Indie andere Schrijvers: Zoologie,* ed. C. J. Temminck, 1–57. Leiden: S. en J. Luchtmans & C. C. van der Hock.

———. 1842. Über die geographische Verbreitung der Saügethiere im Indischen Archipelagus. *Annalen der Erdkunde,* part 1 (March): 251–266; part 2 (April): 289–333.

Mužinić, J., J. F. Bogdan, and B. Beehler. 2009. Julije Klović: The first colour drawing of greater bird of paradise *Paradisaea apoda* in Europe and its model. *Journal of Ornithology* 150: 645–649.

Neill, W. T. 1973. *Twentieth-Century Indonesia.* New York: Columbia University Press.

Owen, R. 1843. *Lectures on the Comparative Anatomy and Physiology of the Invertebrate Animals, Delivered at the Royal College of Surgeons, in 1843.* London: Longman, Brown, Green, and Longmans.

———. 1855a. On the anthropoid apes. *Report of the Twenty-Fourth Meeting of the British Association for the Advancement of Science* 24: 111–113.

———. 1855b. On the anthropoid apes, and their relations to man. *Notices of the Proceedings … of the Royal Institution of Great Britain* 2 (1854–1858): 26–41.

———. 1859. On the classification and geographical distribution of the Mammalia: Being the lecture on Sir Robert Reade's Foundation, delivered before the University of Cambridge, in the Senate-House, May 10, 1859. To which is added an appendix. (Appendix, Part A: "On the extinction of species"; Appendix, Part B: "On the orang, chimpanzee, and gorilla. With reference to the transmutation of species.") London: John W. Parker.

Parley, J. 1977. *The Spontaneous Generation Controversy: From Descartes to Oparin.* Baltimore: Johns Hopkins University Press.

Pascoe, F. P. 1864–1869. *Longicornia Malayana;* or a descriptive catalogue of the species of the three longicorn families Lamiidae, Cerambycidae and Prionidae collected by Mr. A. R. Wallace in the Malay Archipelago. *Transactions of the Entomological Society of London,* 3rd ser., 3: 1–710 and plates.

Paul, D. B. 1988. The selection of the "survival of the fittest." *Journal of the History of Biology* 21: 411–424.

Pope, A. 1734. *An Essay on Man.* London: Printed for J. Wilford. [Much reprinted.]

Prichard, J. C. 1851. *Researches into the Physical History of Mankind,* 4th ed. 5 volumes. London: Houlston and Stoneman. [First edition of 1813 published in 2 volumes.]

Purvis, A. 1995. A composite estimate of primate phylogeny. *Philosophical Transactions of the Royal Society of London B* 348: 405–421.

Raby, P. 2001. *Alfred Russel Wallace: A Life.* Princeton, NJ: Princeton University Press.

Rachels, J. 1986. Darwin's moral lapse. *National Forum (Phi Kappa Phi)* 66(3): 22–24.

Renan, E. 1859. *Essais de morale et de critique.* Paris: Michel Lévy Frères.

Rennie, J. 1831. *The Architecture of Birds.* London: Charles Knight.

———. 1835. *The Faculties of Birds.* London: Charles Knight.

———. 1839. *Natural History of Birds: Their Architecture, Habits, and Faculties.* New York: Harper & Brothers.

———. 1844. *Bird-Architecture,* rev. ed. London: Charles Knight & Co.

Rookmaaker, K., and J. van Wyhe. 2012. In Alfred Russel Wallace's shadow: His forgotten assistant, Charles Allen (1839–1892). *Journal of the Malaysian Branch of the Royal Asiatic Society* 85: 17–54.

Rose, M. D. 1988. Functional anatomy of the Cheiridia. In *Orang-utan Biology,* ed. J. Schwartz, 299–310. Oxford: Oxford University Press.

Rudwick, M. J. S. 1970. The strategy of Lyell's *Principles of Geology. Isis* 61: 5–33.

———. 1972. *The Meaning of Fossils: Episodes in the History of Palaeontology.* London: Macdonald; New York: American Elsevier.

———. 1998. Lyell and the *Principles of Geology.* In *Lyell: The Past Is the Key to the Present,* ed. D. J. Blundell and A. C. Scott, Geological Society Special Publication 143, 3–15. London: Geological Society.

Ruxton, G. D., T. N. Sherratt, and M. P. Speed. 2004. *Avoiding Attack.* Oxford: Oxford University Press.

Sahney, S., and M. J. Benton. 2008. Recovery from the most profound mass extinction of all time. *Proceedings of the Royal Society of London* 275: 759–765.

Scott, W. 1831. *The Pirate.* Waverly Novels, vol. 25. Edinburgh: Robert Cadell; London: Whittaker and Co.

Seba, A. 1734. *Locupletissimi Rerum Naturalium Thesauri,* vol. 1. Amsterdam: Janssonio-Waesbergios & J. Wetstenium, & Gul. Smith.

Shermer, M. 2002. *In Darwin's Shadow; The Life and Science of Alfred Russel Wallace: A Biographical Study on the Psychology of History.* Oxford: Oxford University Press.

Skertchly, S. B. J. 1889. Note on *Ragadia crisia*. *Nature* 40: 10.

Slotten, R. A. 2004. *The Heretic in Darwin's Court: The Life of Alfred Russel Wallace*. New York: Columbia University Press.

Smith, C. H. 1992. *Alfred Russel Wallace on Spiritualism, Man, and Evolution: An Analytical Essay*. Torrington, CT: privately published. http://people.wku.edu/charles.smith/essays/ARWPAMPH.htm.

————. 2013. A further look at the 1858 Wallace-Darwin mail delivery question. *Biological Journal of the Linnean Society* 108: 715–718.

Smith, C. H., and G. Beccaloni, eds. 2008. *Natural Selection and Beyond: The Intellectual Legacy of Alfred Russel Wallace*. Oxford: Oxford University Press.

Smith, F. 1857–1858. Catalogue of the hymenopterous insects collected at Sarawak, Borneo; Mount Ophir, Malacca; and at Singapore, by A. R. Wallace. *Journal of the Proceedings of the Linnean Society (Zoology)*, 2 (6–7): 42–130. [With introductory remarks by A. R. Wallace.]

————. 1873. A catalogue of the aculeate Hymenoptera and Ichneumonidae of India and the eastern archipelago. *Journal of the Linnean Society: Zoology* 11: 285–302. [With introductory remarks by A. R. Wallace.]

Spencer, H. 1851. *Social Statics, Or the Conditions Essential to Happiness Specified, and the First of Them Developed*. London: John Chapman.

————. 1859. Illogical geology. *Universal Review* 2: 57–89.

————. 1862. *First Principles of a New System of Philosophy*. London: Williams and Norgate.

Stauffer, R. C., ed. 1975. *Charles Darwin's Natural Selection: Being the Second Part of His Big Species Book Written from 1856 to 1858*. Cambridge: Cambridge University Press.

Stevens, M. 2005. The role of eyespots as anti-predator mechanisms, principally demonstrated in the Lepidoptera. *Biological Reviews* 80: 573–588.

Stevenson, B. 2009. Samuel Stevens, naturalist (1817–1899). *Micscape Magazine* 166. www.microscopy-uk.org.uk/mag/indexmag.html.

Strick, J. E. 2000. *Sparks of Life: Darwinism and the Victorian Debates over Spontaneous Generation*. Cambridge, MA: Harvard University Press.

Strickland, H. E. 1841. On the true method of discovering the natural system in zoology and botany. *Annals and Magazine of Natural History* 6: 184–194.

Strickland, H. E., and A. G. Melville. 1848. *The Dodo and Its Kindred; or the History, Affinities, and Osteology of the Dodo, Solitaire, and Other Extinct Birds of the Islands Mauritius, Rodriguez, and Bourbon*. London: Reeve, Benham, & Reeve.

Swadling, P. 1996. *Plumes from Paradise*. Port Moresby: Papua New Guinea National Museum; Coorparoo, Australia: Robert Brown and Associates.

Tcherkézoff, S. 2003. A long and unfortunate voyage: Toward the invention of the Melanesia/Polynesia opposition (1595–1832). *Journal of Pacific History* 38: 175–196.

Thomson, J. 1857–1858. *Archives entomologiques, ou Recueil contenant des illustrations d'insectes nouveaux ou rares*. 2 volumes. Paris: Société Entomologique de France.

————. 1859. *Arcana naturae, ou Recueil d'histoire naturelle*. Paris: J.-B. Baillière et Fils.

Tiedemann, F. 1826. *The Anatomy of the Foetal Brain, with a Comparative Exposition of Its Structure in Animals*. Edinburgh: John Carfrae and Son.

Van Wyhe, J., and K. Rookmaaker. 2012. A new theory to explain the receipt of Wallace's Ternate essay by Darwin in 1858. *Biological Journal of the Linnean Society* 105: 249–252.

Walker, F. 1861. Catalogue of the dipterous insects collected at Gilolo, Ternate, and Ceram, by Mr. R. Wallace [*sic*], with descriptions of new species. *Journal of the Proceedings of the Linnean Society (Zoology)* 6: 4–23.

Wallace, A. R. 1847. Capture of *Trichius fasciatus* near Neath. *Zoologist* 5: 1676. [S2]

————. 1850. On the umbrella bird (*Cephalopterus ornatus*), "Ueramimbé, L. G." *Proceedings of the Zoological Society of London* 18: 206–207. [S5]

————. 1852a. On the insects used for food by the Indians of the Amazon. *Transactions of the Entomological Society of London* 2: 241–244. [S10]

————. 1852b. On the monkeys of the Amazon. *Proceedings of the Zoological Society of London* 20: 107–110. [S8]

————. 1852c. Proceedings of natural-history collectors in foreign countries: Letter concerning the ship fire dated 19 Oct. 1852. *Zoologist* 10: 3641–3643. [S7]

———. 1853a. *A Narrative of Travels on the Amazon and Rio Negro, with an Account of the Native Tribes, and Observations on the Climate, Geology, and Natural History of the Amazon Valley.* London: Reeve & Co. [S714]

———. 1853b. On the Rio Negro. *Journal of the Royal Geographical Society* 23: 212–217. [S11]

———. 1854. On the habits of the butterflies of the Amazon Valley. *Transactions of the Entomological Society of London,* n.s., 2 (part VIII): 253–264. [S13]

———. 1855a. Letter from Sarawak [concerning collecting at the Si Munjon Coal Works, Borneo]. *Zoologist* 13: 4803–4807. [S21]

———. 1855b. Letter from Sarawak, Borneo. *The Literary Gazette and Journal of the Belles Lettres, Science, and Art* 2003: 366b–c. [S18]

———. 1855c. On the law which has regulated the introduction of new species [Sarawak Law paper]. *Annals and Magazine of Natural History,* 2nd ser., 16: 184–196. [S20]

———. 1856a. Attempts at a natural arrangement of birds. *Annals and Magazine of Natural History,* 2nd ser., 18: 193–216. [S28]

———. 1856b. A new kind of baby. *Chambers's Journal,* 3rd ser., 6: 325–327. [S30]

———. 1856c. Observations on the zoology of Borneo. *Zoologist* 14: 5113–5117. [S25]

———. 1856d. On the bamboo and durian of Borneo. *Hooker's Journal of Botany and Kew Garden Miscellany* 8: 225–230. [S27]

———. 1856e. On the habits of the orang-utan in Borneo. *Annals and Magazine of Natural History,* 2nd ser., 18: 26–32. [S26]

———. 1856f. On the orang-utan or mias of Borneo. *Annals and Magazine of Natural History,* 2nd ser., 17: 471–476. [S24]

———. 1856g. Some account of an infant "orang-utan." *Annals and Magazine of Natural History,* 2nd ser., 17: 386–390. [S23]

———. 1857a. [Letter extracts to Samuel Stevens from Dobbo, Aru Islands.] *Proceedings of the Entomological Society of London* 1856–1857: 91–93. [S35]

———. 1857b. Letter from Macassar, Celebes. *Zoologist* 15: 5559–5560. [S32]

———. 1857c. On the great bird of paradise, *Paradisea apoda* Linn.; "Burong mati" (Dead bird) of the Malays; "Fanéhan" of the Natives of Aru. *Annals and Magazine of Natural History,* 2nd ser., 20: 411–416. [S37]

———. 1857d. On the natural history of the Aru Islands. *Annals and Magazine of Natural History* 20 (supp.): 473–485. [S38]

———. 1858a. A disputed case of priority in nomenclature. *Proceedings of the Entomological Society of London* 1858: 23. [S42]

———. 1858b. Note on the theory of permanent and geographical varieties. *Zoologist* 16: 5887–5888. [S39]

———. 1858c. On the Arru Islands. *Proceedings of the Royal Geographical Society* 2: 163–170. [S41]

———. 1858d. On the entomology of the Aru Islands. *Zoologist* 16: 5889–5894. [S40]

———. 1858e. On the habits and transformations of a species of *Ornithoptera,* allied to *O. priamus,* inhabiting the Aru Islands, near New Guinea. *Transactions of the Entomological Society of London,* n.s., 4 (part VII): 272–273. [S36]

———. 1858f. On the tendency of varieties to depart indefinitely from the original type [Ternate essay]. *Proceedings of the Linnean Society* 3: 53–62. [S43]

———. 1858g. Proceedings of natural-history collectors in foreign countries: Letter from Amboyna. *Zoologist* 16: 6120–6124. [S44]

———. 1859a. Correction of an important error affecting the classification of the Psittacidae. *Annals and Magazine of Natural History,* 3rd ser., 3 (14): 147–148 [S46]

———. 1859b. [Extract from letter concerning collecting dated October 29, 1858, Batchian, Moluccas; communicated to the Entomological Society of London meeting of March 7, 1859.] *Proceedings of the Entomological Society of London,* 1858–1859: 61; repr., *Zoologist* 17: 6546–6547. [S47]

———. 1859c. Letter from Mr. Wallace concerning the geographical distribution of birds [dated March 1859, Batchian]. *Ibis* 1: 449–454. [S52]

———. 1860a. Note on the habits of *Scolytidae* and *Bostrichidae. Transactions of the Entomological Society of London,* n.s., 5 (part VI): 218–220. [S54]

———. 1860b. Notes on *Semioptera wallacii,* Gray. *Proceedings of the Zoological Society of London* 28: 61. [S55]

———. 1860c. On the zoological geography of the Malay Archipelago [communicated at the Linnean Society meeting of November 3, 1859]. *Journal of the Proceedings of the Linnean Society: Zoology* 4: 172–184. [S53]

———. 1860d. The ornithology of Northern Celebes. *Ibis* 2: 140–147. [S57]

———. 1861. Letters from Ternate, Delli, and Timor. *Ibis* 3: 310–311. [S63]

———. 1862. Narrative of search after birds of paradise. *Proceedings of the Zoological Society of London* 1862: 153–161. [S67]

———. 1863a. Notes on the genus *Iphias;* with descriptions of two new species from the Moluccas. *Journal of Entomology* 2: 1–5. [S74]

———. 1863b. On the identification of the *Hirundo esculenta* of Linnaeus, with a synopsis of the described species of *Collocalia. Proceedings of the Zoological Society of London* 1863: 382–385. [S84]

———. 1863c. On the physical geography of the Malay Archipelago. *Proceedings of the Royal Geographical Society* 7: 217–234. [S78]

———. 1863d. On the proposed change in name of *Gracula pectoralis. Annals and Magazine of Natural History,* 3rd ser., 11: 15–17. [S75]

———. 1863e. Remarks on the Rev. S. Haughton's paper on the bee's cell, and on the *Origin of Species. Annals and Magazine of Natural History,* 3rd ser., 12: 303–309. [S83]

———. 1864. The origin of human races and the antiquity of man deduced from the theory of "natural selection." *Journal of the Anthropological Society of London* 2: clviii–clxx [with an account of related discussion on clxx–clxxxvii]. [S93]

———. 1865a. On the phenomena of variation and geographical distribution as illustrated by the Papilionidae of the Malayan Region. *Transactions of the Linnean Society of London* 25 (part I): 1–71. [A paper read at the Linnean Society meeting of March 17, 1864.] [S96]

———. 1865b. On the varieties of man in the Malay Archipelago. *Transactions of the Ethnological Society of London,* n.s., 3: 196–215. [S82]

———. 1867a. Mimicry, and other protective resemblances among animals. [Review of works by Henry W. Bates, Alfred R. Wallace, Andrew Murray, and Charles Darwin; anonymous, but known to be authored by Wallace.] *Westminster Review,* n.s., 32: 1–43. [S134]

———. 1867b. The philosophy of birds' nests. *Intellectual Observer* 11 (6): 413–420. [S136]

———. 1868. A theory of birds' nests: Shewing the relation of certain sexual differences of colour in birds to their mode of nidification. *Journal of Travel and Natural History* 1: 73–89. [S139]

———. 1869a. *The Malay Archipelago: The Land of the Orang-utan and the Bird of Paradise.* New York: Harper & Brothers. Cited in text as *Malay Archipelago.* [see also S715]

———. 1869b. Notes on the localities given in *Longicornia Malayana,* with an estimate of the comparative value of the collections made at each of them. *Transactions of the Entomological Society of London,* 3rd ser., 3 (part VII): 691–696. [S154]

———. 1869c. Sir Charles Lyell on geological climates and the Origin of Species. [Review of Lyell's *Principles of Geology,* 10th ed., and *Elements of Geology,* 6th ed., anonymous but attributed to Wallace.] *Quarterly Review* 126: 359–394. [S146]

———. 1870. *Contributions to the Theory of Natural Selection.* London: Macmillan & Co. [S716]

———. 1872. The beginnings of life. I & II. [Review of H. C. Bastian (1872), *The Beginnings of Life.*] *Nature* 6: 284–287; 299–303. [S211]

———. 1873. Perception and instinct in the lower animals. *Nature* 8: 65. [S227]

———. 1874. Zoological nomenclature. [Review of D. Sharp 1873, *The Object and Method of Zoological Nomenclature.*] *Nature* 9: 258–260. [S239]

———. 1876. *The Geographical Distribution of Animals; With a Study of the Relations of Living and Extinct Faunas as Elucidating the Past Changes of the Earth's Surface.* 2 volumes. London: Macmillan & Co. [S718]

———. 1878. *Tropical Nature and Other Essays.* London: Macmillan & Co. [S719]

———. 1880a. *Island Life: Or, The Phenomena and Causes of Insular Faunas and Floras, Including a Revision and*

*Attempted Solution of the Problem of Geological Climates.* London: Macmillan & Co. [S721]

———. 1880b. Two Darwinian essays [a review of A. Weismann, *Studies in the Theory of Descent,* part 1, and R. Lankester, *Degeneration: A Chapter in Darwinism*] *Nature* 22: 141–142. [S324]

———. 1882. Review [of *Rhopalocera Malayana* by Distant, 1882]. *Nature* 26: 6–7. [S351]

———. 1883. [Comments by ARW in article entitled "Posthumous Paper by Darwin."] *Times* (London), no. 30997: 10e (December 7, 1883). [S368ab]

———. 1887. Oceanic islands: Their physical and biological relations. [A lecture delivered at the American Geographical Society meeting of January 11, 1887.] *Bulletin of the American Geographical Society* 19: 1–17. [S393]

———. 1889. *Darwinism: An Exposition of the Theory of Natural Selection, with Some of Its Applications.* London: Macmillan & Co. [S724]

———. 1891a. English and American flowers. *Fortnightly Review,* n.s., 50(298) (October 1, 1891): 525–534 (old series, vol. 56) / English and American flowers, II: Flowers and forests of the Far West. *Fortnightly Review,* n.s., 50(300) (December 1, 1891): 796–810 (old series, vol. 56). [S441]

———. 1891b. *Natural Selection and Tropical Nature: Essays on Descriptive and Theoretical Biology.* London: Macmillan & Co. [S725]

———. 1900. *Studies Scientific & Social.* 2 volumes. London and New York: Macmillan and Co. [S727]

———. 1905. *My Life: A Record of Events and Opinions.* 2 volumes. London: Chapman & Hall, Ltd. [S729]

Weismann, A. 1880. *Studies in the Theory of Descent,* part 1: *On the Seasonal Dimorphism of Butterflies.* London: Sampson Low, Marston and Co.

White, A. 1855. *Catalogue of Coleopterous Insects in the Collection of the British Museum. Longicornia II,* vol. 8: 175–412 and plates. London: Taylor and Francis.

White, G. 1834. *A Natural History of Selborne.* London: Allan Bell & Co.

Willich, C. M. 1860. On the angles of dock-gates and the cells of bees. In *Report on the Twenty-ninth Meeting of the British Association for the Advancement of Science; Held at Aberdeen in September 1859,* 10.

Willmer, P. 2011. *Pollination and Floral Ecology.* Princeton, NJ: Princeton University Press.

Wilson, L. G., ed. 1970. *Sir Charles Lyell's Scientific Journals on the Species Question.* New Haven, CT: Yale University Press.

Wood, S. 1995. The first use of the terms "homology" and "analogy" in the writings of Richard Owen. *Archives of Natural History* 22: 255–259.

Woodward, S. P. 1851. *A Manual of the Mollusca: Or, a Rudimentary Treatise of Recent and Fossil Shells Lists.* London: John Weale.

Zollinger, H., and A. Mousson. 1849. *Die Land- und Süsswasser-Mollusken von Java.* Zurich: F. Schulthess.

# Acknowledgments

This book was largely written during a blissful sabbatical year in Germany as a Fellow of the Wissenschaftskolleg zu Berlin, or Berlin Institute for Advanced Study. I am deeply grateful to the Wissenschaftskolleg, Western Carolina University, and the University of North Carolina General Administration for affording me the invaluable luxuries of both time and a living and working environment nonpareil for the completion of this book.

I was profoundly gratified by the readiness with which friends and colleagues endorsed and assisted with the project from inception to completion, showing a generosity of spirit that Wallace would have appreciated. Special thanks go first to Andrew Berry, friend, colleague, indefatigable leader of Harvard's Darwin program in Oxford, and fellow Wallaceophile, without whom this project would not have gotten off the ground. I gratefully acknowledge Ruth Temple and the Council of the Linnean Society of London for permission to publish and annotate the Species Notebook, which resides at the Linnean, and I am appreciative of the kind assistance of the Linnean's librarians, Lynda Brooks, Ben Sherwood, Elaine Charwat, and Andrea Deneau, who provided the high-quality digital images of the notebook, helpfully answered a multitude of questions, and provided ready assistance in my study of the notebook and other Wallace materials at the Linnean.

My work was greatly facilitated by Michael Pearson, who generously sent me his own transcription and photocopies of the recto Species Notebook, done for the Linnean a decade ago, as well as electronic copies of Wallace's Malay Journals. Michael's transcription was an immensely helpful starting point. Anita Murrell intrepidly made her way through a partial second transcription of the recto notebook in July 2011 as a student research intern in the Harvard program that I coteach with Andrew at Oxford each summer.

Leslie Costa, my wife and constant collaborator, meticulously edited my transcription of the recto notebook and generated the complete transcription of the verso side, including its numerous intricate tables. Leslie has truly provided the lion's share of the assistance on the project, and it could not have been completed on time without her considerable and careful efforts.

The manuscript was read by several reviewers, and I am grateful for their keen editorial eyes. Many thanks to Janet Browne and George Beccaloni for their invaluable feedback on the annotations, and to Michael Pearson and Annette Lord for carefully reviewing the transcriptions. Any remaining errors are, of course, solely my responsibility.

I owe much to the leadership and staff of the Wissenschaftskolleg zu Berlin, especially Vera Schulze-Seeger. With regard to this project in particular I express my *tiefste Wertschätzung* to the institute's unfailingly helpful librarians—Sonja Grund, Anja Brockmann, Marianne Buck, and Kirsten Graupner—for their invaluable assistance in procuring myriad books and papers, often obscure, in a timely manner. Thanks, finally, to Thomas Reimer of the Wissenschaftskolleg's library for scanning images, and Kevin McAleer for his assistance with translations.

My editor Michael Fisher at Harvard University Press championed the project from start to finish, and the press's production team was, as ever, a pleasure to work with. Sincere thanks to Susan Wallace Boehmer, Lauren Esdaile, Anne Zarrella, and Christine Thorsteinsson for their help with all aspects of editing and production, and to Julie Ericksen Hagen and Dean Bornstein for their careful efforts with the copyediting and design.

I have benefited in ways both tangible and intangible from a host of friends and colleagues ever willing to discuss Darwiniana and Wallaceana, and I am grateful for the many illuminating conversations I have had over the years with Jim Moore, Janet Browne, Randal Keynes, and especially Andrew Berry and George and Jan Beccaloni. A very special thanks to George for avidly sharing his immense knowledge of the subject, providing all manner of information, comments, and criticisms, and speedily replying to my countless questions. I thank, too, John van Wyhe, Charles Smith, Raghavendra Gadagkar, Jerry Drawhorn, and Gathorne Cranbrook for providing information or sharing ideas, and my fellow Fellows of the Wiko for their camaraderie and perspectives—our *Pausen* and conversations were both welcome tonic and stimulus during my labors on the book.

While working on this project I had the good fortune to make two pilgrimages to Wallace's grave site in Broadstone in the good company of George and Jan Beccaloni, with whom I also explored the area for sites of particular significance to Wallace (including an Indiana Jones–style "exploration" of the woodland that was once Wallace's copse at Old Orchard, his final home, during which a strategic deployment of what was termed my "charming American accent" proved surprisingly helpful). Through their offices I was also privileged to meet Richard Russel (Dick) and Audrey Wallace, and Alfred John Russel (John) and Diana Wallace—Alfred Russel Wallace's grandsons and their delightful spouses. I am deeply appreciative of the Wallaces' warm hospitality and their endorsement of my efforts with their grandfather's remarkable field notebook.

Last but not least, a continent away on the home front, I could not ask for a more professional and dedicated staff than my colleagues at the Highlands Biological Station in North Carolina. Thank you, Tom, Karen, Cynthia, Michelle, Guy, Patrick, Sonya, and Erika: the bliss of my sabbatical leave was complete in the knowledge that the station was in more than capable hands in my absence.

# Note on A. R. Wallace Literary Works

## *Manuscript Facsimile and Transcription Pages*

The facsimile and transcription of Alfred Russel Wallace's Species Notebook (Linnean Society of London manuscript 180) are published with the permission and generous assistance of the Linnean Society of London.

## *Illustration Credits*

# Index

## Note on pagination of index entries:

The index entries pertain to the contents of Alfred Russel Wallace's Species Notebook, not the annotations. Page numbers given correspond to the running pagination of *On the Organic Law of Change*, which presents the recto and verso sides of the Species Notebook with continuous pagination.

Balance in nature, criticism of. *See* Harmony (balance) in nature, criticism of

Baly (Bali), 138, 146; butterflies collected at, 524; inhabitants of, 236

Bamboo, 154, 156

Banana, 268

Batchian (Bacan), 24, 178, 210, 236, 298, 334, 420, 438, 502, 504, 522; butterflies collected at, 522, 524, 526, 528, 530; coal mining on, 502, 504; geyser on, 334; register of insects collected at, 502, 504

Bates, Henry Walter, 468

*Batocera,* 340

*Batrachostomus,* 260

Bats, 52, 136, 146, 150, 210, 316, 338, 514; immense flight in Sarawak, 40

Beck, Mr. van der, assists Wallace in Ceram, 510

Bee Eater, 138

Bees: cells and honeycomb, 374, 376, 378, 380, 382; instinct, 364; putative mathematical precision of, 374, 378, 382

Beetles: collections of, 34, 36, 38, 40, 418, 470–481, 486, 488, 490, 492, 494, 496, 498, 500, 502, 504, 506, 508, 510; crypsis in, 340; drawings of mouthparts of, 442, 446, 448, 450, 452; examples for synonymical catalog, 344; habits, 44, 300; killed by tree sap, 422; cockroaches mistaken for, 324. *See also* Longicornes; Longicorns

Belemnites, 318

*Belideus,* 210

*Belionota stigma,* 438

Bentham, Jeremy, mottoes, 232, 234

Bernstein, Heinrich A., bird collector, address for, 22

Bird & Insect Register of 1858 [ms. WCP4766], 278

Birds: air cells of bones, 252; migration/migratory, 136, 138, 244, 252, 338; nest building, 252, 254, 256, 258, 260, 262, 264, 266, 360; notes on bird architecture, 252, 254, 256, 258, 260, 262, 264, 266; numbers of species of Britain vs. North America, 136; numbers of species by region of Malay Archipelago, 248; plumage, 196; rearing in isolation, untried experiment with, 364; structure of, 92, 250, 252; walking vs. hopping, 144, 146, 244; webbed feet of, 316; White's observations of gait, 244. *See also* Feathers

Birds of Paradise, 24, 496; attracted to fruiting trees in New Guinea, 26; breeding plumage, 170, 174, 176, 198, 204, 298; courtship displays, 174, 176, 198, 298; sketches of, 170, 176, 206, 352, 254; native hunting of, 200, 202; preparation, 202, 204; Wallace's descriptions of, 172, 174, 176, 198, 296, 298, 350, 352, 354, 356, 358. *See also* Paradisea

Bleecker, Pieter, 26, 82

Blimbaig, 268

Blyth, Edward, on types of varieties, 152

Boats: outrigger canoes, 154; praus (prows), 18, 154; rafts, 154; ships, 18, 494, 498

Boisduval, Jean Baptiste, 422, 458, 522, 526, 528

Bontang, Celebes, 479

Borneo, 30, 140, 156, 160, 246, 270, 272, 274, 278, 296, 418, 420, 514, 522, 524, 526, 528, 532; beetles collected in, 418; butterflies collected at, 522, 524, 526, 528; mammals of, 514; Sadong River, 30; Simunjon Coal Works, 30, 32, 34, 36, 54, 84

Bostrichidae, 402, 404, 438, 500

Boswell, James, 140

Bouru (Buru), 270, 522, 524, 526, 528

Bow and arrow, use of by different peoples, 160, 238

Bowring, James, 468

*Brachinus,* 436

Brachyelytra, 406

Branching, evolutionary, parallel and diverging, 102, 104, 136, 390

Brentidae, 402, 422

Broadbill, 370

Buch, Christian Leopold von, 208, 330; "on change of species," 24, 208

Buffon, Comte de, 96

Bugis, 236

*Bulimus,* 248

Buprestidae, 36, 38, 40, 406, 416, 418, 468, 481, 486, 492, 500, 510; note regarding wings of, 416

*Buprestis* sp., 40; *B. gigas,* 36

Burglary, 502

Burong mati, 172, 200

Burong Rajah, 176, 178

Buru. *See* Bouru

Butterflies: Birdwing, 272, 414, 456, 458, 460, 479 (*see also* Orni-thoptera); cabinet design for, 28, 30; collections, 160, 172, 174,

Geology: Darwin's chapter on in *Natural Selection*, 430; effect of geological change on landscape, 322; evidence of uplift, 334, 481; gaps in geological record, 212, 214, 216, 218, 220, 222; geological succession, 100, 102, 104, 106, 218, 318; geological time needed for transmutation, 112, 114, 118, 124; gradual change, 98, 100, 128, 130, 212, 214, 216, 218, 220, 222, 328; Lias formation, 266; limestone, 481; Trias formation, 212, 214. *See also* Coal; Earthquakes; Gradualism; Volcanic islands; Volcanoes

Geotrupidae, 404

Geyser, on Batchian, 334

Gilolo (Halmahera), 246, 296, 420, 522, 524, 526, 528, 530; butterflies collected at, 522, 524, 526, 528, 530; drawing of fly from, 442; grasses, 244, 246; natives of, 296

Glaciers and glaciation, 96, 212, 312, 314

Glaphyridae, 402

*Glenea,* 300, 450

*Gnoma,* 340; *G. subfasciata,* 450

Godart, Jean Baptiste, 286

*Gracula pectoralis,* 270

Gradualism/gradual change, in geology and species, 98, 106, 118, 126, 128, 130, 132, 148, 152, 188, 190, 212, 218, 244, 328. *See also* Geology, gradual change

Grasses: of Gilolo, 244, 246; llanos, 246; plains of, in the tropics, 244, 246; transmutation of wild, 116

Gray, George Robert, 230, 298, 342

Gray, John Edward, 346

Grouse, male displays, 266

Guava, 268

*Gurcinia mangustana,* 268

Guttiferae, 268

*Gymnetis lanius,* 342

Gyrinidae, 404

Habit, 134, 136, 368; of beetles, 44, 300; change in habit but not structure, 134; and instincts, 366, 368. *See also* Orangutan, habits of

Halmahera. *See* Gilolo

Hamilton Gray & Co., address for, 22

*Hammaticherus,* 300, 340; *H. spinicornis,* 424, 448

Hantus, rarity in Lombock, 18

Harmony (balance) in nature, criticism of concept, 126, 128, 308, 310, 320, 322

Haughton, Samuel, 270; Wallace rebuttal of, 374, 376, 378, 380, 382

Hawaii. *See* Sandwich Islands

*Helius* (ship), 410

*Helix,* 248

Helophoridae, 404

Helopidae, 404, 500

Herbst, Johann Friedrich Wilhelm, 406

Hesperiidae (Hesperidae), 477, 488, 532

*Hesthesis,* 418

*Hestia ideopsis,* 532

Heteroceridae, 404

*Heteromera,* 432, 446

Hewitson, William Chapman, 522, 526, 528, 530

Himalayas, 414

Hindu, 158, 228, 236, 240

Histeridae, 404, 500

Homology. *See* Affinity

Homoptera, 477

*Hooker's Journal of Botany,* Wallace papers in, 274

Horiidae, 402

Hornbills, change in habit but not structure of, 134

Horses, of South America, 310

Huc, Evariste Régis, 142, 146

Hugueunis, Mr., 504

Human inhabitants: of Aru Islands, 160, 200, 202; of Flores, 236; of Gilolo, 296; of India, 140, 260, 368; of Malacca, 240; of New Guinea, 238, 240

Human races, in Malay Archipelago: Alfurus, 236, 238, 240, 296; Bugis, 236; Caffre Malegassee, 240; Endamene, 240; Malay, 160, 172, 228, 238, 240, 296; Mongolian, 240; Oceanic, 240; Papuans, 140, 156, 160, 236, 238, 240, 296; Tomore, 236

Humans: accounts of tails in, 156, 210; racial observations, 154, 156, 158, 160, 228, 236, 238, 240, 296. *See also* Instincts, human; Races, human

Hummingbirds, 182, 258, 274

Huxley, Thomas Henry, 362

*Hyades* sp., 473; *H. horsfeldii,* 473

*Hyantis hodeva,* 522; *H. indra,* 522

Hydradephaga, 406, 476, 488